Flow Visualization

Techniques and Examples

Flow Visualization

Techniques and Examples

Editors

A. J. Smits
Princeton University

T. T. Lim
National University of Singapore

ICP Imperial College Press

Published by

Imperial College Press
57 Shelton Street
Covent Garden
London WC2H 9HE

Distributed by

World Scientific Publishing Co. Pte. Ltd.
5 Toh Tuck Link, Singapore 596224
USA office: Suite 202, 1060 Main Street, River Edge, NJ 07661
UK office: 57 Shelton Street, Covent Garden, London WC2H 9HE

British Library Cataloguing-in-Publication Data
A catalogue record for this book is available from the British Library.

First published 2000
Reprinted 2003

FLOW VISUALIZATION: TECHNIQUES AND EXAMPLES

ISBN 1-86094-193-1
ISBN 1-86094-372-1 (pbk)

Printed by FuIsland Offset Printing (S) Pte Ltd, Singapore

PREFACE

Flow visualization is one of the most effective tools in flow analysis, and it has been crucial for improving our understanding of complex fluid flows. In fact, some of the major discoveries in fluid mechanics were made using flow visualization. Professor F.M.N. Brown of the University of Notre Dame wrote:

> ... A man is not a dog to smell out each individual track, he is a man to see, and seeing, to analyse. He is a sight tracker with each of the other senses in adjunctive roles. Further, man is a scanner, not a mere looker. A single point has little meaning unless taken with other points and many points at different times are little better. He needs the whole field, the wide view.

The book is designed to provide source material for those who are intending to carry out flow visualization studies. Although it is written primarily for students and researchers in areas of mechanical, aerospace, and civil engineering, as well as oceanography and physics, we hope that other research workers, including those in medical fields will find the book useful. We hope, too, that the depth and breadth of the book will make it valuable to people who have little or no prior experience in flow visualization as well as to those with considerable experience in this subject. To obtain a complete understanding of the flow behavior, it is usually necessary to complement the flow visualization with quantitative measurements. One of the most exciting advances in flow imaging is that some flow visualization techniques, such as Particle Image Velocimetry (PIV) and Molecular Tagging Velocimetry (MTV), can also provide quantitative results. We have highlighted such dual-use methods in this book.

The text is organized into two major parts. The first part consists of 12 chapters, each dealing with a different technique, or a related set of techniques for flow visualization. The first chapter in this part deals with the interpretation of flow visualization results using critical point theory, and it is a must for everyone as it highlights some of the possible traps and dangers in interpreting flow visualization results. The remaining chapters are devoted to discussion and implementation of particular flow visualization techniques, covering hydrogen bubble, dye and smoke methods, MTV, planar laser imaging, digital PIV, thermochromic crystals, pressure and shear sensitive coatings, methods for compressible flows, three- and four-dimensional imaging, and the interpretation of numerical visualizations of compressible flows with strong gradients. They are all written by recognized experts in flow visualization. We deliberately asked the authors to emphasize the practical aspects of their craft, to help others get started in this field. Extensive references are given for more detailed study. The second part of the text is made up of a collection of flow images taken by leading researchers from around the world. These illustrations give examples of the techniques described in the book, and they were chosen to provide high-quality images of some fascinating fluid flow phenomena.

Flow visualization covers a broad range, and it is certainly impossible for us to include all the topics in a single volume. The choice of the topics must be somewhat controversial and necessitates many arbitrary omissions. We apology for any gaps and omissions in this book.

Finally, we would like to take this opportunity to thank all the authors for sharing their expertise in flow visualization, and their hard work in preparing their particular contributions. It is they who made this book a reality and we hope they are pleased with the final product.

We welcome constructive comments and suggestions.

Alexander J. Smits
Princeton, New Jersey, USA

T. T. Lim
Singapore

CONTENTS

CHAPTER 1

INTERPRETATION OF FLOW VISUALIZATION

A. E. Perry and M. S. Chong[a]

1.1 Introduction

The successful interpretation of fluid flow patterns continues to be an important tool used to investigate and understand the physics of complex three-dimensional eddying motion and turbulence. These flow patterns may be displayed in many ways. They may be photographs of dye or smoke injected into the flow field. They may be long time exposure photographs of particles which have been seeded into the flow. They may be two- or three-dimensional flows. They may be cross-sections of complex three-dimensional flow fields. They may be a single photograph or a sequence of frames. The flow may be steady or unsteady. They may be a vector field measured using some conditional averaging technique with hot-wires or a vector field measured using digital particle-image-velocimetry. These flow patterns may even be artificially created from a numerical computation. Whatever the technique used to generate the flow patterns, one ends up with single or multiple images of flow patterns and it is through the interpretation of these images that one gains an understanding of the physics of the flow field. To be able to successfully interpret these flow patterns requires a thorough under-standing of pathlines, streaklines and streamlines in steady and unsteady flow and a formal classification method to unambiguously describe the flow field.

1.2 Critical Points in Flow Patterns

A flow pattern described by streamlines consists of special points where the streamline slope is indeterminant and the velocity is zero. Such points are called

[a]Dept. of Mechanical and Manufacturing Engineering, University of Melbourne, Parkville, VIC 3052, Australia

1

"critical points" or "stationary points." These points are the salient feature of a flow pattern; given a distribution of such points and their type, much of the remaining flow field and its geometry and topology can be deduced since there is only a limited number of ways the streamlines can be joined and some illustrations of this will be given later.

Of course the properties of the streamline field or velocity vector field seen by an observer depends on the velocity of the observer. If a non-rotating observer is moving with a fluid particle, then there will be a critical point located at the particle, and in the region immediately surrounding the particle and observer, the flow will in most cases be described to first order as

$$u_i = A_{ij}x_j \qquad (1.1)$$

where u_i is the velocity at the position x_j relative to the particle and observer. The quantity A_{ij} is the velocity gradient tensor, that is,

$$A_{ij} = \frac{\partial u_i}{\partial x_j} = \mathbf{A}. \qquad (1.2)$$

Following Chong *et al.* (1989, 1990), the geometry of this streamline pattern can be classified by studying certain invariants of A_{ij} in the characteristic equation

$$\lambda^3 + P\lambda^2 + Q\lambda + R = 0 \qquad (1.3)$$

where P,Q and R are the tensor invariants. These are

$$P = -trace(\mathbf{A}) \qquad (1.4)$$

$$Q = \frac{1}{2}\left(P^2 - trace(\mathbf{A}^2)\right) \qquad (1.5)$$

and

$$R = -det(\mathbf{A}). \qquad (1.6)$$

For incompressible flow, $P = 0$ from continuity and so

$$\lambda^3 + Q\lambda + R = 0. \qquad (1.7)$$

The eigenvalues λ which determine the topology of the local flow pattern depend on the invariants R and Q. In fact the R-Q plane shown in Fig. 1.1 is divided into regions according to flow topology.

The discriminant of A_{ij} is defined as

$$D = \frac{27}{4}R^2 + Q^3 \qquad (1.8)$$

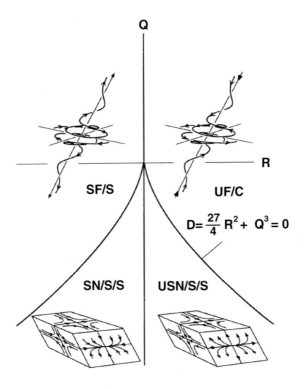

Fig. 1.1. Possible non-degenerate topology in the R-Q plane: stable-focus/stretching (SF/S) (when $D > 0$ and $R < 0$); unstable-focus/contracting (UF/C) (when $D > 0$ and $R > 0$); stable-node/saddle/saddle (SN/S/S) (when $D < 0$ and $R < 0$); and unstable-node/saddle/saddle (UN/S/S) (when $D < 0$ and $R > 0$). Stable means that arrows of time point towards the origin and unstable means that they point away from the origin.

and the boundary dividing the flows with complex eigenvalues from those with all real eigenvalues is

$$D = 0 \qquad (1.9)$$

and is shown in Fig. 1.1.

For $D > 0$ Eqn. 1.3 admits two complex and one real solution for λ. Such points are called foci. If $D < 0$, all three solutions for λ are real and the associated flow pattern is referred as a node/saddle/saddle point according to the terminology adopted by Chong *et al.* (1990). The caption of Fig. 1.1 gives the full description for these regions.

The velocity gradient tensor can be split into two components:

$$A_{ij} = S_{ij} + W_{ij} \tag{1.10}$$

where S_{ij} is the symmetric rate of strain tensor and W_{ij} is the skew symmetric rate of rotation tensor. These are given by

$$S_{ij} = \frac{1}{2}\left(\frac{\partial u_i}{\partial x_j} + \frac{\partial u_j}{\partial x_i}\right) \tag{1.11}$$

and

$$W_{ij} = \frac{1}{2}\left(\frac{\partial u_i}{\partial x_j} - \frac{\partial u_j}{\partial x_i}\right). \tag{1.12}$$

It can be seen that for regions above the $D = 0$ curve, the rotation tensor dominates over the rate of strain tensor and for regions below, the rate of strain tensor dominates. It has been suggested that the core region of a vortex belongs to regions above the $D = 0$ curve. However, the definition of a vortex core has been a subject of much debate. Over the years many workers have been involved in this debate, for example, Truesdell (1954), Cantwell (1979), Vollmers (1983), Dallmann (1983), Chong, Perry & Cantwell (1989, 1990), Robinson (1991), Lugt (1979), Jeong & Hussain (1995), Perry & Chong (1994), and Soria & Cantwell (1994), to mention a few. However, in the study of complex flows, for example, turbulence, it is useful to identify regions of the flow which are "focal" and methods for doing this will shortly be described. Some colloquial terms which better illustrate the physics of the processes involved are as follows: for the upper left part of the Fig. 1.1, the flow pattern could be referred to as a "stretch with a twist", the upper right as a "splat with a twist" or a "squish with a twist", the lower left a "stretch" and the lower right a "splat". These critical points have planes which contain solution trajectories (or streamlines) and these planes will be referred to as eigenvector planes. The upper patterns in Fig. 1.1 possess only one such plane which contains a focus and there exists a real eigenvector about which the trajectories wind in a helix-like manner. The lower patterns possess three such planes which contain solution trajectories. They in general have eigenvector planes which are non-orthogonal. This non-orthogonality always occurs if there exists a rotation tensor. If this is zero, that is, in irrotational flow, then the eigenvector planes are orthogonal. Of course it should be realized that only the first term in a Taylor series expansion has been considered (see Perry & Chong 1986, 1987) and there exist trajectories which osculate to these eigenvector planes close to the critical point but diverge away for large x_j.

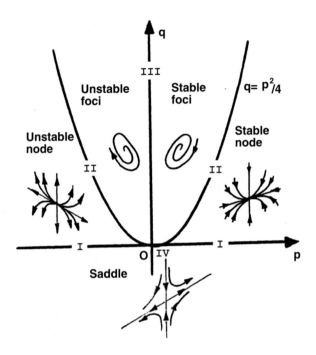

Fig. 1.2. Classification of critical points on the *p-q* chart. Critical points on the boundaries I, II, III, and IV are degenerate. After Perry & Chong (1987).

The lower patterns in Fig. 1.1 shows the projections of the trajectories in the eigenvector planes which pass through the origin.

In general it is extremely difficult to gain an understanding of a three-dimensional critical point no matter how it is displayed and orientated as an image. The best way is to show or at least highlight the trajectories in the eigenvector planes, and so classification of critical points in phase planes becomes very useful. In these planes, simple phase-plane methods can be used to classify the critical-point patterns. The following equation is obtained by defining a new coordinate system in each plane in turn:

$$\begin{bmatrix} \dot{y}_1 \\ \dot{y}_2 \end{bmatrix} = \begin{bmatrix} a & b \\ c & d \end{bmatrix} \begin{bmatrix} y_1 \\ y_2 \end{bmatrix} \tag{1.13}$$

or

$$\dot{\mathbf{y}} = \mathbf{F}\mathbf{y}. \tag{1.14}$$

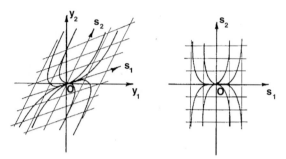

Fig. 1.3. Node in noncanonical form (*left*) . Node in canonical form (*right*). s_1 and s_2 are eigenvectors. After Perry & Fairlie (1974).

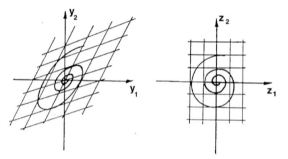

Fig. 1.4. Focus in noncanonical form (*left*). Focus in canonical form (*right*). After Perry & Fairlie (1974).

The two quantities of importance are

$$p = -(a+d) = -trace(\mathbf{F})$$

$$q = (ad - bc) = det(\mathbf{F}). \qquad (1.15)$$

The types of critical points are classified on the p-q chart as shown in Fig. 1.2. Thus nodes, foci or saddles can be obtained. The patterns depends on the region of location of a point defined by p and q on the p-q chart.

If all the eigenvalues are real, either nodes or saddles can be produced. These patterns in general will be in noncanonical form, that is, the eigenvectors in the plane under consideration are nonorthogonal. Fig. 1.3 (*left*) shows a node in noncanonical form, and Fig. 1.3 (*right*) shows a node in canonical form, and s_1

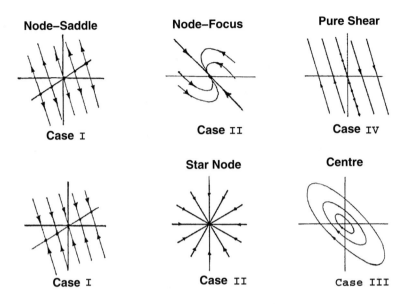

Fig. 1.5. Degenerate critical points, or borderline cases. Case numbers refer to Fig. 1.2. After Perry & Fairlie (1974).

and s_2 are two eigenvector which define the eigenvector plane. This is achieved by distorting the noncanonical pattern by an affine transformation, that is, by a coordinate stretching (with a constant stretching factor) and differential rotation of the coordinates. If the eigenvalues are complex and the (y_1, y_2)-plane contains solution trajectories, a focus is obtained in that plane. Fig. 1.4 (*left*) shows a noncanonical focus, and Fig. 1.4 (*right*) shows a focus in canonical form. When in canonical form, nodes and saddles have solution trajectories that are simple power laws, that is, $y_2 = Ky_1^m$, whereas foci reduce to simple exponential spirals (see Perry & Fairlie, 1974). If the pattern occurs on the boundaries of the p-q chart, that is, when $p^2 = 4q$, $p = 0$, or $q = 0$, then we have "borderline" cases. These are often referred to as "degenerate" critical points and are shown in Fig. 1.5 (see Kaplan, 1958). These patterns rarely occur precisely in practice but may occur momentarily as a flow pattern changes with time from one topological classification to another.

Complicated three-dimensional critical-point patterns can therefore be understood by simply looking at the solution trajectories in each of the eigenvector planes. Furthermore, if these solution trajectories are used in a rendition of a

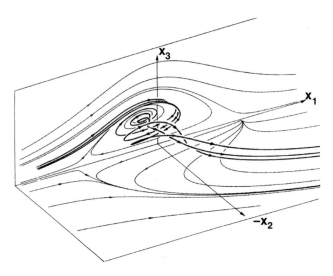

Fig. 1.6. U-shaped separation computed using third order series expansion of u_i in terms of x_j. After Perry & Chong (1986, 1987).

randomly orientated three-dimensional pattern on a computer screen then the pattern will be easily understood. On the other hand, if only trajectories off the eigenvector planes are used, the pattern will be very confusing.

Fig. 1.6 shows a flow pattern for a steady three-dimensional flow separation at a no-slip boundary. In this pattern, only three critical points occur within the field of view. The (x_1, x_3)-plane is a plane of symmetry. The two critical points which occur on the (x_1, x_3)-plane belong to the lower part of Fig. 1.1 and the critical point on the plane of symmetry belong to the top right hand part of Fig. 1.1. However, since we are at a no-slip boundary, the critical points on the (x_1, x_2)-plane require special treatment. If we locate our coordinate system at these critical points in turn we have

$$u_i = x_3 B_{ij} x_j \qquad (1.16)$$

where x_3 is the normal distance from the no-slip surface. The premultiplied x_3 ensures the no-slip condition and in general u_i/x_3 is finite as $x_3 \to 0$ (that is, finite vorticity) but $u_i/x_3 = 0$ at $x_j = 0$. The above could be written as

$$\dot{x}_i = B_{ij} x_j \qquad (1.17)$$

where the dot above the x_i denotes a differentiation with respect to a transformed time τ defined as $d\tau = x_3 dt$. The solution trajectories at $x_3 = 0$ are "limit" trajectories, "limiting" streamlines or skin friction lines. B_{ij} is analogous to A_{ij} and much of the analysis for classification is similar (with minor differences). See Chong *et al.* (1990). For instance, the invariants P, Q and R of B_{ij} produce similar results but P is not zero and the curve separating real from complex solutions becomes asymmetrical at a fixed finite P. Such critical points are called no-slip critical points and the others are referred to as free-slip critical points.

For a more complete treatment of the mathematics of critical points, readers are referred to Perry & Fairlie (1974), Lim, Chong & Perry (1980), Perry & Chong (1987), Chong *et al.* (1990) and Perry (1984).

1.3   Relationship between Streamlines, Pathlines and Streaklines

There is an excellent educational movie made by Kline (1965) where it is shown that in steady flow, streamlines, pathlines and streaklines are identical. The movie also ahows that in unsteady flow, this is no longer true and their relationship becomes most complex.

A streamline is a line drawn in the flow field such that it is tangent to the velocity vectors. In unsteady flow this is also true, giving instantaneous streamlines. Streamlines can never cross except at critical points. Pathlines of various particles cross at any number of points in unsteady flow.

A streakline is the locus of a series of particles which have been released sequentially from a fixed point in the flow. In unsteady flow, streaklines can move normal to themselves.

A good way of illustrating the relationship between these concepts is to consider the unsteady solutions of the Navier-Stokes equations by Perry, Lim & Chong (1979). These are the so called accelerating critical points, one of which is given by

$$u = -b\left(z - \epsilon_z cos(\sqrt{bc}t) + \sqrt{\frac{b}{c}}\epsilon_x sin(\sqrt{bc}t)\right) \qquad (1.18)$$

$$w = c\left(x - \epsilon_x cos(\sqrt{bc}t) - \sqrt{\frac{b}{c}}\epsilon_z sin(\sqrt{bc}t)\right). \qquad (1.19)$$

Equations. 1.18 and 1.19 describe an accelerating and hence unsteady center. There is an analogous family of accelerating saddles and these are related to the dislocated saddles of Perry & Chong (1987). A center is shown as Case III in

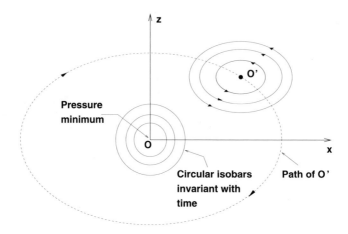

Fig. 1.7. Moving critical point.

Fig. 1.2. The flow is two-dimensional and the streamlines are closed. Fig. 1.7 shows a center at O' which is orbiting O along an elliptical trajectory which is a trajectory of the center shifted to O but with the 'arrow of time' reversed. This generates an unsteady pattern but with a steady pressure field with the pressure minimum at 0 and the isobars are circular. Readers can verify this by substituting Eqns. 1.18 and 1.19 into the Navier-Stokes equations. In Eqns. 1.18 and 1.19, ϵ_x and ϵ_z are arbitrary (they are the initial coordinates of the center O' at time $t = 0$). The constants b and c determine the geometry of the critical point.

Instantaneous velocity vector field directions and the instantaneous stream-line pattern at $t = 0.625T$, where T is the period of orbit of O' about O, are shown in Fig. 1.8. An arbitrary point has be chosen where "dye" is introduced to produce a streakline (the dotted curve is the locus of a series of particles released sequentially from a fixed point F in the flow). Also shown are the path-lines of five arbitrary particles. Note that the pathlines are are tangential to the streamlines and are allowed to cross while streamlines do not cross. This figure also shows that streaklines can also cross themselves. From Figs. 1.8 to 1.10 one can see that the relationship between streamlines, streaklines and pathlines in unsteady flow is most complex. Fig. 1.9 shows the same flow but at time $t = 1.25T$. Of course the streaklines and pathlines would be entirely different if the "dye" is introduced at a different location as shown in Fig. 1.10.

It has been known that particles suspended in a flow and photographed

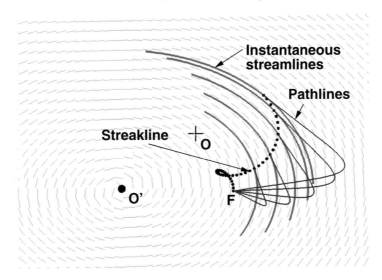

Fig. 1.8. Velocity vector field directions, instantaneous streamlines, streaklines and pathlines in unsteady flow. F is point where "dye" is injected. Here, $b = 1.0$, $c = 0.5$, $\epsilon_x = 0$ and $\epsilon_z = 1.0$ in Eqns. 1.18 and 1.19, and $t = 0.625T$, where T is the period of orbit of O' around O.

at two short time intervals apart give the velocity vector field and if there are enough particles, the instantaneous vector field topology can be recognised without having to compute the velocity field and then integrating the vector field to obtain instantaneous streamlines. This can be illustrated quite clearly by generating an array of dots on a transparency and the same array of dots can be stretched by say $n\%$ in the x_1-direction and shrunk $n\%$ in the x_2-direction with the aid of a computer on another transparency. Fig. 1.11 shows these two transparencies superimposed. A saddle is most obvious, that is, we have simulated a flow field with a pure rate of strain. This is equivalent to superimposing two frames of a movie of a field of particles. To enhance this effect, several images of particles, that is, many movie frames, can be superimposed as shown in Fig. 1.12 (*left*) where 11 frames have been superimposed. An orthogonal saddle can clearly be seen.

If we now rotate each frame relative to the previous frame by a fixed degree of rotation, we would have simulated a rate of rotation plus a rate of strain. Fig. 1.12 (*right*) shows how the flow field changes with different ratio of the rates of rotation to the rate of strain. In Fig. 1.12 (*right*), the ratio of the rate of rotation to the rate of strain is small and the rate of strain dominates and

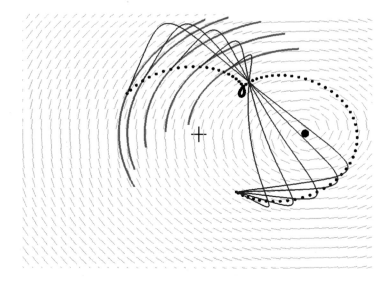

Fig. 1.9. Flow pattern at $t = 1.25T$.

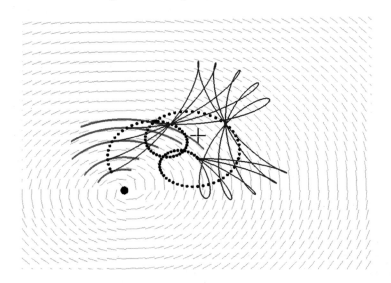

Fig. 1.10. Point where "dye" is injected has been changed. Here, $t = 1.625T$.

we still have a saddle. However, the eigenvectors are no longer orthogonal. By increasing the ratio of the rate of rotation to the rate of strain we are effectively

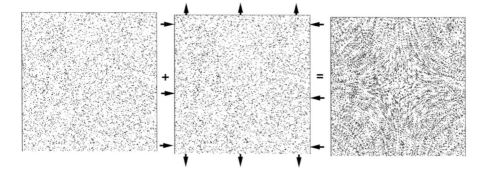

Fig. 1.11. How a saddle is created.

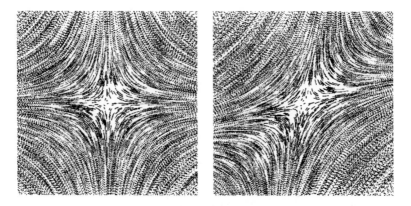

Fig. 1.12. Orthogonal saddle (*left*). Non-orthogonal saddle (*right*).

moving up the q-axis of the p-q chart of Fig. 1.2. Although these pictures do not have streamlines actually drawn on them, it is not difficult to perceive the streamline patterns. Fig. 1.13 show the flow patterns with increasing ratio of rate of rotation to the rate of strain. The degenerate flow pattern shown in Fig. 1.13 (*left*) is close to the origin of the p-q chart, that is, pure shear, Case IV in Fig. 1.5. A further increase in rate of rotation produces a center as shown in Fig. 1.13 (*right*), that is, Case III in Fig. 1.5.

By stretching $n\%$ in the x_2 direction and by $m\%$ in the x_1-direction, nodes are generated if the rate of rotation is small as shown in Fig. 1.14 (*left*). With an increase in the rate of rotation the flow pattern changes to a focus as shown in Fig. 1.14 (*right*).

Fig. 1.13. Degenerate flow pattern (pure shear) (*left*). Center (*right*).

Fig. 1.14. Node (*left*). Focus (*right*).

This principle of obtaining streamlines from short pathlines can be extended very dramatically if we have many frames of a movie or video. By superimposing many frames, instantaneous streamlines show up most clearly and an animation can be produced, as was first done by Perry, Chong & Lim (1982) for the vortex shedding process behind a circular cylinder. Here 40 consecutive frames of a movie made by Prandtl (see Shapiro & Bergman, 1962) were superimposed onto a photographic plate and this was repeated over the vortex shedding cycle. This is analogous to the streamline pattern created by superimposing 20 frames $(1/20^{th}$ of the cycle of the unsteady flow, as shown in Fig. 1.15) at time close to zero for the flow case shown in Fig. 1.8. Also shown is the instantaneous stream-

Fig. 1.15. Relationship between instantaneous streamlines and long time exposure of random particles in unsteady flow. 20 frames superimposed over one-twentieth of the period of orbit of the center. Same flow case as in Fig. 1.8.

lines obtained by integrating the velocity field which is assumed to be "frozen" midway between the first of last superimposed frames. This shows that the instantaneous streamline pattern obtained from a short time exposure of particles corresponds extremely well with the actual streamlines even in unsteady flows. An example of a flow pattern obtained experimentally is shown in Fig. 1.16. This figure shows a short time exposure of particles in the wake of an elliptical cylinder and centers and and saddles are most obvious.

1.4   Sectional Streamlines

A sectional streamline pattern is obtained by integrating the velocity field in a plane where the vectors at that plane have been resolved into that plane. Such patterns can be seen or deduced from time-exposure photographs of clouds of particles illuminated by sheets of laser light. Of course, it is very dangerous to infer the geometry of a three-dimensional critical point from one such plane, unless it is known that we are on an eigenvector plane or a plane of symmetry. Some examples of misinterpretation of flow patterns from sectional streamlines are given in Perry & Chong (1994).

Fig. 1.16. Instantaneous streamlines behind an elliptical cylinder. Reynolds number based on the major axis is 250. After Prandtl & Tietjens (1934).

1.5   Bifurcation Lines

Other salient features of a streamline pattern are bifurcation lines. Bifurcation lines are lines which form asymptotes in the flow field. Fig. 1.17 shows a bifurcation line in mid air. From the work of Hornung & Perry (1984) and Perry & Hornung (1984), the neighbouring trajectories are exponential curves close to the bifurcation lines, and there are two planes which contain these trajectories. In one plane the trajectories converge towards the bifurcation line, and in the other they diverge away if one follows the "arrows of time" indicated in the figure. In the plane orthogonal to the bifurcation line one obtains "sectional" streamlines patterns which are saddles.

Bifurcation lines also occur at no-slip boundaries. The skin friction lines form a family of exponential curves close to the bifurcation line, and the vortex lines at the boundary are orthogonal (Lighthill 1963) and form a family of parabolas as seen in Fig. 1.18. Fig. 1.19 shows the instantaneous skin friction and vortex lines in a turbulent boundary layer obtained by a direct numerical simulation (DNS) of the Navier-Stokes equations (from Chong *et al.* 1998). Kinks in the vortex lines indicate bifurcation lines.

An array of longitudinal vortices aligned in the streamwise direction would

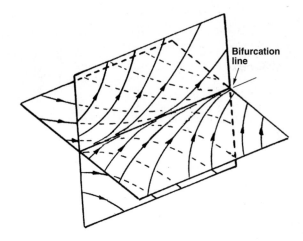

Fig. 1.17. A bifurcation line. After Perry & Chong (1987).

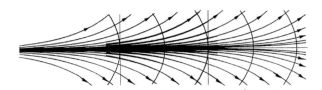

Fig. 1.18. Bifurcation line with skin friction lines and orthogonal vortex lines.

give the bifurcation lines as shown in Fig. 1.20. It is obvious that when viewed in the direction of the arrows, A would appear as a counter-clockwise vortex, B would be a clockwise vortex and C a counter clockwise vortex. Sometimes the principle of identifying the sign of a vortex can be obtained by introducing gravity into the problem. Deducing the flow pattern over an array of grooves was carried out by Perry, Schofield & Joubert (1969). By positioning the array vertically but across the flow as seen in Fig. 1.21 (*top*) and painting the array with a suspension of titanium dioxide with kerosene, the pattern shown is produced. Also shown in Fig. 1.21 (*bottom*) is the deduced two-dimensional mean eddying motions in the cavity. This technique of using bifurcation lines for identifying longitudinal vortices and their sign is most useful over the upper surfaces of delta wings.

$\rightarrow$ Flow direction

Fig. 1.19. Skin friction lines and vorticity lines in a wall-bounded flow. After Chong *et al.* (1998).

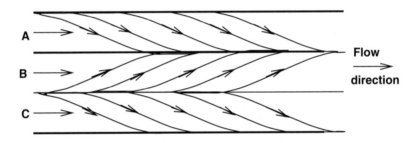

Fig. 1.20. Surface bifurcation lines generated by longitudinal vortices.

1.6   Interpretation of Unsteady Flow Patterns with the Aid of Streaklines and Streamlines

It is well known that in incompressible uniform density flow all vorticity is generated at solid boundaries (see Lighthill, 1963). Vorticity, like dye, moves with the fluid (see Batchelor, 1967). If the dye is injected into the fluid at the location where vorticity is generated, then the dye will mark the vorticity and will continue to do so until such time that viscous diffusion has diffused the vorticity away from the dye. Thus a vortex sheet (that is, a thin shear layer) can be marked by dye.

Fig. 1.22 shows smoke issuing from a tube in an asymmetrical co-flowing wake. The smoke indicates vortex loops since the initial vorticity was marked by the smoke. The flow has been investigated in considerable detail by Perry & Lim (1978), Perry, Lim & Chong (1980), Perry & Tan (1984) and Perry

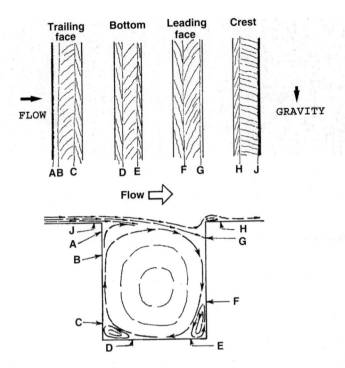

Fig. 1.21. Surface f low patterns around '*d*' type roughness. After Perry, Schofield & Joubert (1969).

& Chong (1987). The smoke can be thought of as a bundle of streaklines. Fig. 1.23 shows the instantaneous velocity vector field obtained by hot-wire measurements and the streamline pattern down the plane of symmetry as seen by an observer moving with the eddies or smoke. A distribution of saddles and nodes can be seen. Trajectories which are connected to saddle points have been highlighted and these are called "separatrices" since they divide the flow into distinct regions. These become eigenvectors at the saddle points. Smoke and dye, like vorticity patterns are invariant to the velocity of the observer but the velocity and streamline fields depend very much on the observer velocity. Fig. 1.24 shows the same pattern with a 10% change in velocity. The row of saddles and nodes above the eddy structure has turned into a bifurcation line. If the velocity of the observer is aligned such that he/she is moving with the vortex loops, the streamline pattern time variation is a minimum and the foci line up with the loops of smoke and parts of the smoke tend to align with the

Fig. 1.22. Externally illuminated single-side wake pattern passing a hot-wire probe. After Perry & Chong (1987).

separatrices. This tendency is also observed in two-dimensional von Kármán vortex streets which consist of an array of saddles and centers. However, here the flow is three-dimensional and the spiralling in the foci indicates vortex stretching.

Quite often a pattern can be understood by knowing its vortex skeleton (Perry & Hornung, 1984). By taking strobed laser sheet cross-sections of the smoke, the crude vortex skeleton as shown in Fig. 1.25 was obtained. By application of the Biot-Savart law a pattern with the same topology as given in Fig. 1.23 is obtained. Quite often it is asked, "What is the vortex skeleton of a turbulent boundary layer." The next section describes this.

When studying DNS data of turbulent flows, particularly for wall-bounded flows, it is quite perplexing to decide what quantities to consider and how to display them. Some years ago it was the dream of many workers in the field to display the vortex lines of a turbulent boundary layer, jet or wake from DNS data, but when this became possible it was quickly realized that such displays are a complete mess and almost impossible to interpret. The vorticity lines looked like a complete tangled mess of wire. Blackburn, Mansour & Cantwell (1996) examined channel flow data by identifying regions of the flow where the rotation tensor denominated over the rate of strain tensor. This was done by mapping out isosurfaces of the discriminant D in Eq. (1.8) and a value above

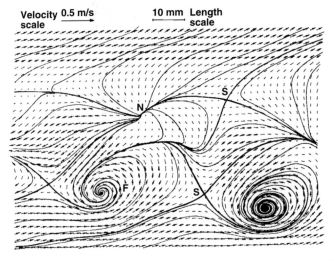

Fig. 1.23. Typical instantaneous (phase-averaged) velocity vector field for smoke pattern shown in Fig. 1.22. After Perry & Chong (1987).

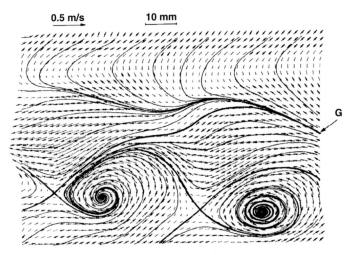

Fig. 1.24. Flow pattern given in Fig. 1.23 with 10% change in convection velocity. G is a bifurcation line. After Perry & Chong (1987).

zero was chosen. It is found that these isosurfaces enclose a region of flow which is focal, and Fig. 1.26 shows these isosurfaces. It is also found that these isosurfaces enclose reasonably well-ordered and concentrated lines of vorticity,

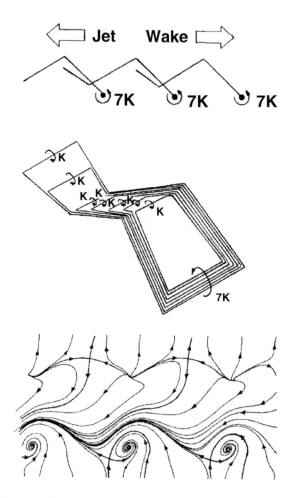

Fig. 1.25. Side view of vortex skeleton for single-sided structure (*top*). Oblique view of a typical cell. K denotes a unit of circulation (*middle*). Computed velocity field using the Biot-Savart law (as seen by an observer moving with the eddies) (*bottom*). After Perry & Chong (1987).

and the picture resembles the sides of the long conjectured ∩-like vortices in wall bounded flow. This approach was applied by Chong *et al.* (1998) to a zero pressure gradient boundary layer and a layer which undergoes separation and reattachment, and the focal regions were clearly seen. In this latter work it is

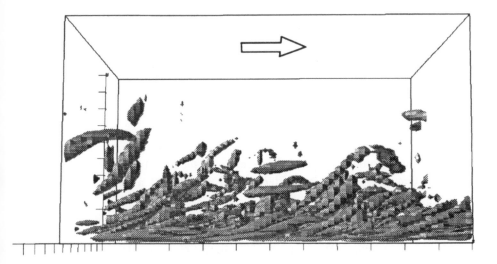

Fig. 1.26. Isosurfaces of constant D. After Blackburn, Mansour & Cantwell (1996).

shown that if the isosurfaces are chosen with D slightly above zero, most of the enstrophy in the flow is accounted for and that these focal regions retain their identity for a considerable time as they convect downstream. One can see from Fig. 1.26 that as we trace one of these worm-like structures in the streamwise direction, they are first aligned along the wall longitudinally before bending up off the wall. This may explain the skin friction bifurcation lines seen in Fig. 1.19. These might well be the footprints of attached vortex loops (for example, see Perry & Chong, 1982). Head & Bandyopadhyay (1981) proposed such structures from flow visualisations with smoke introduced at the wall.

1.7 Concluding Remarks

It can be seen that the study and interpretation of flow patterns is aided greatly by the application of the mathematics of critical point theory. This has become extremely useful in recent times where it is possible to produce large databases for the description of flow patterns from laboratory measurements and numerical computations. Exciting new developments are beginning to emerge, particularly in the study of turbulence.

1.8 References

Batchelor, G. K. 1967. *An Introduction to Fluid Dynamics.* Cambridge University Press.

Blackburn, H. M., Mansour, N. N. & Cantwell, B. J. 1996. Topology of fine-scale motions in turbulent channel flow. *J. Fluid Mech.*, **310**, 269–292.

Cantwell, B. J. 1979. Coherent turbulent structures as critical points in unsteady flow. *Archiwum Mechaniki Stosowanej* (Archives of Mechanics), **31**, **5**, 707–721.

Chong, M. S., Perry, A. E. & Cantwell, B. J. 1989. A general classification of three-dimensional flow fields. *Proceedings of the IUTAM Symposium on Topological Fluid Mechanics*, Cambridge, ed. H. K. Moffatt and A. Tsinober, 408–420.

Chong, M. S., Perry, A. E. & Cantwell, B. J. 1990. A general classification of three-dimensional flow fields. *Phys. of Fluids*, **A4** (4), 765–777.

Chong, M. S., Soria, J. Perry, A. E., Chacin, J., Cantwell, B. J. & Na, Y. 1998. Turbulence structures of wall-bounded shear flows found using DNS data. *J. Fluid Mech.*, **357**, 225–247.

Dallmann, U. 1983. Topological structures of three-dimensional flow separations. *DFVLR Rep. IB 221-82-A07*, Gottingen, West Germany.

Head, M. R. & Bandyopadhyay, P. 1981. New aspects of turbulent structure. *J. Fluid Mech.*, **107**, 297–337.

Hornung, H. G. & Perry, A. E. 1984. Some aspects of three-dimensional separation. Part I. Streamsurface bifurcations. *Z. Flugwiss. Weltraumforsch*, **8**, 77–87.

Jeong, J. & Hussain, F. 1995. On the identification of a vortex. *J. Fluid Mech.*, **285**, 69–94.

Kaplan, W. 1958. *Ordinary Differential Equations.* Addison-Wesley, Reading, Mass.

Kline, S. J. 1965. FM-48 Film loop. National Committee for Fluid Mechanics film.

Lighthill, M. J. 1963. Attachment and separation in three-dimensional flow. In *Laminar Boundary Layers*, ed. L. Rosenhead, 72–82, Oxford University Press.

Lim, T. T., Chong, M. S. & Perry, A. E. 1980. The viscous tornado. Proceedings of the 7th Australasian Hydraulics and Fluid Mechanics Conference, Brisbane, 250–253.

Lugt, H. J. 1979. The dilemma of defining a vortex. In *Recent Developments in Theoretical and Experimental Fluid Mechanics*, ed. U. Muller, K.G. Roesner and B. Schmidt, Springer, 309–321.

Perry, A. E. 1984. A study of degenerate and non-degenerate critical points in three-dimensional flow fields. *Forschungsber. DFVLR-FB 84-36*, Gottingen, Germany.

Perry, A. E. & Chong, M. S. 1994. Topology of flow patterns in vortex motions and turbulence. *Appl. Scientific Research*, **54** (3/4), 357–374.

Perry, A. E. & Chong, M. S. 1986. A series expansion study of the Navier-Stokes equations with applications to three-simensional separation patterns. *J. Fluid Mech.*, **173**, 207–223.

Perry, A. E. & Chong, M. S. 1987. A description of eddying motions and flow patterns using critical-point concepts. *Ann. Rev. Fluid Mech.*, **19**, 125–155.

Perry, A. E. & Fairlie, B. D. 1974. Critical points in flow patterns. *Adv. Geophys.*, **18B**, 299–315.

Perry, A. E. & Hornung, H. G. 1984. Some aspects of three-dimensional separation. Part II. Vortex skeletons, *Z. Flugwiss. Weltraumforsch*, **8**, 155–160.

Perry, A. E. & Lim, T. T. 1978. Coherent structures in co-flowing jets and wakes. *J. Fluid Mech.*, **88**, 451–63.

Perry, A. E. & Tan, D. K. M. 1984. Simple three-dimensional motions in coflowing jets and wakes. *J. Fluid Mech.*, **141**, 197–231.

Perry, A. E., Chong, M. S. & Lim, T. T. 1982. The vortex shedding process behind two-dimensional bluff bodies. *J. Fluid Mech.*, **116**, 575–578.

Perry, A. E., Lim, T. T. & Chong, M. S. 1979. Critical point theory and its application to coherent structures and the vortex shedding process. *Report FM-11*, Mechanical Engr. Dept., University of Melbourne.

Perry, A. E., Lim, T. T. & Chong, M. S. 1980. The instantaneous velocity fields of coherent structures in coflowing jets and wakes. *J. Fluid Mech.*, **101**, 33–51.

Perry, A. E., Schofield, W. H. & Joubert, P. N. 1969. Rough wall turbulent boundary layers. *J. Fluid Mech.*, **37**, 383–413.

Prandtl, L. & Tietjens, O. G. 1934. *Applied Hydro- & Aeromechanics.* Dover.

Robinson, S. K. 1991. Coherent motions in the turbulent boundary layer. *Ann. Rev. Fluid Mech.*, **23**, 601–639.

Shapiro, A. H. & Bergman, R. 1962. Experiments performed under the direction of L. Prandtl, (Gottingen). FM-11 Film loop. National Committee for Fluid Mechanics.

Soria, J. & Cantwell, B. J. 1994. Topological visualisation of focal structures in free shear flows. *Appl. Scientific Research*, **53**, 375–386.

Truesdell, C. 1954. *The Kinematics of Vorticity.* Indiana University Press.

Vollmers, H. 1983. Separation and vortical-type flow around a prolate spheroid. Evolution of relevant parameters. AGARD Symposium on Aerodyn. of Vortical Type Flow in Three-Dimensions, Rotterdam, *AGARD-CP342*, 14.1–14.14.

CHAPTER 2

HYDROGEN BUBBLE VISUALIZATION

C. R. Smith, C. V. Seal, T. J. Praisner and D. R. Sabatino[a]

2.1 Introduction

During the past 40 years hydrogen bubble visualization has greatly facilitated the fundamental understanding of a wide variety of fluid dynamic phenomena, including boundary layers, turbulence, aerodynamics, separated flows, and wakes, to mention but a few. For example, much of our appreciation for the flow structure of turbulent boundary layers can be attributed to the initial examination and discoveries made using hydrogen bubble visualization (Kline *et al.*, 1967; Kim *et al.*, 1971). As an illustration of a hydrogen bubble visualization, Fig. 2.1 shows the characteristic low-speed streak pattern that Kline and his colleagues discovered to be the dominant flow pattern occurring adjacent to the wall beneath a turbulent boundary layer.

Hydrogen bubble visualization provides a relatively simple and low-cost flow visualization technique which consists of creating material sheets and time lines of very small hydrogen bubbles by a process of electrolysis in a water flow. When properly illuminated, these material sheets/lines allow detailed visualization of the flow field (Schraub *et al.*, 1965; Kline *et al.*, 1967; Lu & Smith, 1985). The hydrogen bubbles are generated using a very fine (25–50 μm) conductive wire stretched between two conductive supports as one electrode of a DC electrolysis circuit. The other terminal of the electrical circuit is usually a metal or carbon electrode located elsewhere in the water flow. By establishing the wire as the *negative* electrode, small hydrogen bubbles form on the wire, and are subsequently swept off by the flow and carried downstream, which enables the visualization. The bubbles are typically of the order one half to one wire diam-

---

[a]Dept. of Mechanical Engineering and Mechanics, Lehigh University, Bethlehem, PA 18015, U. S. A.

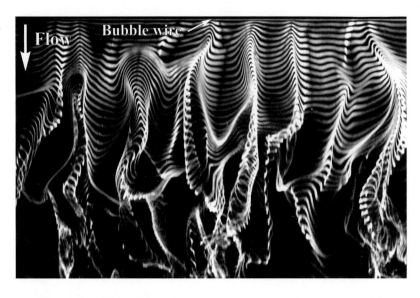

Fig. 2.1. Turbulent low-speed streak pattern visualized by horizontal bubble time-lines in the near-wall region of a turbulent boundary layer. Time-line generation frequency of 30 Hz. $Re_x = 2.2 \times 10^5$ and $Re_{\delta*} = 746$, with wire at $y^+ = 5$. From Praisner & Sabatino, private communication.

eter so that the rise rate of the bubbles is essentially negligible compared to the local velocity. Note that using the wire as a *positive* electrode will result in the generation of oxygen bubbles from the wire. Generally, the generation of oxygen bubbles is undesirable since the molecular structure of water will yield a bubble generation rate that is only one-half that of hydrogen. Additionally, oxygen gas seems to form larger bubbles than hydrogen for the same diameter wire, which creates grainy appearing images, as well as buoyancy problems.

One of the advantages of hydrogen bubble visualization is its versatility. Hydrogen bubble probes can normally be located essentially anywhere in a flowfield and in any orientation, with negligible to limited interference with the local flow. This versatility allows more varied and creative visualizations of a given flow than are possible with other techniques. Another advantage is simplicity and cost effectiveness. A hydrogen bubble visualization system can be constructed from off-the-shelf components at modest cost. And its simplicity of operation allows extensive visualization studies to be done relatively quickly and with minimal set-up time. Additionally, hydrogen bubble visualization can produce quantitative data (Schraub *et al.*, 1965; Lu & Smith, 1985) using a pulsed voltage to generate time lines of bubbles, coupled with image capture/processing.

The primary limitation of hydrogen bubble visualization is that it is only effective for relatively low Reynolds number water flows. Also, the hydrogen bubbles tend to dissipate within a moderate distance downstream of the probe, limiting the region that can be effectively visualized.

While hydrogen bubble visualization can provide spectacular and insightful visualizations of complex flows, it can also be a rather frustrating technique to employ in practice. This is due to both the relatively high degree of trial and error involved in effectively employing this type of visualization, and the fragility of the small diameter wires employed. However, once familiar with the use of the technique, it can be a powerful tool for exploring the manifold complexities of fluid behavior.

2.2 The Hydrogen Bubble Generating System

Schraub *et al.* (1965) provide a detailed description of the hydrogen bubble technique, including details for configuring an appropriate power supply, techniques for obtaining quantitative data, and uncertainty analyses. The present chapter reviews some of the basic details for implementing a system, and then focuses on the employment of the technique, including some of the experimental art form required for effective applications.

The basic requirement for a bubble generation system is a variable voltage D.C. power supply, with a voltage range of at least 50-70 volts and a current capacity of approximately 1 amp. While Schraub *et al.* (1965) provide a schematic for an appropriate system, modern technology provides many alternatives for creative designs, and markedly more compact and powerful variable D.C. power supplies. Note that the longer the length of a generating bubble wire, or if multiple wires are operated simultaneously, the higher the voltage requirement for the power supply. For example, the system employed in our laboratory has a range of 0-300 volts at a maximum current of 2 amps, and is generally used to power single-wire probes with wires 150-250 mm in length. Typical operating characteristics to obtain appropriate visualization is 150 volts at approximately 0.5 to 1 amp, depending on the wire length and diameter, and the amount of electrolyte dissolved in the water (see below). Note that other, more powerful systems, with capabilities of 0-250 volts and up to 8 amps have been employed successfully for operation of multiple wire probe "rakes" (Magness *et al.*, 1990).

When connected directly to an appropriate bubble wire probe (see Section 2.3), the current flow from the DC power supply will stimulate a steady

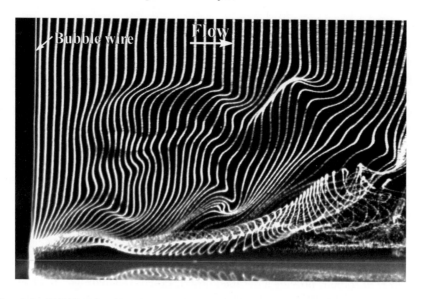

Fig. 2.2. Bubble time-line visualization normal to a surface using a vertical bubble wire (with a hair support) within a turbulent bundary layer. Time-line generation frequency of 30 Hz. $Re_x = 2.2 \times 10^5$ and $Re_{\delta*} = 746$, with wire at $y^+ = 5$. From Praisner & Sabatino, private communication.

electrolytic process at the generating wire, which will result in the production of a continuous sheet of hydrogen bubbles. The motion and deformation of this generated sheet of bubbles will act as a material sheet, moving and deforming with the corresponding motion and deformation of the local fluid behavior. Often, observation of this material sheet deformation alone is sufficient to assess the local fluid behavior. However, the bubble generation process can also be periodically interrupted, or "pulsed," to create a series of "time lines" of hydrogen bubbles which allow either the qualitative or quantitative assessment of local velocity behavior. An example of such bubble time lines is shown in Fig. 2.2, illustrating the flow behavior in a turbulent boundary layer. This pulsing process is achieved by using a square wave generator to gate the voltage signal from the DC power supply via a power MOS transistor. Depending on the characteristics of the square-wave generator employed, both the frequency of time line generation and the duty cycle (that is, the portion of the cycle that bubbles are actually generated) can be controlled. The resultant time lines of bubbles will appear as a segmented material sheet, with the spacing between the time lines proportional to the local velocity. These time line visualization patterns can thus be used to qualitatively assess local velocity behavior, with quantitative velocity behavior obtainable by careful use of image acquisition and processing (for example, Paxson & Smith, 1983; Lu & Smith, 1985, 1991). However, it is important to realize that the bubbles generated in the wake of a bubble wire move slightly slower than the local fluid velocity due to the wake defect of the wire. Lu & Smith (1991) describe a correction method to quantitatively account for this wake defect effect.

Since electrolysis generates the hydrogen bubbles, the attainment of high quality hydrogen bubble visualization depends on the presence of a sufficiently conductive solution to achieve electrolysis at a minimum applied voltage. Generally, plain tap water does not contain sufficient dissolved electrolytes to facilitate a good electrolytic process, and requires the addition of salts to the water to establish an effective electrolyte solution. We have found that our laboratory tap water requires the addition of 0.12 grams of sodium sulfate per liter of water to create an electrolyte concentration which facilitates effective hydrogen bubble visualization. Although addition of other electrolytes (including table salt) is possible, these do not perform as effectively as sodium sulfate. The addition of a small amount of hydrochloric acid is another method for promoting an electrolytic solution, but the acid solution can rapidly degrade the bubble probes, causing them to fail more frequently.

Note that establishing the appropriate concentration of sodium sulfate or other electrolyte requires some trial and error for a particular water source. When the electrolyte concentration is too low, the generated bubble sheet will be more diffuse, and a higher voltage setting will be necessary to achieve adequate bubble concentration. In contrast, if the electrolyte concentration is too high, bubbles will generate at lower voltage levels, but will often form larger diameter bubbles, which can create buoyancy problems. Additionally, the higher electrolyte level can precipitate corrosion problems with the exposed metals comprising the bubble probe and within the flow channel. It should be apparent from the above discussion that the characteristics of a hydrogen bubble generating system are not particularly critical in order to achieve effective visualization.

It should also be apparent that an electrical system to effectively power this electrolytic visualization process is quite powerful and potentially dangerous, due to the presence of high voltage/current in a conducting electrolytic medium. One must be extremely careful when employing hydrogen bubble generating systems, since contact with the wire probe, the positive electrode, or the water flow can result in a potentially life threatening electric shock.

2.3 Bubble Probes

One of the advantages of hydrogen bubble visualization is the capacity for generating hydrogen bubbles almost anywhere within the flow field via various types of positioning probes. An initial comment regarding the construction of hydrogen bubble probes is that one should not be discouraged if the first probes constructed do not perform effectively. One of the unfortunate truisms of this type of visualization is that the bubble wire probes often fail (normally the generating wire breaks, not the main probe structure), so that one generally has ample opportunity to practice and improve their probe construction skills.

The particular design of a probe will be a function of the flow geometry being examined and the aspect of the flow one wishes to visualize. For example, if the behavior adjacent to a surface is to be examined, such as flow in the wall region of a turbulent boundary layer, it would be appropriate to employ a horizontal wire that can be located parallel to the surface and traversed both vertically and laterally. However, if one wants to assess flow or velocity behavior normal to a surface, then a vertical-wire probe is the appropriate choice. Fig. 2.3 shows examples of generic designs for both horizontal and vertical-wire probes.

Generally, hydrogen bubble probes consist of a fine conductive wire (usually 25–50 μm platinum) strung taut between two metal, conducting supports (for example, brass rod/tubing is an effective probe construction material). It is important that the wire be under tension between the supports to remove any slack, and is free of any kinks[b] such that a clean, flat sheet of bubbles, free of initial distortions, is generated. However, too much tension on the wire will result in accelerated, and sometimes immediate, failure of the wire. Determining the correct amount of tension is one of the more "artistic" aspects of probe construction and usually requires trial and error to develop an appreciation for the proper tension and the appropriate construction method. One method is to first solder an initial end of the wire to the tip of one wire support. The wire is then held under gentle tension (to keep the wire taut) using the fingers of one hand, and soldered to the second wire support using the opposing hand. An alternative method is to solder one end of the wire to the tip of the first wire support, gently bend the second wire support inward toward the first wire support, lay the wire over the tip of second support (with as little slack as possible), and solder the wire in place. Releasing the wire supports then places

---

[b]Because the wire is usually spooled, it will sometimes tend to form kinks, not unlike a garden hose. Such kinks can be problematic when studying small-scale flow phenomenon such as near-wall boundary layer flow because they usually provide sites for larger bubbled to form.

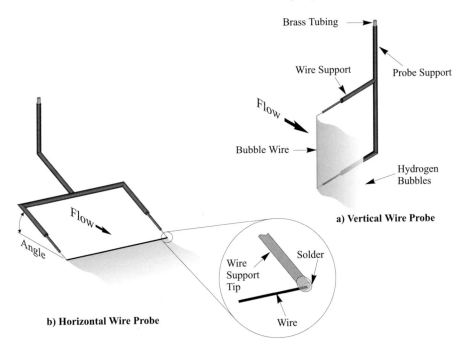

Fig. 2.3. Generic examples of hydrogen bubble probe designs. (a) Vertical wire probe configuration; (b) Horizontal wire probe configuration.

the wire in tension. Note that if the wire needs to be located very close to a solid surface, such as was required for the image shown in Fig. 2.1, then care needs to be taken to assure that the wire is soldered as closely to the tips of the wire supports as possible.

Regarding effective wire materials, platinum, steel, stainless steel, aluminum, and tungsten have all been used, with varying degrees of effectiveness (Schraub *et al.*, 1965). As a result of the electrolyte added to facilitate the electrical conductivity of the water, oxidation is a problem with both steel and aluminum, which results in rapid failure (although aluminum wire gives a very nice bubble quality while it lasts). Stainless steel and tungsten are strong, but yield generally poor quality bubbles, with tungsten also being particularly expensive. The best compromise for bubble quality, strength, and price is platinum. Being a noble metal, platinum will not corrode or react, can be soldered very effectively, and has appropriate conductivity to generate an effective, bubble-generating electric field.

To assure that hydrogen bubbles form only on the wire, the rod or tubing comprising the probe supports, and the soldered connections must be electrically insulated (in the absence of electrical insulation, electrolysis will cause bubbles to form on all exposed, conducting probe surfaces). This insulation process has generally been accomplished using both shrink-fit insulating tubing, and commercially available liquid tape. The shrink-fit tubing works well for the majority of the tubing comprising the wire supports, and must be applied before soldering the wire in place. The liquid tape is required at the tips of the wire supports, where the elevated temperatures created during the soldering process can cause shrink fit tubing to melt. Typically, liquid tape is used to insulate the tips of the probe (and the initial portions of the generating wire, if appropriate). However, skill is required to assure an adequate insulating layer of the coating, without creating a large, obtrusive build-up of material on the probe tips.

An important parameter in probe design is the distance between the wire supports. Because the wire supports are often subject to vortex shedding, the spacing between the supports should be sufficient to prevent the shedding from influencing the flow of interest in the central region of the bubble wire. However, as the spacing between wire supports increases, the wire-support structure will become more delicate, and prone to wire breakage due to induced probe vibration caused by vortex shedding from the supports. Wide spacing of the wire supports normally necessitates use of larger tubing (for structural rigidity), which can exacerbate the shedding problem. A longer active wire length also increases the potential for wire breakage, as well as makes the maintenance of a uniform bubble sheet more difficult, due to the lengthwise diminution of the electrolytically active electric field around the bubble wire. A successful probe will provide a balance for these conflicting spatial considerations.

Another important factor to consider when constructing a horizontal wire probe (Fig. 2.3b) is the angle between the wire supports of the probe and the surface. Since such probes are employed predominantly for plan view visualization, if the angle is too large, the cross member of the probe can interfere with the line of sight to the bubble sheet. If the angle is too small, the cross member may be below the water level of a channel flow, which can result in additional vortex shedding from the cross member, creating further structural and stability problems with the probe.

When configuring a bubble probe, it is important to consider the scale of the phenomenon to be visualized so that the wire may be sized to minimize induced flow disturbances due to either the wire or the associated probe supports. For example, turbulent boundary layer flows usually require use of approximately 25 μm wire, since a sheet of very small bubbles is necessary for visualization of the generally small scales often associated with such flows. However, the smaller the wire diameter, the more fragile the bubble wire probe, and the greater potential for wire breakage. For larger-scale flows it may be more desirable to use larger diameter wire, which is stronger and more durable (the tensile strength is proportional to the diameter squared). However, the size of the generated bubbles is directly proportional to the diameter of the bubble wire employed, and the smaller the generated bubbles, the better the perceived quality of the visualization. Generally, wire diameters greater than 50 μm will provide markedly poorer quality visualization, creating bubble sheets that are grainy in appearance and subject to significant buoyancy effects.

One characteristic problem with the electrolytic process used to generate the bubble sheets is that the charged electrodes also attract other dissolved ions in the water flow. This electrical attraction causes the sustained build-up of foreign material on the surface of the wire, which results in a corresponding degeneration of the visualization process, generally characterized by the formation of larger, more buoyant bubbles. When this material build-up on the wire becomes obtrusive, the wire must be "cleaned." The most effective cleaning method is a momentary reversal the electrolytic polarity by incorporating a polarity switch into the power supply circuitry. Reversing the polarity of the power supply for approximately 2-3 seconds normally facilitates cleaning of the wire. Since a reversal of polarity with the system under power will create an electrical surge, this polarity switching can only be performed during bubble generation if the operating voltage is below approximately 50 volts. Otherwise, the system will blow a fuse in the power supply or trip a circuit breaker in conventional electrical systems. If the operating voltage is above 50 volts (which is generally the case), a reversal of the electrolytic polarity requires that the operating voltage first be manually reduced below 50 volts. The polarity can then be safely reversed to "clean" the wire, followed by the subsequent return of the operating voltage to the original polarity and level. In any event, when the polarity is reversed, the electric field in the wire is reversed which drives off the charged material adhering to the wire. Normally this cleaning process can be accomplished within 6-10 seconds. Once cleaned, returning the electrolytic circuit to the original polarity and voltage level typically restores the original visualization quality of the generated bubble sheets, with the best visualization quality being obtained immediately following the cleaning of the wire.

2.4   Lighting

Although hydrogen bubbles can generally be observed with the naked eye, proper illumination is required to create clear, definitive visualization images that can be photographically recorded and analyzed. Portable high-wattage (1000 W works well) photographic lamps, which are available through most photographic supply stores, have proven to be effective and economical light sources. Slide projectors are also very effective as general light sources. Normally, hydrogen bubbles are most effectively illuminated using angled back-lighting of the bubble sheet. The brightness of the illuminated bubbles is a function of the angle formed between the axis of illumination and the line-of-sight of the camera. As a guideline, Schraub *et al.* (1965) recommend an angle of about 115°. However, our experience has been that determination of the optimum lighting angle is somewhat of an art form, and depends on the test section being viewed, the illumination and visual access, the visual background, and the type of image recording system, to name only a few of the contributing factors. Optimization of the illumination again requires significant trial and error. However, a rule of thumb is that when photographing from the side or at shallow oblique angles under general lighting, illumination from the bottom is normally most effective. When photographing from steep oblique angles or in plan view, illumination from the side or normal to the viewing direction is generally more effective.

Note that whatever the viewing direction, the presence of a sharply contrasting background is important to assure appropriate image quality. Normally, high image contrast is obtained by painting background surfaces matte black, if possible, or by configuring temporary black backgrounds (usually black pasteboard) behind the area to be visualized. However, there is often a trade-off between establishing an appropriately contrasting background and providing adequate transparent access for the illumination source. Establishing the appropriate background often requires another series of compromises of viewing and illumination angles in order to optimize a visualization image.

It should go without saying that the quality of a visualization greatly depends on maintaining clean and clear water, which generally necessitates a combination of continuous filtering (with 5 μm or smaller filters) and chlorination. In fact, the chlorine used to prevent organic growth in the water often has a positive impact on bubble quality since it helps facilitate the ion concentration in the water. Generally, we have found that 0.3 ppm chlorine works well in controlling algae growth. Note that development of any degree of algae or dirt in the water will rapidly render a "cloudy" appearance to visualization images, greatly reduce image sharpness and contrast, and exacerbate the coating of the wire by foreign materials, as discussed in Section 2.3.

Besides general illumination via photographic lamps, it may be desirable to selectively illuminate parts of the bubble sheet. For example, relatively thin light sheets (useful for illuminating cross-sections of a bubble sheet) can be created by selectively masking the containment surfaces of a water channel. However, optical properties and illumination characteristics generally limit light sheets generated by such masking processes to a minimum thickness of about 5 mm (particularly when using inexpensive, unfocussed light sources). Such finite thickness light sheets yield visualization cross-sections for which the images are integrated across the light sheet thickness, such that the images are often ill defined. If a well-defined, visualization cross-section is desired, a laser, in conjunction with appropriate generating optics such as a cylindrical lens or scanning mirror, can be used to create thin light sheets on the order of 1 mm in thickness. Laser illumination offers a number of interesting lighting possibilities. For example, a few appropriately-oriented mirrors can be employed to create a horizontal laser light sheet which can be accurately positioned to visualize behavior adjacent to a flat surface, such as for turbulent boundary layer flows.

2.5   Unique Applications

As previously mentioned, a unique aspect of hydrogen bubble visualization is the capability of positioning a bubble wire at selected locations and in specific orientations within a flowfield. This can often be accomplished using standard vertical or horizontal probes mounted in conventional traversing mechanisms. However, more unique and novel approaches may often be warranted to facilitate the desired visualization. For example, orienting a hydrogen bubble wire normal to a surface and employing pulsing of the voltage to the wire creates "time-lines" of hydrogen bubbles which reveal the behavior of the streamwise velocity profiles adjacent to a surface. However, the use of a conventionally constructed vertical probe of the type described in Section 2.3 and illustrated in Fig. 2.3a will introduce unacceptably large disturbances adjacent to a surface due to the presence of the lower wire support. To circumvent this problem, Lu & Smith (1985) utilized a vertical wire which was anchored to an upper wire support, passed down through a 0.8 mm hole in a flat plate, and was secured to the opposite side of the plate. One problem encountered using this wire mounting technique is that a resident bubble will often form at the juncture where the wire passes through the plate, due to the gas generated within the hole in the

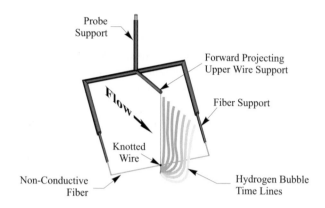

Fig. 2.4. Vertical bubble wire probe employing non-conductive fiber for lower support.

plate. The presence of this bubble effectively acts as a small, but continually increasing, "bump" on the surface, creating a local wake disturbance. However, periodic removal of the bubble using a soft paintbrush will temporarily restore the visualization quality at the surface. An additional drawback with mounting the wire through the surface is that the wire is fixed at a particular location.

Another way to visualize streamwise velocity profiles above a flat plate is to modify the lower wire support to be as non-obtrusive as possible. Fig. 2.4 is an example of how this has been done effectively using a non-conductive fiber as a wire support member. Using this approach, a very fine, transverse non-conductive fiber is connected under tension between parallel, vertical supports (generally by gluing the fiber to the tips). The vertical bubble wire is then tied very carefully around the center of the fiber, placed in moderate tension, and soldered to the upper wire support. A portion of wire is left extending below the fiber, which is carefully manipulated to be contiguous with the axis of the vertical wire above the fiber, and trimmed to 5-10 mm in length. Due to the low drag on the wire near the surface, the extended end of the wire will remain essentially in line with the vertical wire, and allow the end of the wire to be traversed until it is in contact with a solid boundary. Thus, visualization of velocity profiles can be done all the way to the surface. Fig. 2.2 is an example of such a vertical visualization of velocity profiles in a turbulent boundary layer using a fiber support probe similar to that illustrated in Fig. 2.4. The limitations of this type

---

[c] Note that due to the small wire diameter and malleable nature of platinum, such a process of wire-tying works much more effectively than adhesives or other methods.

of probe are that the fiber will create a minimal wake in the visualization profile, a bubble may develop near the knotted portion (this can be periodically removed using a soft artist's paintbrush), and one must use a fiber that is sufficiently strong, and yet very thin. After some trial and error, it was found that a long, human hair provided the best source of a strong, thin, non-conductive fiber support.

One of the more interesting applications of hydrogen bubble visualization is for assessment of three-dimensional, out-of-plane motion by observation of a generated bubble sheet in either oblique or end-view. Oblique viewing can be done using either vertical or horizontal wires, with the wires being traversed through a sequence of positions to reveal the three-dimensional character of the flow (Acarlar & Smith, 1987; Fitzgerald *et al.*, 1991; Seal, 1997). Normally, oblique views can be effectively illuminated, viewed, and photographed, and are very useful for both developing models of flow processes or assessing local behavior to plan for the employment of more quantitative instrumentation, such as Laser Doppler Anemometry (LDA) or Particle Image Velocimetry (PIV). Fig. 2.5 is an excellent example of an oblique visualization sequence of the highly three-dimensional transitional breakdown of a laminar flow approaching a bluff body junction. This visualization, done with a horizontal hydrogen bubble probe located upstream of the plate-body junction, illustrates the development of very compact, discrete "hairpin-shaped" vortices caused by the three-dimensional destabilization and breakdown of the laminar boundary layer impinging on this junction region.

In contrast, end-on viewing of hydrogen bubbles under general lighting can be difficult, with images often having a very diffusive appearance since bubble sheet behavior is integrated over an extended distance. Thus, to achieve sharp images of the out-of-plane motion it is necessary to either view individually generated time lines of hydrogen bubbles (for example, Schwartz & Smith, 1983; Paxson & Smith, 1983), or illuminate only a portion of the bubble sheet using cross-stream illumination. This cross-stream light sheet can be created using an appropriate slit with general lighting, or a laser sheet (see Section 2.4). The laser sheet lighting is quite effective, and can clearly illustrate the cross-stream deformations of an impinging sheet. Fig. 2.6 is an excellent example of an end view visualization using hydrogen bubbles for examination of the trailing vortex from a delta wing by Magness *et al.* (1990). This study utilized end views (via a downstream mirror) of multiple bubble sheets generated by a parallel grid of bubble wires, illuminated with a cross-stream laser sheet. In a further extension of the use of multiple bubble wires, Grass *et al.* (1993) employed a crossed grid of

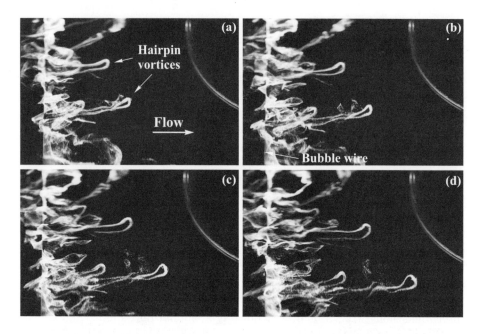

Fig. 2.5. Temporal sequence of hydrogen bubble images within a transitional juncture flow illustrating the generation of hairpin vortices. Time betwen images is 0.2 s. $Re_x = 2 \times 10^5$ and $Re_{\delta*} = 784$. From Praisner & Sabatino, private communication.

wires (132 nodes) to not only visualize the flow, but to extract three-dimensional velocity field results using digital analysis of stereoscopic images. This remarkable feat helped to shed significant light on the near-wall flow structure of a turbulent boundary layer.

Note that whether using single time lines or cross-stream light sheet illumination, the degree of cross-stream deformation will be dependent on how far removed either the bubble time line or the light sheet is from the generating wire. Thus, any end-view image will reflect the *integrated* effect of the local flow field on the bubble sheet during transit to the point of imaging, and cannot be construed as the instantaneous behavior of the bubble line or sheet.

An additional caution when employing multiple wires in close proximity to each other is that the electric field surrounding each of the wires will stimulate sympathetic current flow in the other adjacent wires. During the simultaneous operation of all the wires, this generally is unnoticeable. However, if one or more wire is made inactive, while the others remain in operation, the induced

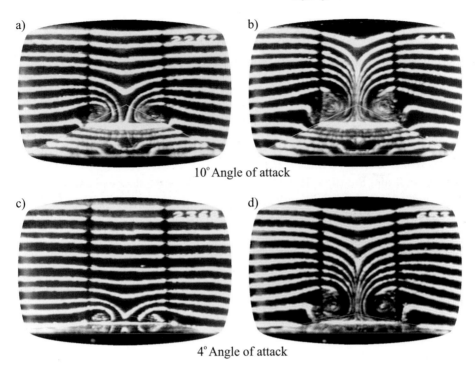

10° Angle of attack

4° Angle of attack

Fig. 2.6. End view of bubble sheets generated by a wire grid upstream of a pitching delta wing. Plane normal to the viewing direction is illuminated by a laser sheet. $Re_c = 3.8 \times 10^4$. (From Magness *et al.*, 1990).

sympathetic current flow can cause the supposedly inactive wires to produce bubbles, which may interfere with the desired visualization.

2.6 References

Acarlar, M.S. and Smith, C.R. 1987. A study of hairpin vortices in a laminar boundary layer. Part I. Hairpin vortices generated by a hemisphere protuberance. *J. Fluid Mech.*, **175**, 1–42.

Fitzgerald, J.P., Greco, J.J., and Smith, C.R. 1991. Cylinder end-wall vortex dynamics. *Phys. of Fluids A*, **3** (9), 2031.

Grass, A.J., Stuart, R.J., and Mansour-Tehrani, M. 1993. Common vortical structure of turbulent flows over smooth and rough boundaries. *AIAA Journal*,

3 (5), 837–847.

Kim, H.T., Kline, S.J. and Reynolds, W.C. 1971. The production of turbulence near a smooth wall. *Journal of Fluid Mechanics*, **50**, 133–160.

Kline, S.J., Reynolds, W.C., Schraub, F.A. and Runstadler, P.W. 1967. The structures of turbulent boundary layers. *Journal of Fluid Mechanics*, **95**, 741–773.

Lu, L.J. and Smith, C.R. 1985. Image processing of hydrogen bubble flow visualization for determination of turbulence statistics and bursting characteristics. *Expts. in Fluids*, **3**, 349–356.

Lu, L.J. and Smith, C.R. 1991. Use of quantitative flow visualization data for examination of spatial-temporal velocity and bursting characteristics in a turbulent boundary layer. *J. Fluid Mech.*, **232**, 303–340.

Magness, C., Utsch, T., and Rockwell, D. 1990. Flow visualization via laser-induced reflection from bubble sheets. *AIAA Journal*, **28** (7), 1199–1200.

Paxson, R.D. and Smith, C.R. 1983. A technique for evaluation of three-dimensional behavior in turbulent boundary layers using computer augmented hydrogen bubble-wire flow visualization. *Expts. in Fluids*, **1**, 43–49.

Schraub, F. A., Kline, S. J., Henry, J., Runstadler, P. W., and Little, A. 1965. Use of hydrogen bubbles for quantitative determination of time-dependent velocity fields in low-speed water flows. *Journal of Basic Engineering*, June.

Schwartz, S.P. and Smith, C.R. 1983. Observation of streamwise vortices in the near-wall region of a turbulent boundary layer. *Physics of Fluids*, **26** (3), 641–652.

Seal, C.V. 1997. *The Control of Junction Flows*. Ph.D. Dissertation, Dept. of ME/Mech., Lehigh University.

CHAPTER 3

DYE AND SMOKE VISUALIZATION

T. T. Lim[a]

3.1 Introduction

The observation of fluid motion using smoke and dye is one of the oldest visualization techniques in fluid mechanics, dating back to the time of Leonardo da Vinci. The technique is inexpensive and easy to implement. Above all, it offers significant insight into the phenomena occurring in complex fluid flows. In fact, some of the major discoveries in fluid phenomena were made using this simple technique. A classic example is the experiment by Osborne Reynolds in which dye injection method was used to show the transition from a laminar flow to turbulent flow in a pipe (Reynolds, 1883). A more recent example is the investigation by Head & Bandyophadyay (1981) in which smoke injection was used to show the existence of hairpin or Λ-shaped vortices in a turbulent boundary layer. This discovery would not have been possible with point-by-point measurements using hot-wire or laser Doppler anemometry. Of course, visual observations alone do not provide the complete answer regarding flow mechanisms, and observations need to be complemented with quantitative investigations so that the observed phenomena can be described quantitatively. With the recent advances in computer imaging technology, some flow visualization techniques can now provide quantitative results. A good example of this is the Particle Image Velocimetery (PIV) which has gained immense popularity in recent years as a measurement tool (see Chapter 6). Other visual methods which can provide limited quantitative data include the hydrogen bubble (Chapter 2) and smoke-wire techniques (this chapter).

Here, we describe smoke and dye visualization techniques which are often

[a]Dept. of Mechanical and Production Engineering, National University of Singapore, 10 Kent Ridge Crescent, SINGAPORE 119260

used to study fluid motion. The terms "smoke" and "dye" are used in a loose sense to include both $TiCl_4$ and electrolytic precipitation techniques since they also seed the flow with fumes or particles which help to mark the fluid motion. The focus of this chapter is on the practical application of these techniques, and not so much on the literature review of the subject, since many excellent reviews already exist (see, for example, Clayton & Massey, 1967; Werlé, 1973; Merzkirch, 1987a; Gad-el-Hak, 1988; Freymuth, 1993, and Mueller, 1996). To make the chapter reasonably self-contained, I have also included a brief discussion on photographic tools and techniques, which are inseparable parts of flow visualization techniques. The discussion is intended for reader who is new to photography, and contains some useful hints on how to produce high-quality flow visualization images.

3.2   Flow Visualization in Water

3.2.1   Conventional dye

Of all the flow visualization techniques, dye visualization is perhaps the easiest to carry out. Most often, food dye is used because it is safe to handle, and is available in most supermarkets. Although the choice of color is a matter of personal preference, it has been found that reds, blues and greens generally produce better picture contrast than the rest.

In general, food dye sold in the supermarket comes in a concentrated solution, and it has a specific gravity greater than one. Unless it is made neutrally bouyant, the dye would not follow the flowfield as intended, and this can lead to serious misinterpretation of flow visualization results. To make the dye neutrally bouyant, a small quantity of alcohol, such as methanol or ethanol, is normally added to the dye solution. The exact amount of alcohol needed entails some degree of trial and error because commercial grade alcohols do not all come with the same degree of purity. Once the dye/alcohol mixture is neutrally buoyant, it is then diluted with the operating fluid from the tunnel. This practice is to ensure that the temperature difference between the dye/alcohol mixture and the working fluid is kept to a minimum, because a large temperature different can lead to other undesirable buoyancy effects. The extent to which the dye should be diluted depends very much on the application, and requires some degree of personal judgement, but a concentrated dye may obscure the salient features of the flow, while a "thin" dye may lead to poor contrast in the flow image.

3.2.2 Laundry brightener

This household product has proven to be an excellent tracer in water, and it has been widely used by the research group at the University of Melbourne (see Lim & Nickels, 1992; Kelso *et al.*, 1996; Lim, 1997). The liquid has a specific gravity surprisingly close to 1, and in most cases does not require the addition of alcohol to make it neutrally bouyant. It is marketed in Australia under its commercial name, Reckitt's Bluo. For optimal results, it should be diluted with an equal amount of operating water. An added advantage of this liquid is that it does not contaminate the water as quickly as the traditional food dyes, thus extending the useful running time of the experiment. Unfortunately, it is only available in one color, blue.

3.2.3 Milk

Milk is another indicator which is often used to visualize liquid flows in much the same way as smoke is used to visualize gaseous flows. It is preferred by many because of its high reflective properties which help to improve the contrast of flow images. Although it is normally used in its natural white state, food dye is sometimes added to it in applications where differentiating various parts of the flowfield is important. Another reason cited for choosing a milk/dye mixture over a dye solution alone is because the fat content in the milk helps to retard the diffusion of the dye. This has been found to be beneficial in studies involving high shear flows where the dye diffuses easily. However, the milk must be completely flushed from the injection system at the end of the experiment: any milk left in the system may curdle, and block the injection ports or slots. Moreover, aggregates of curdled milk may enter the operating fluid and degrade the flow quality.

3.2.4 Fluorescent dye

Under normal lighting, a dilute fluorescent dye solution appears almost transparent, but when illuminated with a laser light source of an appropriate wavelength the dye fluoresces strongly. This unique property has attracted many to use it to visualize the internal structures of fluid flow by illuminating it with a thin sheet of laser light. Some of the common fluorescent dyes used in flow visualization include Fluorescein, Rhodamine-B and Rhodamine-6G. When illuminated with the output of an argon ion laser, fluorescein displays a green color, while Rhodamine-B and Rhodamine-6G give dark red and yellow colors, respectively.

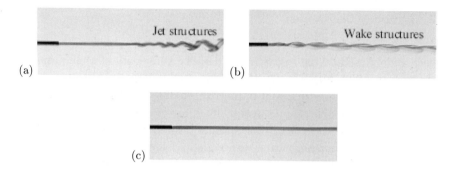

Fig. 3.1. Dye injection using a right-angle probe. The flow is from left to right. (a) Jet structure caused by the probe's exit velocity higher than the freestream ; (b) Wake structure caused by the exit velocity lower than the freestream; (c) Smooth dye filament indicating a correct exit velocity.

3.2.5   Methods of dye injection

There are various methods of introducing dye into the flow. Most commonly, it is released through a dye probe which is usually fabricated using either a hypodermic needle, or stainless steel tubing of 1.5 to 2.0 mm in diameter. The advantage of this technique is that the probe can be moved easily within the flow, and the dye can be released at the location of interest. However, its greatest drawback is the disturbance that the probe may cause to the flow field. To minimize this effect, the probe is often located some distance upstream from the point of observation. The dye is usually supplied to the probe by either a gravity-feed, or a pressurized reservoir. Although a gravity-feed system is easier to implement, a pressurized reservoir can provide a more consistent flow rate, and does not depend on the height of the dye level. In either case, the dye exit velocity must be equal to the local flow velocity in order to minimize the disturbance to the flow field. When the exit velocity is too high, a jet flow is produced, which generates "mushroom-like" structures (see Fig. 3.1). Similarly, when the exit velocity is too low, wake structures are formed, which appear as a series of interconnecting vortex loops. When a correct exit velocity is reached, the dye should appear as a smooth filament.

Another common technique of releasing dye into the flow is through dye ports or slits which are usually fabricated as part of the model (for example, see Fig. 3.2). When deciding the locations of the ports or slits, one must bear in mind that dye-lines are streaklines, and streaklines only display a spatially integrated

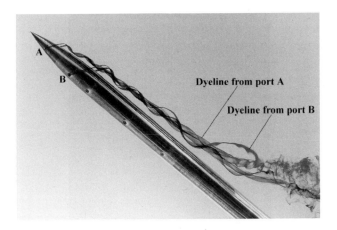

Fig. 3.2. Picture showing dye lines of flow past a tangent ogive cylinder at high angle of attack. The flow is from left to right. Note the dependence of the streakline pattern on the location where the dye is released (Luo *et al.*, 1998). Figure also shown as Color Plate 1.

view of the flow structures. This is because dye distorts as it travels downstream. Accordingly, the streakline pattern seen at some distance downstream of a test-model is a result of the accumulated distortion which can be traced all the way back to the point of release. In other words, the streakline pattern at a given location is a function of the location where the dye is released. This behavior of dye is clearly demonstrated in Fig. 3.2 where the flow past a tangent ogive cylinder at high angle of attack is shown. Here, it is obvious that the dye released from port A follows a different path from that released from port B, even though the two ports are physically quite close to each other. Also, when releasing dye through dye ports, one must ensure that its exit velocity is kept to a minimum since a large exit velocity can significantly alter the flow behavior. Moreover, if dye is intended to mark vorticity, it must be released at the location where the vorticity is generated

Another method of releasing dye into the flow is to coat a test model with a concentrated solution of dye and alcohol. By allowing the alcohol to evaporate, a thin layer of saturated dye crystals is formed on the surface of the model. When the model is towed at a low speed, or placed in a test section with a slow moving stream, dye is washed from the surface of the model, allowing the flow structure to be observed. A slight variation of this technique is the dye-layer technique used by Gad-el-Hak (1986). Here, a concentrated alcohol/dye mixture

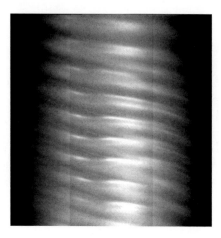

Fig. 3.3. Visualization of forced Taylor-Couette flow using rheoscopic fluid (Sinha & Smits, 1999).

is painted on cotton strings which are stretched on a rack. The size of the string is dictated by the freestream velocity, but it should be small enough so that the wakes behind the strings are laminar. When the rack is immersed in the tank and towed at low speed, the dye which is washed away from the strings forms several thin sheets. In the presence of a weak stable saline stratification, the dye layers remain queiscent until disturbed by a moving model. This technique is particularly suited to visualizing flow separation. The main drawback of the coating technique is its short running time, which is dictated by the amount of dye on the surface of the model or string.

3.2.6   Rheoscopic fluid

This fluid was invented by Paul Matisse originally for use in his artwork. It has since been used successfully to visualise Taylor-Couette flows (Fig. 3.3), Rayleigh-Bénard convections, and surface flow patterns. The fluid contains a suspension of microscopic crystalline guanine platelets with an average size of $6 \times 30 \times 0.07$ μm and a density of 1.62 gm/cm^3. Their slow settling velocity (approximately 0.1 cm/hour in undisturbed water) and high reflective index ($\propto 1.85$) makes them highly suitable for flow visualization applications (see Matisse & Gorman, 1984). With an appropriate choice of working fluid, such as glycerin/water mixtures and liquid perchloroethylenethe, the platelets can be suspended in the fluid for an even longer period of time. When the platelets

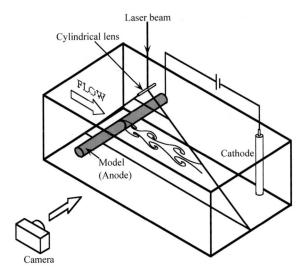

Fig. 3.4. Typical experimental setup of the electrolytic precipitation technique.

experience a shear, they align with the direction of the local shear stress, so that only those platelets with their "faces" oriented toward the observer will direct light to the observer, thus making them appearing as white, while those oriented in any other direction will appear darker. The fluid is non-toxic and non-reactive, and it can be purchased from the Kalliroscope Corporation in Massachusetts.

3.2.7 Electrolytic precipitation

This technique applies only in water and is best suited to applications where the velocity is low, typically from 0.1 to about 5 cm/s. The operating principle is based on the electrolysis of water, and the setup is similar to that used in the hydrogen bubble technique discussed in Chapter 2 except that instead fine insoluble metallic particles are produced at the anode. The particles are white in color with an average diameter of about $1 \mu m$, and when illuminated with a light source they appear like a white smoke. The technique is simple to implement and does not require the use of special metals or chemicals.

Figure 3.4 shows a typical experimental setup used in the electrolytic precipitation technique (see also Taneda *et al.*, 1977; Taneda, 1977). Most often, the test model is used as an anode, and the cathode is located at some distance downstream where it does not disturb the flow. Although metals such as copper, iron, lead, tin, and brass may be used as an anode, solder is found to produce the "best" smoke. In most cases, brass is used because of its rigidity, and a thin layer of solder is usually coated onto the brass surface to facilitate smoke generation. The extent of the solder covering the model depends very much on the application. In two-dimensional flow, a narrow strip of solder covering the body around the center-plane of the model is usually sufficient. In three-dimensional flow, a much larger area, possibly covering the whole model, may be required. In any case, those parts of the model which are not coated with solder must be insulated with a thin layer of non-conducting material such as epoxy resin, to prevent unwanted smoke generated from the metal in contact with the electrolyte.

As for the cathode, the material used does not matter appreciably, although brass is commonly used. However, the shape of the cathode, and in particular the surface area in contact with the electrolyte, is an important parameter which determines the quantity of the smoke. Some researchers have gone to the extent of using an aluminium honeycomb, such as that commonly used as a flow straightener in wind or water tunnels. The large surface area of the honeycomb has been found to improve the smoke quality considerably.

Other factors which affect the quality of smoke include the distance between the two electrodes, the electrolyte concentration, and the strength of the electric current. For best results, it is recommended that the distance between the anode and cathode should be less than 1 meter. If it is used in a towing tank, the distance between them should remain constant. As to the amount of additive electrolyte required, it depends to a great extent on the type of water used. If "hard" water is used, then additional electrolyte may not be needed. However, if the water is "soft," additional electrolyte is required to improve the electrolysis process. The most effective electrolyte for this purpose is sodium chloride (that is, common salt). If the anode is made of tin, then NH_4NO_3 can also be used. The quantity of electrolyte required depends on the amount of current used. In most cases, approximately 5 kilogram of table salt per cubic meter of water is enough to produce smoke of sufficient quality. This amount of salt has a negligible effect on the overall density of the water. Distilled water is not recommended in this application even with the addition of electrolyte.

In most applications, a voltage of about 10 V (DC) and a current of about 10 mA is sufficient. However, as the flow velocity increases, the current must also be increased in order to maintain an acceptable smoke quality. When the freestream velocity is above 5 cm/s, the quantity of smoke produced may not be

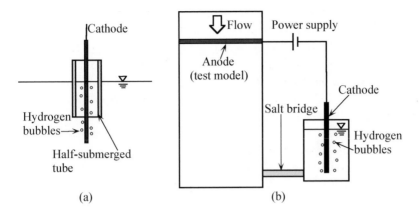

Fig. 3.5. Techniques to eliminate undesirable hydrogen bubbles from the cathode. (a) Half submerged tube technique used by Honji *et al.* (1980). (b) Salt bridge technique used by the author and his colleagues.

sufficient for flow visualization.

While this technique minimizes the disturbance to the flow field compared with the dye injection method, it has a number of disadvantages. One of them is the deterioration of the anode (that is, the model) with time due to chemical errosion, and the anode must be cleaned regularly with fine sand paper, or replaced. Another disadvantage is the generation of hydrogen bubbles at the cathode which can degrade the quality of the images. One way of addressing the problem is by confining the bubbles to rise through a half-submerged tube as Honji *et al.* (1980) have done. However, this method works well only in a horizontal water channel. For applications involving a vertical water tunnel, it is much better to use a "salt bridge" as shown in Fig. 3.5. A salt bridge is commonly used in electrochemical processes, and its primary function is to close the electrical circuit between two containers of solution (see, for example, Castellan, 1972; Atkins, 1982). It is usually made of a saturated salt, such as potassium chloride, in a gel form enclosed in a tube. A salt bridge can be made by heating agar-agar with water until it is completely dissolved. Salt is then added until the solution is saturated. The mixture is then transferred into a glass tube with its ends stuffed with cotton wool. By using a salt bridge, the hydrogen bubbles are confined in a container located outside the test-section. However, it requires substantially higher supply voltage in order to produce "smoke" of a sufficient quality.

Fig. 3.6. Kármán vortex street behind a circular cylinder at $Re = 140$ visualized using the electrolytic precipitation technique.(Taneda, 1982).

Figure 3.6 shows the well-known Kármán vortex street behind a circular cylinder visualized by Taneda (1982) using the electrolytic precipitation method.

3.3   Flow Visualization in Air

3.3.1   Smoke tunnel

One of the most important items of equipment in flow visualization involving air is a low turbulence wind tunnel. A well-designed smoke tunnel should have a turbulence level in the test-section, preferably, on the order of 0.02%. The most commonly used tunnel in flow visualization is a non-return or indraft suction type, where air is drawn through a large settling chamber consisting of a honeycomb and several screens, sometimes as many as 12, followed by a large contraction before the air enters the test-section. With this design, the smoke can be exhausted to the outside of the building. Figure 3.7 shows various components of a typical smoke tunnel. The purpose of the honeycomb is to breakup the large-scale air turbulence entering the tunnel, although some tunnels do not have them (for example, see Mueller, 1996). The function of the screens is to further reduce the turbulence level before entering the contraction section. For optimal results, the screens should be arranged in the order of decreasing

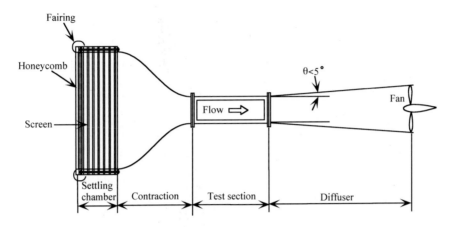

Fig. 3.7. Schematic of a typical smoke tunnel setup.

mesh size. The function of the contraction, apart from further decreasing the streamwise component of turbulence, is to ensure that the velocity profile at the entrance of the test-section is uniform (that is, a "top hat" profile). The contraction ratio, which is defined as the inlet to outlet area ratio, in most smoke tunnels is normally larger than those in conventional wind tunnels, and ranges from 9 to as high as 96. Of all the components making up the wind tunnel, the contraction section is perhaps the most crucial. Special attention must be paid to its design, as too sharp a contraction can cause the flow to separate, leading to poor quality flow, while too gentle a contraction can lead to an undesirable increase in the thickness of the wall boundary layer. The methods of designing a contraction vary considerably, and interested readers should consult the articles by Cohen & Ritchie (1962), Chmielewski (1974), Morel (1975), and Mikhail (1979). Comprehensive design rules pertaining to the design of wind tunnels are given by Mehta (1977), Mehta & Bradshaw (1979), and Rae & Pope (1984).

3.3.2   Smoke generator

Smoke for flow visualization purposes may be generated by burning tobacco, wood, and wheat straw, or by vaporizing hydrocarbon oils. Regardless of the source of the smoke used, it is essential that the smoke meets the following criteria:

1. It must be able to track the flowfield accurately. In other words, the smoke

particles must be sufficiently small so that their motion reflects the motion of the flow.

2. It must not significantly affect the flowfield under investigation.

3. It must posses high reflective properties.

4. It must be nontoxic.

In term of the reliable control of the quantity of smoke, vaporizing hydrocrabon oils is perhaps the best.

Most smoke generators are manufactured for the entertainment industry. A number of them can be used for flow visualization studies. A good example of this is the portable smoke generator manufactured by Symtron Systems Inc, in New Jersey. It is marketed as Smokemaster, and comes self-contained in a compact carrying case. It has a fast warm-up time of about 60 s from cold start and produces a high quantity of smoke with a particle size of about 0.5 μm. Most importantly, the manufacturer claims that the smoke is nontoxic, non-irritating, enviromentally safe and nonflammable under normal operating conditions.

A smoke generator such as that designed by F.N.M. Brown (1971) at the University of Notre Dame may also be built in-house. This particular generator operates by heating kerosene on flat electrical strip heaters, and the smoke produced is forced through the generator with a blower or by compressed air. Detailed information on its design can be found in Merzkirch (1987b) and Mueller (1996).

3.3.3   Smoke-wire technique

Here, smoke is produced by vaporizing oil from a fine wire heated by an electric current. The smoke-wire technique is similar to the hydrogen-bubble technique in water where hydrogen bubbles are produced from a fine wire by an electro-chemical process (see Chapter 2). The smoke-wire technique was originally developed to measure velocity profiles in boundary layers, and it has also been used successfully to visualize complex three-dimensional flows such as separation bubbles, flow structures in turbulent free shear flows and boundary layers, a jet in a cross-flow, as well as Kármán vortex street behind a circular cylinder (see for example, Bastedo & Mueller, 1986; Cimbala *et al.*, 1988; Fric & Roskho, 1994).

Compared to the more elaborate smoke generator discussed above, this technique is relatively inexpensive to implement. In principle, it requires only a fine metal wire, mineral oil and a power source. Most metals with sufficient strength and electrical resistivity can be used, but the three most commonly used wires are made of stainless steel, nichrome and tungsten. The size of the wire is dictated to some extent by the flow speed. For low speed applications, it is better to use a smaller diameter wire because a smaller wire produces smoke which is sharper. At higher speeds (a few meters per second), a larger diameter wire is recommended because its larger surface area can maintain a higher smoking rate, and since the wires are typically stretched taut, it will be able to accommodate the required tension at a higher temperature. Another factor which must be considered when deciding the wire size is the Reynolds number. To mimimize the flow disturbance, the Reynolds number based on the wire diameter should be less than 20. For most applications, the optimal size is about 0.1 mm in diameter.

There are a variety of oils which can be used to produce smoke filaments, including paraffin, kerosene, lubricating oil, silicon oil and model train oil. Paraffin is perhaps the most effective. To ensure that the smoke is produced uniformly along the length of the wire, it is essential that the wire is coated evenly with the oil. The coating may be applied by a gravity feed method, or manually with an applicator, or automatically with "wipers" or brushes. The gravity feed technique is easy to set up, but it is not as effective as the manual coating technique, and good results are seldom obtained. The manual technique allows better control of the oil thickness, but it is tedious and troublesome, and the test-section wall must be removed repeatedly to coat the wire.

To solve these problems, automatic oil coating systems have been designed. Among the various designs published in the literature, the one by Liu & Ng (1990) is the perhaps the best. A schematic of their design is shown in Fig. 3.8. The system is made up of two parts: a wire drive system and a control circuit (shown in Fig. 3.9). The wire drive system consists of a fine wire which traverses vertically through the center of the section. One end of the wire is connected to a pulley driven by a stepper motor, and the other end is attached to a weight to provide the necessary tension. Located outside the tunnel on its top and bottom surfaces are two small paint brushes connected to electro-magnetic solenoid actuators. During operation, the stepper motor turns first one way, say in the clockwise direction, to pull the wire down, while at the same time the upper solenoid is activated to push the brush containing mineral oil against the wire. This enables the wire to be coated evenly with the oil while it is traversing through the brush. The length of the wire pulled by the stepper motor is equal to the height of the test-section. Once a sufficient length of wire has been pulled

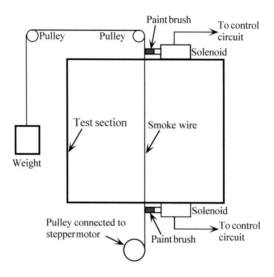

Fig. 3.8. Schematic of an automatic oil coating system used by Liu & Ng (1990).

through, the brush is retracted and the heating of the wire begins. The amount of smoke produced and its duration depends to a great extent on the voltage and current applied to the wire. A higher current vaporizes smoke quickly, but a lower current produces smoke which is too faint to photograph. For most applications, a 24V DC power supply with the current of about 0.5 to 0.8 amp for a stainless wire of 0.1 mm diameter is more than enough to produce smoke of sufficient quality. On the next run, the stepper motor rotates in the counter-clockwise direction while the bottom solenoid is actuated to coat the wire. The whole operations is fully controlled by the electronic circuit shown in Fig. 3.9. The function of each electronic component is discussed in details by Liu & Ng (1990). The advantage of this system is that the brushes are located outside the tunnel, and hence there is no disturbance to the flow in the test-section. However, the electronic circuit shown in Fig. 3.9 has a minor drawback because it can only control the movement and the heating of the wire. There is no provision for synchronization with the camera and the lighting. This shortcoming can be easily rectified by incorporating a couple of extra timers and a solenoid driver in the circuit. There are numerous alternative control circuits described in the literature to perform a similar function (see, for example, Torii, 1977; Nagib, 1977; and Batill & Mueller, 1981).

Although the system shown in Fig. 3.8 is designed for operating a vertical

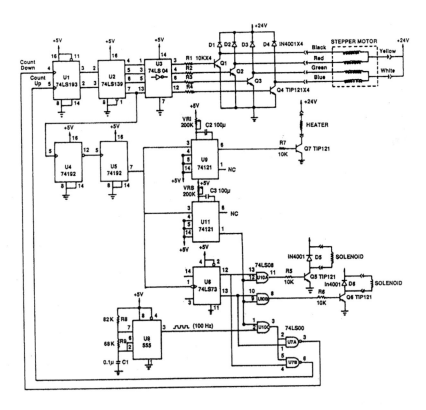

Fig. 3.9. Control circuit used by Liu & Ng (1990) to generate smoke from a wire.

wire, it can easily be adapted for a horizontal wire by rearranging the motor and the pulley system.

3.3.4 Titanium tetrachloride

The use of titanium tetrachloride ($TiCl_4$) to generate smoke or fumes for flow visualization in air can be traced to Simmons & Dewey (1931). Since then, it has been applied to the study of accelerating flow around an aerofoil (Freymuth, 1985), and vortex flows (Visser *et al.*, 1988), just to name two examples. The technique makes use of the fact that when $TiCl_4$ is exposed to moist air, it develops dense white hydrochloric acid fumes and minute particles of titanium dioxide according to the reaction:

$$TiCl_4 + 2H_2O = TiO_2 + 4HCl \qquad (3.1)$$

TiCl$_4$ is inexpensive and can be purchased commercially. However, hydrochloric acid fumes are toxic and can pose serious health hazard. Experiments should therefore be carried out in a well-ventilated environment, and if possible, the fumes should be exhausted to the outside of the laboratory. There are a number of methods of introducing the "smoke" into the tunnel. One is to apply the liquid on the surface of model to be tested using a small diameter pipette made of either stainless steel or brass. If the model is small enough, it can be immersed into the liquid. Due to its toxicity, all contact with eyes and skin should be avoided. In case of accident, wash with plenty of water. A comprehensive description of the correct method of using this technique is given by Freymuth *et al.* (1985).

A somewhat safer and convenient method of introducing the fumes into the wind-tunnel was devised by Visser *et al.* (1988), by taking advantage of the fact that $TiCl_4$ has a low vapor pressure and vaporises easily under standard atmospheric condition. In this technique, a pressurized inert gas such as nitrogen is passed through a chemical bottle containing the $TiCl_4$ liquid and forces vaporised $TiCl_4$ into the tunnel through a probe (see Fig. 3.10). When the vapor comes into contact with moist air in the tunnel, dense white fumes is produced. With this method, the user has minimum contact with the liquid, and the smoke can be released at any point in the wind-tunnel. In addition, the smoke can be turned on and off at will.

3.4　Photographic Equipment and Techniques

3.4.1　Lighting

Proper lighting is one of the most important aspects of photography. The two most common sources of lighting used in flow visualization are the *conventional light source*, and *the laser*. Conventional light sources include spot-lights, quartz-iodine lights, tungsten-halogen lamps, mercury lamps, an electronic flash and stroboscopic light, and they are used to visualize external features of the flow. In contrast, lasers are frequently used to visualize the internal structure of the flow.

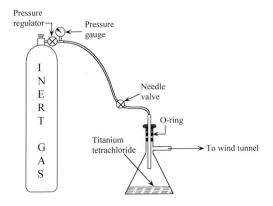

Fig. 3.10. A safer technique for generating smoke from TiCl₄ (after Mueller, 1996).

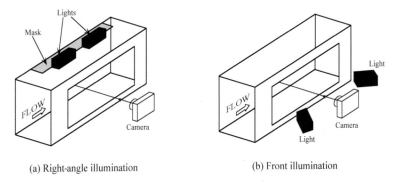

(a) Right-angle illumination   (b) Front illumination

Fig. 3.11. Lighting arrangements commonly used to illuminate external flow features.

External illumination

With external illumination, the locations of the sources of light in relation to the subject can strongly affect the quality of the photographic images. Two common lighting arrangements are shown in Fig. 3.11. The choice of the arrangement depends very much on the application. For example, in studies involving smoke and water tunnels with dark test-sections, the right-angle arrangement is preferable in terms of obtaining high-quality images. With the front-illumination arrangement the light reflecting from the front and back of the tunnel can adversely affect the image quality. Nevertheless, with trial and error, reasonable image

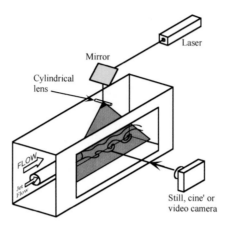

Fig. 3.12. Schematic showing the generation of a laser sheet using the cylindrical lens technique.

quality can be also be obtained with this arrangement. With the right-angle illumination, further improvement in the photographic quality can be achieved by limiting the light falling onto the back of the test-section. A narrow strip of cardboard or mask is normally used for this purpose (see Fig 3.11). However, if the test-section has a white background as in most water tunnel studies using dye, the cardboard should not be used because the shadow cast by the cardboard onto the test-section wall can degrade the photographic quality. Moreover, with a white background, it is generally better to use front-lighting, or a combination of both front and right-angle lighting since light falling on the white background helps to enhance contrast.

In most flow visualization studies where the flow speed is not very high, conventional lighting is usually sufficient for most cameras (still, cine, or video). However, there are times when the flow speed is too high and the shutter speed of the *still* camera is too slow to capture sharp images. Under these conditions, flash photography may be necessary. There are many types of commercial flash units to choose from, some of which have a flash duration shorter than 20 μs. The advantage of using flash is that it enables high speed fluid motion to be arrested without compromising on the depth of field.

Internal illumination

To visualize the internal features of a flow, light sheets are used. In the past, a narrow sheet of light could be obtained by allowing flood lighting to pass through two narrow slits arranged some distance apart. However, the technique is not particularly effective because no more than 10% of the original illumination can be produced for a beam thickness of 1 to 2 mm. With the advent of lasers, a thin sheet of light can now be formed very easily using a cylindrical lens, as shown in Fig. 3.12. Glass rods are often used. The spreading angle of the sheet depends on the diameter of the lens: the smaller the diameter, the larger the spreading angle. For most visualization applications, lenses of between 2 to 10 mm diameter are sufficient.

Another method of generating a laser light sheet is to use an oscillating mirror mounted on an optical scanner (see, for example, Gad-el-Hak, 1986). The technique is found to produce a more uniform light intensity because glass rods usually contain imperfections. However, to achieve a uniform sheet with the oscillating mirrors, the frequency of the oscillation must be equal to at least the inverse of the camera shutter speed.

Once the laser sheet is formed, it is ready to be used for "sectioning" the flowfield, usually by aligning the plane of the light sheet perpendicular to the line of view. For most studies, a 4 to 5 W CW laser provides sufficient light intensity.

3.4.2   Camera

Still cameras, cine cameras, and video cameras are all popular for flow visualization studies. The Single Lens Reflex (SLR) camera is the most popular still camera design because it eliminates the parallex error normally associated with direct vision finder cameras. Most importantly, it enables a visual check on the depth of field of the subject. Depth of field is defined as a distance between the nearest and farthest parts of the subject matter which can be brought to acceptable focus.

Although most SLR cameras take 35 mm film, several models are made in medium and large format sizes (see Table 3.1). Large format cameras are used mainly by professional photographers and will not be discussed here. Of the remaining two formats, 35 mm cameras are the most often used in flow visualization, partly because the film can be processed very quickly in most photographic shops. However, in terms of photographic quality, medium format cameras are significantly better. Moreover, the larger film size allows better

Format and film size	Camera
Small format (35 mm film) *Negative size:* 24 × 36 mm	Pentax, Nikon, Olympus, Canon, Leica, Minolta
Medium format (120 and 220 roll film) *Negative size:* 60 × 45 mm, 60 × 70 mm, 70 × 70 mm	Contax 645, Mamiya RB67, Bronica, Hasselblad 200 and 500 series
Large format *Negative size:* 90 × 120 mm to 4 × 5 in.	Linholf, Speed Graphics

Table 3.1. Different camera formats.

reproduction for enlargement. With the 35 mm camera, the film size is 3 to 6 times smaller than the medium format films (Table 3.1). In addition, some medium format cameras come with the option of an interchangeable Polaroid back, which enables results to be seen instantly. Polaroid produces a wide range of instant films, ranging from color film (for example, T-669, T-679, and T-689) to black and white film (for example, T-665 and T-667).

In recent years, digital camera technology has improved considerably in terms of its resolution and it has gained considerable popularity in the research community. One of the advantages of digital cameras is that the image can be viewed almost instantaneously, without the delay involved in developing the film. Digital cameras are grouped under three general categories based on the resolution of the image they generate, namely VGA, XGA and megapixel cameras. VGA cameras generally have a resolution of about 300,000 pixels/image, while XGA cameras have approximately 800,000 pixels/image. Megapixel cameras can capture images of more than 1 million pixels/image. For flow visualization, megapixel cameras are the only camera that can provide reasonably sized photo-realistic pictures.

Although digital cameras have revolutionized the task of taking pictures, they still cannot fully emulate film-based cameras. For example, current digital images are not suitable for enlargement because the image sharpness will be considerably reduced.

While still cameras are excellent for capturing "instantaneous" pictures of streakline patterns made by dye and smoke, the only way to capture the flow motion is to use a motion picture camera or a video camera. Motion picture cameras are grouped according to their shutter speed, as well as film format. A normal-speed camera offers a shutter speed ranging from 5 to 75 fps (frames/s), while a high-speed version can go as high as 1600 fps. Non-framing cameras can achieve megahertz rates, albeit for only a limited number of images. As for the film format, it ranges from 8 mm to 70 mm. For many flow visualization investigations, 16 mm normal-speed cameras such as the Bolex H16 Reflex and Arriflex 16 SR3 cameras, are adequate. If a high-speed camera is required, the Kodak Ektapro high-gain camera, a Wollensak WF-3 Fastex 16 mm camera or a Nac E-10 Electronic Eye high speed camera, with a framing rate of up to 8000 fps for a full frame capture, may be used. However, when using a high speed camera, the illumination needs to be increased considerably since the exposure time for each frame is proportionally reduced.

For film-based cameras, the long processing time has always been the main drawback, and as a result, more people are turning to video recording systems. Normal-speed video system comes with a standard framing rate of 25 fps for the PAL format, and 30 fps for the NTSC format. For the high-speed version, the framing rate can be as high as 1000 fps. At such high speeds, unless the illumination power is increased substantially, an image intensifier may be needed. One such intensifier is the IMCO Intensifier Lens System, but it is relatively expensive. Moreover, the resolution of the images captured by a high-speed video camera is generally poorer than that of the comparable film-based system. A time-code card, which can be used to display the frame-number on a monitor, is an important accessory. This feature is most useful during slow motion flow analysis.

Digital video cameras are now also gaining in popularity. Most digital video cameras today have an image quality far exceeding the analog formats such as VHS and Hi8 used by traditional video cameras. In addition, most of them can also be used to capture still-images, thus doubling as a still camera. However, when using a digital video camera to record moving images, be prepared to have a large computer disk space ready to store the images. If needed, the images can be encoded on CD-ROM for storage.

Image quality and framing rate can be traded off to some extent. Princeton Scientific Instruments produces a framing camera that can capture 32 digital images at megahertz rates with a resolution of about 256^2 pixels (see Chapter 5).

3.4.3 Lens

The lens is one of the most important elements in photography, and probably the most expensive part of a camera. The quality of the photographic images

Typical lens set	Small Format (35 mm film)	Medium Format (roll film)	Large Format (4 × 5 in film)
Wide angle lens	28 mm	50 mm	90 mm
Normal lens	50 mm	90 mm	150 mm
Long focal lens	100 mm	150 mm	280 mm

Table 3.2. Examples of typical lens sets for different format cameras.

depends to a great extent upon the properties of the lens. Lenses can be classified into the following categories:

1. **Wide angle lens:** These lenses generally have an angle of view of over 70°, and they are particularly useful in cramped working conditions. This class of lens has an inherent image distortion near the edges of the focal plane. The wide-angle focal length is approximately equal to the short side of the negative format. For a 35 mm camera, it works out to be approximately 28 mm (see Table 3.2).

2. **Normal angle lens:** This term usually refers to a lens which has an angle of view that ranges from 45° to 50°. The focal length of the lens in this category is approximately equal to the diagonal side of the negative format. For 35 mm camera, it is about 50 mm.

3. **Long focus Lens:** These lenses are normally used to produce large images when the camera is at an unavoidably large distance away from the subject. The angle of view is usually 35° or less. The focal length of a long focus lens is approximately twice the long side of the negative format.

4. **Zoom lens:** These lenses are designed to have a variable focal length. Some of the commonly used zoom lenses include 70–150 mm f-4.5, and 80–200 mm f-3.5.

5. **Macro lens:** These lenses are specially designed to deliver optimum resolution at short distances (that is, to capture life size images) as well as normal-range work. The most common macro focal lengths are 50 mm and 100 mm.

In addition to the focal length, lenses are also specified according to their speed. The term "speed of a lens" is defined as the largest aperture at which a lens can be used, and it is usually marked on the rim of the lens mount as "$f/$" followed by a number. The letter f is an abbreviation for the term "factor," and the number which follows is calculated by dividing the focal length of the lens by the diameter of the effective lens aperture. Therefore, the smaller the f-number, the greater the light beam entering the lens, and hence, the brighter the photographic image. It follows that "speed of a lens" is a measure of the maximum light passing power of a lens when the aperture is set at its largest size. The advantage of a fast lens is that it reduces the required exposure time, which is a big advantage in action photography. For most lenses the sequence of f-number aperture control is as follows:

$$f/1.4; \ 2.8; \ 4; \ 56; \ 11; \ 16; \ 22; \ 32$$

The largest f-number signifies the smallest aperture size, hence the dimmest illumination. Each change to a lower f-number in this sequence indicates a doubling of the illumination. All lenses with the same f-number setting transmit the same amount of light.

Apart from controlling the amount of light, the size of the aperture also influences the depth of field of a camera. In general, the smaller the aperture size, the greater the depth of field. As a practical tip, it is always better to set the aperture at its maximum size, because the short depth of field allows one to be more critical during focusing.

One of the important factors to consider when taking pictures is the correct setting of the shutter speed. Choosing the most suitable speed for a particular application depends on the total amount of light available, the aperture required, and whether the subject is stationary or moving. If the subject is moving, one must know how fast it is moving. Some of these requirements can conflict with each other. In flow visualization work, action-stopping the flow motion is the most important requirement. Therefore, it is always better to first select the minimum shutter speed based on the flow speed in a particular application. Once the shutter speed has been determined, the aperture must be selected on the basis of the illumination available. But if the depth of field is also an important consideration, the illumination needs to be substantially increased. A completely different way of maintaining a good depth of field while still ensuring that the flow appears stationary is to move the camera at the same speed as the flow. However, this may not be practical in many applications.

3.4.4   Film

Photographic film comes in two types: slide or "reversal" films, and print or "negative" films. The choice depends on the application. Both the reversal and negative films come in a variety of standard sizes to suit different format cameras. Of the three formats discussed earlier, 35 mm film is undoubtedly the most popular because it can be processed commercially with a short turn-around time. Apart from the size, films are also specified according to their speed. For color films, the speed can range from as low as ISO 25 to as high as ISO 1000 (ISO, for International Standard Organization, is a new speed scale which replaces the old ASA, for American Standard Association, scale). For black and white film, the speed can go as high as ISO 3200 for regular film, and ISO 20,000 for Polaroid film. A high-speed film can capture light faster than a slow film, and therefore requires less lighting to get a proper exposure. Faster speed films, however, are generally grainier and render subjects less sharply. Generally, the lower the film speed, the smaller is the size of the grains, and the greater is the sharpness of the photographic image. In addition, lower speed film requires more light for proper exposure. For many flow visualization studies, film speeds between ISO 400 and 800 is sufficient to offer a good balance of fine grain and high definition.

Before using a film, it is important to also know its characteristics since that will help to improve the quality of the photograph. For example, if tungsten lighting is used to illuminate a subject, then a tungsten film should be used. If a normal daylight film is used with tungsten lighting, the image will appear with a slight orange tinge. This is because a tungsten filament emits more red than blue light. Although this problem can be addressed by using a *80B* filter which absorbs more red than blue, the filter will cause a reduction in the amount of light falling onto a film. This must therefore be compensated by increasing either the camera aperture or the illumination level.

3.5   Cautionary Notes

Dye and smoke flow visualizations have made many significant contributions to our fundamental understanding of fluid flow. However, like most experimental techniques, flow visualization has its limitations and pitfalls, which can lead to a misinterpretion of the results. These limitations/pitfalls can be attributed to a number of factors:

First, when investigating fluid flow where strong vortex stretching is present such as during vortex interactions, one must be aware that the time evolution of vorticity is not identical to that of a passive scalar, including both the dye and smoke particles. The difference in their behavior can best be understood with reference to their respective transport equations. For vorticity, the transport equation is governed by

$$\frac{\partial}{\partial t} = -\left(\mathbf{V} \cdot \nabla\right) + \left(\cdot \nabla\right)\mathbf{V} + \nu\nabla^2 \qquad (3.2)$$

where $= \nabla \times \mathbf{V}$ is the vorticity and ν is the kinematic viscosity. The first term on the right hand side represents the advection of vorticity by the local mean velocity, and the second term is related to vortex stretching by the local strain, while the last term represents diffusion of vorticity due to viscosity. For the passive scalar, the transport equation is

$$\frac{\partial S}{\partial t} = -\left(\mathbf{V} \cdot \nabla\right)S + \kappa\nabla^2 S \qquad (3.3)$$

where κ is the diffusivity of the material S. The passive scalar transport equation contains the same advection and diffusion terms as the vorticity equation, but it lacks the corresponding stretching term. It is the absence of the stretching term which is responsible for the difference in their behavior. The extent of their difference is governed by the relative importance of the stretching and the advection terms in equation 3.2. If the stretching term is small in comparison to the advection term, the passive scalar will advect and diffuse in the same way as the vorticity, provided the Schmidt number (ν/κ) is unity. On the other hand, if the stretching term is dominant, as is commonly encountered in the study of vortex dynamics, then the difference between them can be significant. This is because when a vortex filament stretches, the vorticity is intensified, while the density of a passive scalar is decreased. During intense vortex stretching, the concentration of the passive scalar may reduce to an extent that the presence of the passive scalar in the flow field may not be obvious. In other words, the absence of passive scalar may not necessarily indicate the absence of vorticity. This behavior is clearly demonstrated by the a numerical study of Kida *et al.* (1991) in which the time evolution of passive scalar quantity is compared with that of the vorticity for the case of two vortex rings colliding at an angle. Their analysis clearly shows that during the initial stage of the vortex ring interaction, the effect of vortex stretching is small, and the scalar quantity closely follows the vorticity. However, at a later time when the effect of stretching becomes significant, the difference between them is quite distinct as can be clearly seen

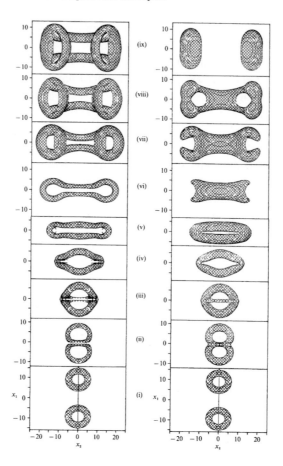

Fig. 3.13. Numerical simulation of the collision of two vortex rings. The diagrams show perspective views of the iso-surfaces of the vorticity norm (left) and a passive scalar (right). (i) $t = 0s$, (ii) $t = 1s$, (iii) $t = 1.5s$, (iv) $t = 2s$, (v) $t = 3s$, (vi) $t = 5s$, (vii) $t = 10s$, (viii) $t = 15s$, and (ix) $t = 21s$. (Reprinted with permission from Journal of Fluid Mechanics, Kida *et al.*, 1991).

in Fig. 3.13. This is despite the assumption made in the computation that the Schmidt number (ν/κ) is unity.

Second, under normal circumstances, if a passive scalar is released at the location where vorticity is generated, the passive scalar will coincide with the vorticity at all times provided the Schmidt number is unity. However, when dye or smoke is used as a tracer, the Schmidt number is typically of $O(1000)$.

This implies that the dye/smoke will only follow the vorticity during the initial stage of the flow development because at a later time viscous diffusion will cause the vorticity to diffuse away from the dye. For turbulent flows, the difference may not be significant because the operative non-dimensional parameter is the turbulent Schmidt number which is always close to unity (see Kelso *et al.*, 1997).

Third, the streakline pattern displayed by smoke/dye filaments shows only a spatially integrated view of the flow pattern. This is because dye/smoke is distorted by the local flowfield as it travels downstream. As indicated earlier, the streakline pattern seen at some distance downstream of a test-model is a result of the "accumulated" distortion which the dye filaments have undergone all the way from the point of introduction, so that the streakline pattern depends strongly on the location where dye/smoke is released. This effect was clearly demonstrated by Cimbala *et al.* (1988) in a smoke-wire experiment where they showed that the smoke released at different locations downstream of a circular cylinder in cross-flow displayed entirely different wake patterns. This suggests that to obtain a true picture of the flow pattern the smoke or dye must be introduced as close to the location of observation as possible.

Finally, in unsteady flow, a streakline pattern is not the same as streamline pattern. Therefore, it is *wrong* to interpret streaklines as equivalent to stream-lines as some researchers have done in the past. The only condition that the two are the same is when the flow is steady. The relationship between them in an unsteady flow is clearly demonstrated in Chapter 1.

3.6   References

Atkins, P.W. 1982. *Physical chemistry*. 2nd edition,Oxford University Press.

Bastedo, W.G. and Mueller, T.J. 1986. Spanwise variation of laminar separation bubbles on wings at low Reynolds numbers. *J. Aircraft*, **23**, 687-694.

Batill, S.M. and Mueller, T.J. 1981. Visualization of transition in the flow over an airfoil using the smoke-wire technique. *AIAA J.*, **19**, 340–345.

Brown, F. M. N. 1971. *See The Wind Blow*. University of Notre Dame.

Castellan, G.W. 1972. *Physical Chemistry*. 2nd ed., Addison-Wesley Publishing Co.

Chmielewski, G.E. 1974. Boundary considerations in the design of aerodynamic contractions. *Journal of Aircraft*, **11**, 435–438.

Cohen, M.J. and Ritchie, N.J.B. 1962. Low speed three-dimensional contraction design. *Journal of Royal Aero. Soc.*, **66**, 231–236.

Cimbala, J.M., Nagib, H.M. and Roshko, A. 1988. Large structure in the far wakes of two-dimensional bluff bodies. *J. Fluid Mech.*, **190**, 256–298.

Clayton, B.R., and Massey, B.S. 1967. Large structure in the far wakes of two-dimensional bluff bodies. *J. Sci. Instrum.*, **44**, 2–11.

Freymuth, P., Bank, W. and Palmer, M. 1985. Use of titanium tetrachloride for visualization of accelerating flow around airfoils. *In Flow Visualization III.*, ed. W.J. Yang, 99–105, Hemisphere, New York.

Freymuth, P. 1985. The vortex patterns of dynamic separation: A parametric and comparative study. *Prog. Aerospace Sci.*, **22**, 161–208.

Freymuth, P. 1993. Flow visualization in fluid mechanics. *Rev. Sci. Instruments.*, **64**, 1–18.

Fric, T.F. and Roshko, A. 1994. Vortical structure in the wake of a transverse jet. *J. Fluid Mech.*, **279**, 1–47.

Gad-el-Hak, M. 1986. The use of dye-layer technique for unsteady flow visualization. *Trans ASME.*, **108**, 34–38.

Gad-el-Hak, M. 1988. Visualization techniques for unsteady flows: An overview. *J. Fluids Eng.*, **110**, 231–243.

Head, M.R. and Bandyopadhyay, P. 1981. New aspects of turbulent boundary layer structure. *J. Fluid Mech.*, **107**, 297–338.

Honji, H., Taneda, S. and Tatsuno, M. 1980. Some practical details of electrolytic precipitation method of flow visualization. *Reports of Research Institute for Applied Mechanics, Kyushu University.*

Kelso, R. M., Lim, T.T. and Perry, A.E. 1996. An experimental study of a round jet in cross-flow. *J. Fluid Mech.*, **306**, 111–144.

Kelso, R. M., Delo, C. and Smits, A.J. 1997. The structure of the wake of a jet in cross-flow. *Phys. Fluids*, **306**, 111–144.

Kida, S., Takaoka, M. and Hussain, F. 1991. Collision of two vortex rings. *J. Fluid Mech.*, **230**, 583–646.

Lim, T.T. and Nickels, T.B. 1992. Instability and reconnection in head-on collision of two vortex rings. *Nature*, **357**, 225–227.

Lim, T.T. 1997. A note on the leapfrogging between two coaxial vortex rings at low Reynolds numbers. *Phys Fluids*, **9**, 239–241.

Liu, C.Y. and Ng, K.L. 1990. A low-cost mini smoke tunnel with automatic smoke wire fueling mechanism. *Int J. Mech. Eng. Education.*, **18**, 85–91.

Luo, S.C., Lim, T.T., Lua, K.B., Chia, H.T., Goh, E.K.R. and Ho, Q.W. 1998. Flowfield around ogive/elliptic-tip cylinder at high angle of attack. *AIAA J.*, **36**, 1778–1787.

Matisse, P. and Gorman, M. 1984. Neutrally bouyant anisotropic particles for flow visualization. *Phys Fluids*, **27**, 759–760.

Mehta, R.D. 1977. The aerodynamic design of blower tunnels with wide-angle diffusers. *Prog Aerospace Sci.*, **18**, 59–120.

Mehta, R.D. and Bradshaw, P. 1979. Design rules for small low speed wind tunnels. *Aeronautical J.*, **83**, 443–449.

Merzkirch, W. 1987a. Techniques of flow visualization. *NATO AGARD Report 302.*

Merzkirch, W. 1987b. *Flow Visualization.* Academic Press, New York.

Mikhail, M.N. 1979. Optimum design of wind tunnel contractions. *AIAA J.*, **17**, 471–477.

Morel, T. 1975. Comprehensive design of axisymmetric wind tunnel contractions. *J. Fluids Engineering*, **97**, 225–233.

Mueller, T. J. 1996. Flow visualization by direct injection. *In Fluid Mechanics Measurements.*, ed. R.J. Goldstein, 367–450, Washington, DC:Taylor & Francis.

Nagib, H.M. 1977. Visualization of turbulent and complex flows using control sheets of smoke streaklines. In *Proc. of the Intern. Symp. on Flow Visualization.*, 257–263, Tokyo, Japan.

Rae, W. H. Jr. and Pope, A. 1984. *Low-Speed Wind Tunnel Testing.* 2nd ed., John Wiley & Sons, New York.

Reynolds, O. 1883. An experimental investigation of the circumstances which determine whether the motion of water shall be direct or sinuous and the laws of resistance in parallel channels. *Phil. Trans Royal. Soc. London*, **174**, 51.

Simmons, L.F.G. and Dewey, N.S. 1931. Photographic records of flow in the boundary layer. *Reports and Memoranda, Aeronautical Research Council 1335*, London, 1931.

Sinha, M. and Smits, A.J. 1999. Quasi-periodic transitions in axially forced Taylor-Couette flow. In *Proc. of the 11th Intern. Couette-Taylor Workshop.*, ZARM, University of Bremen, Germany.

Taneda, S., Honji, H. and Tatsuno, M. 1977. The electrolytic precipitation method of flow visualization. In *Proc. of the Intern. Symp. on Flow Visualization.*, ed. T. Asanuma, 133–138, Tokyo, Japan.

Taneda, S. 1977. Visual study of unsteady separated flows around bodies. *Prog. Aerospace Sc.*, **17**, 287–348.

Taneda, S. 1982. Kármán vortex street behind a circular cylinder at $Re=$ 140. In *An Album of Fluid Motion*, 56, ed. M. Van Dyke, The Parabolic Press.

Torii, K. 1977. Flow visualization by smoke-wire technique. In *Proc. of the Intern. Symp. on Flow Visualization.*, ed. T. Asanuma, 251–156, Tokyo, Japan.

Visser, K.D., Nelson, R.C. and Ng, T.T. 1988. Method of cold smoke generation for vortex core tagging. *J. Aircraft*, **25**, 1069–1071.

Werlé, H. 1973. Hydrodynamic flow visualization. *Ann. Review of Fluid Mech.*, **5**, 361–382.

CHAPTER 4

MOLECULAR TAGGING VELOCIMETRY

Walter R. Lempert[a]

4.1 Introduction

The advent of the laser as a relatively common tool for flow visualization has stimulated the development of velocimetry techniques based on the use of photo-sensitive molecules. While a variety of different molecules have been employed for this purpose, they share the common attribute that laser excitation is used to produce a defined pattern of long-lived tracers which are embedded within the flow field. This process is often referred to as "tagging." After a suitable time delay, a CCD (or other) camera is used to obtain an image of the displaced pattern (often termed "interrogation"). The observed displacement divided by the elapsed time is assumed to constitute a measurement of vector velocity field. Dependent upon the details of the optical processes and/or the nature of the tracer, the technique has alternately been referred to as Laser-Induced Photochemical Anemometry (Falco & Nocera, 1993), Flow Tagging Velocimetry (Lempert, 1995) and Molecular Tagging Velocimetry (Gendrich, Koochesfahani & Nocera, 1997).

As will be shown, the properties of available photo-sensitive materials vary quite significantly, albeit it in somewhat subtle ways. The purpose of this chapter is to provide a framework sufficiently detailed to enable potential users to effectively match the diagnostic to their particular measurement environment. To avoid confusion, we shall adopt the terminology of Gendrich & Koochesfahani (1996), and use the term Molecular Tagging Velocimetry to encompass the variety of time-of-flight velocimetry techniques which are based on photo-sensitive molecules.

[a]Departments of Mechanical Engineering and Chemistry, The Ohio State University, Columbus, OH 43210, U. S. A.

4.2   Properties of Photo-Sensitive Tracers

4.2.1   Photochromic dyes

While not the primary focus of this chapter, we begin with a brief mention of photochromic molecules. The use of photochromic dyes for liquid phase velocimetry was introduced by Popovich & Hummel (1967). A photochromic material is one in which the absorption of a photon induces a temporary change in the absorption spectrum of the absorbing molecule. In general, photochromic tracers used in fluid studies are initially transparent in the visible region of the spectrum. Upon absorption of a single photon, typically but not necessarily, in the ultraviolet (UV) region of the spectrum, the molecule becomes absorbing over a wide region of the visible. Upon subsequent back illumination with white light, the fluid containing the activated tracer takes on a dark (generally blue) color which is actually a manifestation of the high absorbance at green to red wavelengths.

The principal advantage of photochromic dyes is the relatively low cost of both the tracer and the required instrumentation. The tagging step can often be performed using relatively inexpensive nitrogen lasers ($\lambda = 0.337\mu$m) and the interrogation performed with common white light flash sources. The use of photochromic dyes has been described in some detail by Fermigier & Jennfer (1987).

4.2.2   Phosphorescent supramolecules

The principal disadvantage of photochromic tracers is that the absorption-based interrogation produces images of inherently limited contrast. In order to circumvent this, new molecular tagging techniques based on optically emitting materials have recently been developed. Phosphorescence, which is similar to fluorescence, refers to spontaneous radiative emission from relatively long lifetime "metastable" electronic states of molecules, which are populated as a result of optical absorption. As a somewhat arbitrary rule, materials which exhibit radiative decay lifetimes of order 1 ms or greater are termed phosphorescent, whereas materials with shorter lifetimes are termed fluorescent. Molecular Tagging Velocimetry based on phosphorescent tracers uses a single laser to prepare the metastable excited state by ordinary absorption. The spontaneous radiative emission serves as the interrogation, and does not require a second optical source. While the concept is quite simple, the challenge has been the synthesis of suitably long life-time molecules which are soluble in non-organic solvents and exhibit tolerable emission yields in the presence of water or oxygen.

Recently, Nocera and co-workers (Ponce, 1993; Hartmann, 1996) have presented what they term "phosphorescent supramolecules," which consist of an active lumophore (1-Bromonapthalene) which is bound to a site on the interior of a cup-shaped glucosyl-modified cyclodextran (Gβ-CD) molecule. Upon addition of alcohol, a protective "lid" is formed which prevents quenching of the lumophore from dissolved oxygen and/or water. Fig. 4.1 shows the chemical structures and representative emission spectra for aqueous solutions which are 10^{-5} M in Bromonapthalene and 10^{-3} M in Gβ-CD, with and without the presence of added alcohol. The excitation wavelength is 308 nm, corresponding to that from a XeCl excimer laser. The feature in the vicinity of 325 nm, which appears both with and without the protective alcohol, corresponds to ordinary fluorescence and has a lifetime of order 9 ns. The feature in the range 480–650 nm appears only in the presence of alcohol and has a measured decay lifetime in the range ~0.10 to 5.0 ms, dependent upon the concentration and specific type of alcohol (Ponce, 1993). It should be noted that the intensity axis on the right hand side of Fig. 4.1 has been expanded relative to the left hand side.

Practical implementation of MTV using phosphorescent supramolecules requires some consideration of composition in order to match the emission decay time to the flow velocity regime. Clearly the emission life time must be long enough to permit sufficient displacement to occur. For example, a fluid element tagged in a flow field with a local mean velocity of 10 cm/s would experience 500 μm of displacement after 5 ms. Such a displacement is readily measurable, but is near the low end of that which is generally employed using MTV. Obviously, displacement, and, correspondingly, measurement accuracy, increases as flow velocity and/or delay time is increased. More quantitatively, it is easy to show that the relative detected signal, S, is given by:

$$S = I_o \tau e^{-\Delta t/\tau} \left[1 - e^{-\tau_{exp}/\tau} \right] \qquad (4.1)$$

where τ is the phosphorescence decay time, Δt is the tag-interrogate delay time, t_{exp} is the interrogation exposure time, and I_o is a constant which incorporates many parameters such as optical absorption, tracer concentration, phosphorescence quantum yield, and optical collection efficiency. Well-designed experiments will have delay times which are long enough to permit sufficient displacement (on the order of 20 detector pixels), but not too long compared to τ. Similarly, the exposure time should be adjusted to be short compared to flow times, but also not too short compared with τ. Fortunately, the required chemicals are not prohibitively expensive, so that weak signals can be boosted

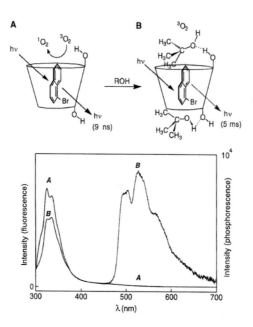

Fig. 4.1. Chemical structure and emission spectra for 1-BrNp in CDs with (B) and without (A) added alcohol (from Gendrich, 1997).

by simply increasing the solution concentration. It must be noted, however, that as of this writing, there was no known commercial supplier of phosphorescent supramolecule tracers.[b]

4.2.3   Caged dyes

An alternative MTV approach based on caged dye Photo-Activated Fluorophore (PAF) tracers has recently been presented (Lempert, 1995; Harris, 1996). PAF's are nominally fluorescent dyes that have been rendered non-fluorescent by strategic attachment of a chemical "caging" group. The chemical caging group is photolytically cleaved upon absorption of a single photon of ultraviolet light. After photolysis, the fluorescent dye is recovered and can be tracked indefinitely using ordinary laser sheet imaging techniques. In effect, as illustrated in Fig. 4.2, the

---

[b] Interested parties should contact Prof. Daniel Nocera, Dept. of Chemistry, Massachusetts Institute of Technology, Cambridge, MA (USA).

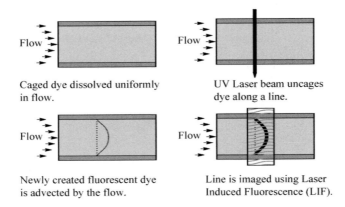

Caged dye dissolved uniformly in flow.

UV Laser beam uncages dye along a line.

Newly created fluorescent dye is advected by the flow.

Line is imaged using Laser Induced Fluorescence (LIF).

Fig. 4.2. Schematic illustration of use of caged dye PAF's for molecular tagging.

tagging laser is used to photo-chemically create a user-defined pattern of ordinary fluorescent dye. The convection of the "locally seeded" fluid elements are subsequently interrogated using a second laser. In all work reported to date, the tagging was performed using the third harmonic of a Nd:YAG laser at 0.355 μm. Interrogation utilizes visible lasers, most commonly argon-ion, flashlamp-pumped dye, and the second harmonic of Nd:YAG (continuous wave or pulsed).

Most MTV measurements reported to date have utilized some form of caged fluorescein dye. Fig. 4.3 shows absorption spectra of both the caged and uncaged form of the dye. It can be seen that the uncaged dye has a rather broad absorption feature centered at approximately 350 nm, which is characteristic of benzene type compounds. After uncaging, the spectrum is identical to that of ordinary disodium fluorescein, which is a very commonly employed material for passive scalar measurements. As can be seen in Fig. 4.3, the uncaged dye has a very strong absorption in the vicinity of 490 nm., which is well matched to the argon-ion laser.

Caged dye tracers have two principal advantages in comparison to long life-time phosphorescent materials. The first is that the uncaging is permanent, so that arbitrarily long time intervals, limited only by mass diffusion, can be employed between tagging and interrogation. This provides the capability to perform measurements in exceedingly low speed flows. For example, Harris (1996) has reported quantitative measurements in electrohydrodynamic flows with mean velocity of order 2-4 μm/s. The second advantage is that the uncaged

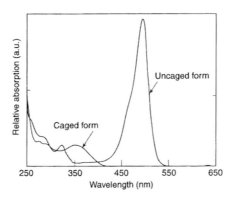

Fig. 4.3. Absorption spectrum of caged and uncaged form of fluorescein PAF (from Lempert, 1995).

dye exhibits exceedingly high signal levels. We can see this more explicitly by considering the following simple expression for the rate of absorption of interrogation photons per unit volume of fluid:

$$\frac{dA}{dt} = \left(\frac{I}{h\nu}\right)\varepsilon c \tag{4.2}$$

where I is the intensity of the laser (W·s^{-1}·area^{-1}), (ε is the molar extinction coefficient (liter·mol^{-1}·cm^{-1}), c is the molar concentration of uncaged dye, and the factor $h\nu$ converts from energy to photons. Since the fluorescence quantum yield is of order 1.0 (Drexhage, 1990), the photon absorption rate is approximately equal to the fluorescence emission rate. We can therefore divide Eqn. 4.2 by the number density of uncaged dye molecules and substitute $\varepsilon \sim 10^5$ liter^{-1}·mol^{-1}·cm^{-1} with the result that the photon emission rate per uncaged dye molecule is given by:

$$\text{Photons/Molecule} \approx 400\,I \tag{4.3}$$

where I is the laser intensity in W/cm^2. Eqn. 4.3 is valid as long as the laser intensity is less than the so-called "saturation" intensity, which for fluorescein dye is of order 3×10^5 W/cm^2 (Chen, 1967). Substitution of this value into Eqn. 4.3 shows that as many as 10^8 photons/sec can be radiated from a single uncaged dye molecule. Physically this corresponds to optical "recycling" of the tracer during a single interrogation event, which occurs rapidly due to the fast (order 4.5×10^{-9} s) excited state radiative lifetime (Chen, 1967).

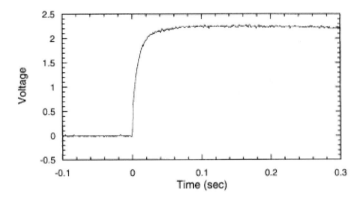

Fig. 4.4. Fluorescence rise time of caged Q-rhodamine in methanol (from Lempert, 1998).

There are, however, some significant disadvantages associated with caged dye tracers. Fundamentally, the most significant is the finite kinetic rate of the photochemical cage breaking step. This is illustrated in Fig. 4.4 which shows the fluorescence rise time of a caged rhodamine PAF tracer in methanol (Lempert, 1998). The data was obtained by loosely focusing the third harmonic output from a pulsed Nd:YAG laser into a cuvette containing approximately 2 mg/liter of the PAF. Simultaneously, the uncaged dye was interrogated using the 0.514 μm output of a CW argon-ion laser. At $t = 0$, the Nd:YAG laser was fired and it can be seen that the visible fluorescence signal evolves to order 70% of its maximum after approximately 10 ms. If the time axis in Fig. 4.4 is expanded it is found that the fluorescence evolves to $\sim$25% of its maximum at a time less than 1 ms (Lempert, 1998). The effect of the finite rise time is to establish a minimum time delay between tagging and interrogation, which can be particularly important in high Reynolds number flows. To date, the minimum reported delay time which has been used in caged dye PAF studies is 200 μs (Lempert, 1995). This corresponds to $\sim$5% of the kinetic e^{-1} rise time.

The second principal disadvantage of caged dye PAF's is the fact that the materials can only be used once, since the uncaging is irreversible. This is compounded by the high cost of the tracer itself. PAF's were originally developed by the biological sciences community for cellular related studies which typically require exceedingly small (order milligram) quantities of tracer. As will be seen in the next section, the high brightness of the uncaged dye results in very modest concentration requirements, but the cost can still be prohibitive.[c]

4.3   Examples of Molecular Tagging Measurements

In this section we will give a brief survey of representative reported MTV measurements, with emphasis on experimental detail. The prime purpose of this section is to provide the reader with a real sense of the range of flow environments in which MTV studies have been performed and to aid in design of potential measurements in their laboratories.

4.3.1   Phosphorescent supramolecules

Fig. 4.5 shows a schematic diagram of a vortex ring/wall interaction experiment (Gendrich, 1997) in which Gβ-CD supramolecules were used to obtain a series of planar velocity fields in the near-wall region. The output of a 100 mJ/pulse XeCl excimer laser was split into two equal power beams, each of which was further split into approximately 20 low energy, parallel beams, using a custom built beam blocker. The two sets of parallel tagging beams are brought incident to the flow at right angles, forming a grid pattern. The beam blocker is an aluminum plate with a set of approximately 1 mm wide slots cut into it. While a significant fraction of the total laser energy is thrown away, the beam blocker provides a simple and flexible method for grid formation. Typical individual beams have a diameter of order 250 μm and a single pulse energy of about 1 to 2 mJ.

Figs. 4.6 and 4.7 show a typical pair of interrogation images and the corresponding velocity field. The time delay between the images is 8 ms and the axis of symmetry of the counter rotating vortex pair is indicated by the dashed line. The interrogation images were captured using a pair of relatively low cost CCD video cameras, although for time delays exceeding $\sim 5\tau$, more expensive, intensified cameras are sometimes required. The concentration of the Gβ-CD and 1-BrNp were 2×10^{-4} M (0.2 g/liter) and 10^{-5} M, respectively, and the measured $1/e$ radiative decay time was 3.7 ms. The solvent is 0.05 M cyclohexanol in water.

An important feature of Molecular Tagging Velocimetry is its inherent capability in three-dimensional flow. Unlike Particle Image Velocimetry (PIV), or

---

[c]A variety of caged dye PAFs are available commercially from Molecular Probes, Inc., Eugene, OR, U. S. A.

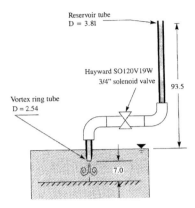

Fig. 4.5. Schematic diagram of vortex ring/wall interaction study (Gendrich, 1997).

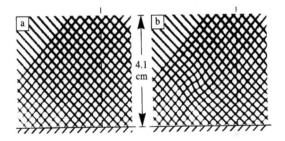

Fig. 4.6. Representative interrogation image pair for vortex ring/wall interaction study. Left image acquired 1 μs after tagging and represents original grid. Right image acquired 8 ms later (Gendrich, 1997).

other scattering-based particle tracking techniques, there is no particular difficulty associated with obtaining data in cross sectional planes perpendicular to a principal flow axis. The reason for this is that in MTV the tagging optics define which fluid elements will be tracked. Upon activation, the tagged fluid, at least in principle, can convect to any location within the flow field prior to interrogation. Fig. 4.8 shows an example set of vector data obtained in a highly three-dimensional periodically forced wake flow (Koochesfahani, 1996). Note that the magnitude of the vector velocity in the v-w plane is as high as 40% of the mean velocity in the principal (x) direction. The absolute free stream mean velocity is 10 cm/s. Other optical considerations, such as tracer concentration, laser power, and time delay, are comparable to that used to obtain the images

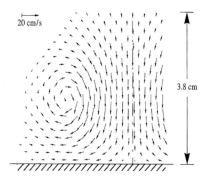

Fig. 4.7. Vector velocity field derived from the image pair in Fig. 4.6 (Gendrich, 1997).

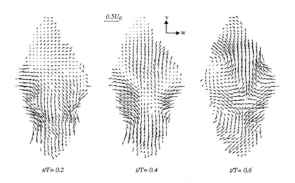

Fig. 4.8. Instantaneous MTV vectors in periodically forced wake flow at three times within the cycle. Mean flow direction is out of page (Koochesfahani, 1996).

in Fig. 4.6.

4.3.2   Caged dye tracers

Caged dye tracers have been employed in a variety of flow problems including vortex breakdown in a cylinder with a single rotating endwall (Harris, 1996), transition and turbulence in Taylor–Couette flow (Biage, 1996), measurement of internal circulation in droplets and electrohydrodynamic flows (Harris, 1996), spreading of surface tension driven flows (Dussaud, 1998), swelling of polyelectrolyte gels (Achilleos, 1998), scalar mixing in turbulent pipe flow (Guilkey, 1996), and convection in microchannel flow (Paul, 1998).

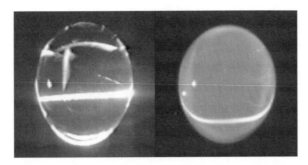

Fig. 4.9. Tag (left) and interrogate (right) image pair obtained from free falling water droplet using caged fluorescein PAF. Time delay is 29.5 ms (from Harris, 1996).

Fig. 4.9 shows a tag/interrogate image pair obtained from a free falling water droplet (Harris, 1996). The image on the left is due to elastic scattering from a single tagging pulse. The image on the right shows the interrogated line segment 29.5 ms later.

The droplets in Fig. 4.9 are approximately 5 mm in diameter and were formed from a common laboratory burette that was gravity fed from a reservoir containing a 0.20 mg/liter (6.7×10^{-8} M) solution of 3000 molecular weight caged fluorescein. A single 1 to 2 mJ tagging pulse was focused with a 20 cm lens, producing a tagged line approximately 100 μm thick. The interrogation was performed with a single 10 to 20 mJ pulse from a flashlamp-pumped dye laser which was formed into a sheet approximately 3 cm high by 300 μm thick. The pulse duration of the dye laser is 1 to 2 μs, effectively instantaneous with respect to fluid time scales. The fluorescence from the interrogation laser was focused with approximately unity magnification onto an ordinary CCD video camera (Cohu Model 4810). An OG-515 colored glass filter was employed to block the "blue" elastic scattering and transmit the "green–red" fluorescence.

Fig. 4.10 shows the velocity profile obtained from the image pair in Fig. 4.9. The lower line indicates the apparent velocities derived directly from the raw data. This data is obtained by taking vertical slices through the image and applying least-squares type fitting routines to locate the center of the grey scale intensity at each horizontal pixel location. This procedure gives the apparent result that all of the fluid is traveling downward with respect to the reference frame of the droplet. This unphysical result illustrates the significant influence of droplet curvature on the apparent velocity profiles. In effect, the droplet itself is acting as a spherical lens, introducing significant distortion.

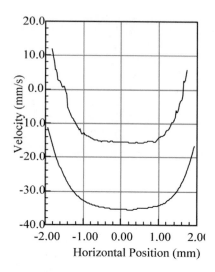

Fig. 4.10. Droplet velocity profile calculated with (upper) and without (lower) ray tracing correction (from Harris, 1996).

Several techniques have been reported for removing this optical distortion, including rather sophisticated approaches which remove both geometrical and intensity distortion (Zhang & Melton, 1994). Application of a simpler ray-tracing procedure, described in Section 4.4, show velocities becoming positive near the sides of the droplet, consistent with internal circulation. It can also be seen that the region very close to the droplet edge is obscured. This is due to total internal reflection.

As an example of a physically larger and higher Reynolds number flow field, Fig. 4.11 shows a set of representative images obtained from a flow produced by rotating concentric cylinders (Taylor-Couette flow, Biage, 1996). The cylinder height was 102 cm and the inner and outer cylinder radii were 3.14 cm and 8.18 cm, respectively. The images in Fig. 4.11 were obtained at a Reynolds number of 1.7×10^4, based on the inner/outer cylinder gap spacing, and were part of a study which obtained data for Reynolds numbers in the range 154 to 3.5×10^5. Sets of images such as those of Fig. 4.11 were used to obtain instantaneous velocity profiles and spectral density functions.

In a similar study, Harris (1997, 1999) has studied the flow produced in a cylinder with a single rotating endwall (Escudier, 1984; Brown & Lopez, 1990).

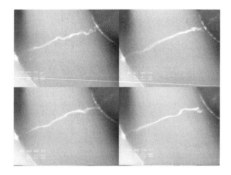

Fig. 4.11. Representative caged dye images obtained from a Taylor-Couette flow at Re = 1.7 x 10^4.

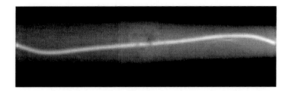

Fig. 4.12. Bottom view of interrogated line in cylinder flow with single rotating endwall. Reynolds' number is 1410 and time delay after tagging is 0.40 s (from Harris, 1997).

Fig. 4.12 shows an experimental image obtained in a plane orthogonal to the principal axis of the cylinder. A comparison between data and numerical computation is illustrated in Fig. 4.13. This comparison was performed directly in the Lagrangian reference frame by "writing" a line into the computation and allowing it to evolve in a series of real time steps. Note that in Fig. 4.13, the displacement and radial position have been non-dimensionalized by the cylinder radius.

As a final example, Fig. 4.14 shows a time sequence of three images of flow induced by a single p-xylene droplet spreading on a water surface (Dussaud, 1998). These images were obtained by writing a pair of vertical lines into a 16 cm diameter tank filled with a 0.5 mg/liter solution of caged fluorescein in water. A single drop of p-xylene was then deposited onto the quiescent water surface with a microsyringe. The droplet rapidly spreads over the surface, forming a thin film of volatile fluid. The sublayer flow pattern induced during the spreading of the volatile film was captured by viewing the interrogated image from the side.

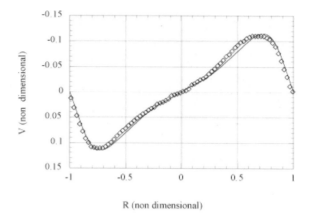

Fig. 4.13. Lagrangian frame comparison between data of Fig. 4.12 and computation. Line was written into computation and allowed to evolve in series of real time steps (from Harris, 1997).

4.4   Image Processing and Experimental Accuracy

In this section we will briefly review MTV image processing techniques and provide some estimates of experimental accuracy limits. It is straightforward to show that the relative statistical uncertainty in velocity is given by the simple expression:

$$\frac{\sigma_\nu}{\nu} = \sqrt{\left(\frac{\sigma_x}{x}\right)^2 + \left(\frac{\sigma_t}{t}\right)^2} \tag{4.4}$$

where x corresponds to displacement and t to time. In most experiments, the contribution due to time is exceedingly small and can be ignored. The statistical uncertainty in velocity, therefore, is determined by the accuracy with which fluid element displacement can be determined. As we will see, displacement accuracy of order $\pm 1\%$ is generally attainable without enormous difficulty.

4.4.1   Line processing techniques

The simplest MTV experiments are those involving single lines, or sets of parallel lines. In this case, the experiment is generally configured such that the lines are written perpendicular to the principal flow axis. The measured quantity is then the component of velocity parallel to this axis, although motion in other directions produces measurement ambiguity. This has been discussed in detail

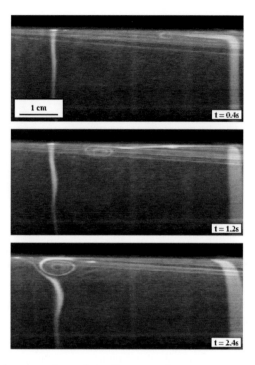

Fig. 4.14. PAF images of p-xylene film spreading on water surface, obtained 0.4 s (upper), 1.2 s (middle), and 2.4 s (lower) after deposition of single droplet (from Dussaud, 1998).

by Hill & Klewicki (1996) who have shown that for a specified (x, y)-location, the relative uncertainty in the u (principal or x-axis) component of velocity due to finite v (y-component of velocity) is given by:

$$\frac{\triangle u}{u} = \triangle t \left(\frac{v}{u} \frac{\delta u}{\delta y} \right) \qquad (4.5)$$

Alternatively, the effect of finite v can be thought of as producing an ambiguity in the y-position of the fluid element whose u component of velocity is being measured. In either case, some a priori knowledge of the flow field is clearly important.

Keeping in mind the ambiguity implicit in Eqn. 4.5, extraction of velocity is basically a matter of determining fluid element displacement. For the simplest case of single lines, least-squares fitting approaches are generally utilized to

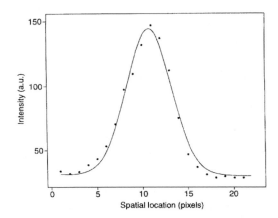

Fig. 4.15. Typical digitized intensity trace (dotted) and least squares fit (solid) from a single lateral pixel location along a caged dye interrogation line (from Lempert, 1995).

determine the center of grey-scale intensity in the y-direction (Lempert, 1995; Hill & Klewicki, 1996). As an illustrative example, Fig. 4.15 shows a typical digitized vertical "slice" through the intensity profile at a single lateral pixel location along a horizontal line obtained using caged fluorescein dye. The solid curve is the best least-squares fit of the digitized intensity profile to the sum of a constant baseline and presumed Gaussian spatial profile.

The fitting routine utilizes four variable parameters, corresponding to baseline, intensity normalization, line width, and line center. Estimate of the statistical uncertainty in each parameter is obtained by assuming that all of the residual between the data and the fit is attributable to statistical scatter. The procedure is to assume that the uncertainty in each data point is equal to the normalized rms residual in the overall fit and then to apply standard error propagation techniques (Bevington, 1969). While these assumptions are clearly not absolutely true, the procedure provides a reasonable estimate of the accuracy with which the center of intensity can be located. For the data of Fig. 4.15, the procedure yields an uncertainty (2σ) in the center of intensity of ± 0.15 pixels. This corresponds to approximately 2.5% of the $\sim$6 pixel full width at half maximum, which is not atypical. Assuming that the initial position can be determined with equal accuracy, then 1% measurement precision can be obtained by allowing the fluid to displace through $\sim$20 pixels. This assumes that the displacement uncertainty is given by $\sqrt{2}\times 0.15$ pixels. In reality, the initial position can often be determined with higher accuracy by averaging several images.

Hill & Klewicki have presented a similar approach, which uses a combination of smoothing and least-squares curve fitting to analyze lines obtained using long lifetime phosporescent molecules. They report somewhat higher uncertainties (order 0.3 to 0.4 pixels). They have also extended the technique to analyze multiple parallel lines.

4.4.2 Grid processing techniques

By writing a pattern of intersecting lines, such as that in Fig. 4.6, two-dimensional velocity vectors can be determined which are free of the ambiguity associated with Eqn. 4.5. In this section we briefly summarize two types of image processing techniques which have been developed specifically for the analysis of MTV grid data.

The most straightforward image processing approach is to extend the single line least-squares fitting to two dimensions. Hill & Klewicki (1996) have presented an algorithm in which a region of interest (ROI) is defined around each intersection point. The ROI is defined by four approximate points, two on each of the intersecting lines. Least squares fitting is used to precisely locate the line center of each of these points. A pair of straight lines are then defined by connection of each of the two pairs of points determined from the least squares procedure. The final grid point is defined by the intersection of the resulting two lines.

Gendrich & Koochesfahani (1996) have developed an alternative MTV grid image processing algorithm based on spatial cross-correlations. The technique is similar to that often employed for Particle Image Velocimetry. A rectangular window, termed the source window, is selected in the vicinity of each line crossing in the original "tagging" image. This source window is vector displaced throughout a larger "roam" window in the displaced image. The vector displacement is determined based on the maximum spatial correlation between the two images. Gendrich & Koochesfahani report typical uncertainty of ± 0.10 pixel, based on 95% confidence.

4.4.3 Ray tracing

We conclude this section by briefly considering the effects of curved surfaces, such as optical windows, or the fluid itself. Fig. 4.16 illustrates a simple droplet ray tracing procedure outlined by Harris (1996). The basic idea is to computationally transfer the intensity from the CCD image plane back to the original

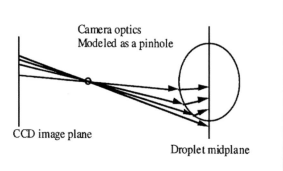

Fig. 4.16. Two dimensional illustration of simplified ray tracing procedure. Actual computations are done in three dimensions.

fluid object plane. The image optics are modeled as a simple pinhole, and intensity is translated in a straight line from a given CCD pixel, through the pinhole, until the curved surface is reached. Snell's law is then used to calculate a three-dimensional refraction angle. The translation is then continued until the plane containing the original tagged line and the principal flow axis is reached (in Fig. 4.16 this plane is assumed to contain the droplet centerline). The procedure is repeated for each CCD pixel. For small curved objects, such as droplets, the systematic error can be significant due both to imprecisely defined curvature and the lack of precise knowledge of the tagging position. In his droplet studies, Harris (1996) concluded that uncertainty of only a few percent in the droplet radius was sufficient to generate uncertainties of up to $\pm 20\%$ in the absolute velocity. This was more than an order of magnitude larger than the reported uncertainty in the relative velocity profiles.

If stereoscopic interrogation images are available, then both images can be ray-traced simultaneously. The corrected object location corresponds to the point where the two transferred fluid elements intersect (Harris, 1999).

4.5   References

Harris, S. R., Lempert, W. R., Hersh, L., Burcham, C. L., Saville, D. A., Miles, R. B., Gee, K., and Haughland, R. P., 1996, "Quantitative measurements of internal circulation in droplets using flow tagging velocimetry," AIAA J. 34, pp. 449-454.

Achilleos, E. C. and Kevrekidis, I. G. 1998. Private communication.

Bevington, P. R. 1969. *Data Reduction and Error Analysis for the Physical Sciences*. McGraw Hill.

Biage, M., Harris, S. R., Lempert, W. R. and Smits, A. J. 1996. Quantitative velocity measurments in turbulent Taylor-Couette flow by PHANTOMM flow tagging. *8th Int. Symposium on Applications of Laser Techniques to Fluid Mechanics*, Lisbon, Portugal.

Brown, G. L. and Lopez, J. M. 1990. Axisymmetric vortex breakdown. Part 2. Physical mechanisms. *J. Fluid Mech.*, **221**, 553–576.

Chen, R.F., Burck, G.G. and Alexander, N. 1967. Fluorescence decay times: proteins, coenzymes, and other compounds in water. *Nature*, **156**, 949–951.

Drexhage, K. H. 1990. Structure and properties of laser dyes. In *Topics in Applied Physics*, **1**, ed. F. P. Schafer, Springer-Verlag, Berlin, 1.

Dussaud, A., Troian, S. M. and Harris, S. R. 1998. Fluorescence visualization of a convective instability which modulates the spreading of volatile films. *Phys. of Fluids*, **10**, 1588–1596.

Escudier, M. P. 1984. Observations of the flow produced in a cylindrical container by a rotating endwall. *Expts. in Fluids*, **2**, 189–196.

Gendrich, C. P. and Koochesfahani, M. M. 1996. A spatial correlation technique for estimating velocity fields using Molecular Tagging Velocimetry (MTV). *Expts. in Fluids*, **22**, 67–77.

Gendrich, C. P., Koochesfahani, M. M. and Nocera, D. G. 1997. Molecular tagging velocimetry and other novel applications of a new phosphorescent supramolecule. *Expts. in Fluids*, **23**, 361–372.

Guilkey, J. E., Gee, K. R., McMurty, P. A. and Klewicki, J. C. 1996. Use of caged fluorescent dyes for the study of turbulent passive scalar mixing. *Expts. in Fluids*, **21**, 237–242.

Falco, R. E. and Nocera, D. 1993. Quantitative multi-point measurements and visualization of dense liquid-solid flows using Laser-Induced Photochemical Anemometry. In *Particulate Two-Phase Flow*, ed. M. C. Rocco, Butterworth-Heinemann, Boston, 59–126.

Fermigier, M. and Jenffer, P. 1987. In *Flow Visualization IV*, ed. C. Veret, Hemisphere Publishing Corporation, Washington, DC., 153–158.

Hartmann, W. K., Gray, M. H. B., Ponce, A., Nocera, D. G. and Wong, P. A. 1996. Substrate induced phosphorescence from cyclodextrin lumophore host-guest complexes. *Inorganica Chimica Acta*, **243**, 239–248.

Harris, S. R., Lempert, W. R., Hersh, L., Burcham, C. L., Saville, D. A., Miles, R. B., Gee, K. and Haughland, R. P. 1996. Quantitative measurements of internal circulation in droplets using flow tagging velocimetry. *AIAA J.*, **34**,

449–454.

Harris, S. R., Miles, R. B. and Lempert, W. R. 1997. Comparisons between flow tagging measurements and computations in a complex rotating flow. *Paper No. 97-0852*, 35th AIAA Aerospace Sciences Meeting, Reno, NV.

Harris, S. R. 1999. *Quantitative Measurements in a Lid Driven, Cylindrical Cavity using the PHANTOMM Flow Tagging Technique.* Ph.D. Thesis, Princeton University.

Hill, R. B. and Klewicki, J. C. 1996. Data reduction methods for flow tagging velocity measurements. *Expts. in Fluids*, **20**, 142–152.

Koochesfahani, M. M, Cohn, R. K, Gendrich, C. P. and Nocera, D. G. 1996. *8th Int. Symposium on Applications of Laser Techniques to Fluid Mechanics*, Lisbon, Portugal.

Lempert, W. R., Magee, K., Ronney, P., Gee, K. R. and Haughland, R. P. 1995. Flow tagging velocimetry in incompressible flow using PHoto-Activated Nonintrusive Tracking Of Molecular Motion. *Expts. in Fluids*, **18**, 249–257.

Lempert, W. R., Lee, D., Harris, S. R., Miles, R. B. and Gee, K. 1998. Miniaturization of caged dye flow tagging velocimetry for microgravity droplet diagnostics. *Paper No. AIAA-98-0512*, 36th AIAA Aerospace Sciences Meeting, Reno, NV.

Paul, P.H, Garguilo, M. G. and Rakestraw, D. J. 1998. Imaging of pressure- and electrokinetically driven flows through open capillaries. *Anal. Chem.*, **70**, 2459–2467.

Ponce, A., Wong, P. A., Way, J. J. and Nocera, D. G. 1993. Intense phosphorescence triggered by alcohols upon formation of a cyclodextrin ternary complex. *J. Phys. Chem.*, **97**, 11137–11142.

Popovich, A. T. and Hummel, R. L. 1967. Light-induced disturbances in photochromic flow visualization. *Chem. Eng. Sci.*, **29**, 308–312.

Zhang, J. and Melton, L. A. 1994. Numerical simulations and restorations of laser droplet-slicing images. *Appl. Opt.*, **33**, 192–200.

CHAPTER 5

PLANAR LASER IMAGING

Richard B. Miles[a]

5.1 Introduction

The availability of high-power, narrow linewidth, frequency-tunable lasers has opened up an array of new flow visualization techniques which are based either laser-induced fluorescence or on light scattering. For most of these applications, the laser beam is expanded into a thin light sheet, which is passed through the flow field. A camera is placed orthogonal to the direction of the light propagation, along the normal to the plane formed by the sheet of light. The laser beam is either directly scattered from molecules or particles in the flow or is absorbed by molecules which subsequently fluoresce. In either case, the camera sees a cross-sectional image of the flow field, and the intensity of the light pattern from that image can be used for either qualitative or quantitative measurements of flow field properties.

Often, the flow field is seeded with either an atomic or a molecular species which acts as the fluorescer, or with particles which follow the flow and act as scattering centers. The various approaches that rely on atomic or molecular fluorescence from species in the flow have come to be known collectively as *Planar Laser-Induced Fluorescence* (PLIF) (Hanson, 1988; Vancruyningen *et al.*, 1990). Early work in this field used fluorescence from sodium or iodine seeded into flow fields, but, more recently, has concentrated on the use of nitric oxide, acetone, or oxygen. The great advantage of these laser-induced fluorescence approaches is that the signal levels are generally strong since the excitation is through a resonant interaction with a particular atomic or molecular species. In many cases, however, the fluorescence is reduced in strength by collisional

---

[a]Dept. of Mechanical and Aerospace Engineering, Princeton University, Princeton, NJ 08544, U. S. A.

93

quenching, molecular dissociation, or intramolecular energy transfer. In particular, collisional quenching adds significant complexity to the measurement of quantitative properties such as density and temperature since the quenching rate is a function of these variables. If the laser linewidth is narrow enough, and if the frequency of the laser can be accurately controlled, then the laser-induced fluorescence approach can also be used to measure velocity through the Doppler shift associated with the absorption line feature. PLIF may be extended to flow tagging (Miles & Lempert, 1997), either through a localized excitation of some of the molecules, such as oxygen, or by a localized laser-induced chemical reaction such as the formation of ozone or OH. In this case, two laser systems are required, one to do the tagging, and the second to do the interrogation.

Direct light scattering from particles or molecules also has the potential of giving the velocity as well as imaging flow structure, and, in the case of molecular scattering, temperature and density. The most straightforward application of direct light scattering is the identification of flow structure by imaging of the light from the region of interest. If the scattering is due to molecules only, then the scattering intensity can be related to the density of the molecules, and density field images can be acquired. If particles are present, then the light scattered cannot be quantitatively associated with the density field, but may, nevertheless, produce information regarding flow structure and fluid motion. Utilization of a double-pulsed laser system together with particles for *Particle Imaging Velocimetry* (PIV) will be discussed in the following chapter and is not pursued here. Rayleigh scattering from the molecules in the flow is generally very weak and is often overwhelmed by background scattering from windows and walls. In this case, the addition of a tenuous fog of nanoscale particles can be used to enhance the Rayleigh signal for measurements of velocity and flow structure.

By combining scattering from either the molecules or particles together with atomic, molecular, or interference filters, velocity images of the flow can be acquired (McKenzie, 1996; Forkey *et al.*, 1996a; Seasholtz *et al.*, 1997). In contrast to the PLIF and PIV techniques, these images contain out-of-plane velocity information, and three such images can be used to reconstruct a three-dimensional velocity field using only a single laser sheet for illumination. In many circumstances, these same atomic and molecular filters can be used to block out background scattering from windows and walls to enhance image quality.

5.2 Planar Laser-Induced Fluorescence

For imaging with laser-induced fluorescence, the laser itself must be capable of being tuned into resonance with an optical absorption line associated with an atomic or molecular species in the flow. In general, this is accomplished using a narrow linewidth tunable dye laser, or, in the case of oxygen, a frequency-narrowed argon-fluoride laser. The strength of the laser-induced fluorescence is determined by the fraction of the illumination laser light absorbed times the fraction of that light subsequently re-emitted. This can be expressed as the product of the intensity, I_L, of the illumination laser, the atomic or molecular absorption cross section, σ, the fluorescence efficiency factor, η, and the collection efficiency of the detection system, ζ:

$$P = \frac{I_L}{\hbar \omega} \sigma \eta \zeta \qquad (5.1)$$

where P is the number of photons detected per second from a single absorbing molecule, ω is the radial frequency of the emission, and $\hbar$ is Plancks constant divided by 2π. Since fluorescence is incoherent, the total number of photons scales directly with the number of absorbing molecules in the observation volume.

The fluorescence efficiency is determined by the ratio of the radiation rate, A (s^{-1}), to the total rate at which excited atoms or molecules are lost or deactivated:

$$\eta = \frac{A}{A + Q + D + I} \qquad (5.2)$$

Various competing deactivation processes include quenching, Q, dissociation, D, and internal nonradiative deactivation, I. Usually, the radiation rate for "allowed transitions" is on the order of 10^5 s^{-1} for molecules, and 10^7 s^{-1} for atoms. If any of the other deactivation rates are faster, the fluorescence signal is reduced. Since quenching is a collisional effect, if the quenching term dominates, then the fluorescence efficiency becomes a function of the gas mixture, gas pressure, gas temperature, and the particular state excited.

Atoms and molecules that have transitions in the visible region of the spectrum tend to be reactive and are, therefore, not naturally found in low temperature flows. Before frequency-tunable ultraviolet lasers were available, much work was done using flows seeded with such reactive gases as sodium (Zimmermann & Miles, 1980) and iodine (Hiller & Hanson, 1990). The very corrosive character of iodine requires that the surface of the test vessel or wind tunnel be protected by materials such as Teflon (Eklund *et al.*, 1994). Sodium is used in very low concentrations, so the protection of walls is not a significant issue. Sodium reacts with oxygen, so it is useful in non-oxygen bearing flows such as nitrogen or helium, or at high temperature such as in engines or engine exhaust where some atomic sodium is present. As an example, sodium-seeded nitrogen and sodium-seeded helium may be used to study boundary layer structure in supersonic and hypersonic flow fields (Erbland *et al.*, 1998). In this case, a frequency-doubled Nd:YAG laser drives a frequency-narrowed tunable dye laser. The dye laser is tuned to the D2 transition of sodium at 0.5896 μm. The laser pulse lasts approximately 10 ns, and the fluorescence lifetime of the sodium upper state is 16 ns $(A = 1/(2\pi\tau) = 10^7 \text{ s}^{-1})$, so images are "frozen" in time and give an instantaneous view of the turbulent boundary layer structure. Quenching of sodium is negligible in helium, and, since dissociation does not occur and there are no non-radiative internal transitions, the fluorescence efficiency is unity. Quenching does occur in nitrogen, but it is not strong enough to significantly reduce the signal level. Since the sodium is injected only into the boundary layer, the outer edge of the boundary layer is easily detected. Examples of such images taken in a Mach 8 wind tunnel with both nitrogen and helium injection are shown in Fig. 5.1. It is interesting to note that the boundary layer structure is very different when helium is injected, as compared to when nitrogen is injected. The boundary layer structure observed in the nitrogen injection case is virtually identical to the naturally occurring boundary layer and has large-scale ejections of hot wall fluid into the free stream and large-scale incursions of free stream fluid toward the wall, as one would expect. The helium boundary layer, on the other hand, looks almost laminar.

For imaging the flow field, laser-induced fluorescence from nitric oxide, acetone, or from oxygen may be used. In the case of nitric oxide, a frequency-tripled Nd:YAG laser drives a tunable dye laser which operates in the vicinity of 0.452 μm. This laser is then frequency-doubled to 0.226 μm to overlap the molecular resonance of nitric oxide. These transitions are called the Gamma Band and promote the molecule from the ground vibrational state $[X(\nu'' = 0]$ to the lowest vibrational state of the first electronic band $[A(\nu' = 0]$. There are numerous rotational lines associated with this absorption band, and their relative strength depends on the rotational temperature of the molecule. Various types of quantitative images can be acquired using nitric oxide fluorescence. Unfortunately, due to the rather long lifetime of the A state (217 ns) (McDermid & Laudenslager, 1982), collisional quenching significantly affects the brightness of the fluorescence. This means that for quantitative measurements of nitric oxide concentration, the quenching rates must be factored in. The predominant

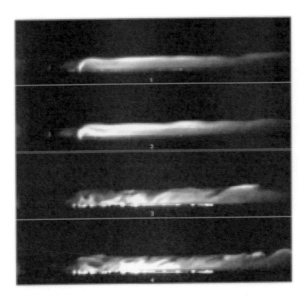

Fig. 5.1. Planar Laser-Induced Fluorescence of a Mach 8 flat plate boundary layer with normal sonic injection of sodium seeded gas. Streamwise view, with flow from left to right. Top panels show helium injection with momentum flux ration of 0.13. Bottom panels show injection of nitrogen at a momentum flux ratio of 0.11.

quencher in air is oxygen, which is approximately 1600 times more efficient at quenching NO than is nitrogen. Other important quenchers include water vapor, CO_2, and nitric oxide itself (Greenblatt *et al.*, 1987).

Quenching is of less importance for velocity and temperature measurements since those are made using various differencing schemes, so the quenching, for the most part, cancels out. For example, the measurement of temperature is made by comparing the fluorescence from the excitation of two rotational lines, each having a population fraction related to the temperature. When these fluorescent signals are divided, density and quenching terms drop out (assuming both upper states are equally quenched), and what remains is only a function of temperature (Lee *et al.*, 1993; Lachney & Clemens, 1998). Similarly, velocity measurements are made by comparing fluorescence levels from forward-propagating and backward-propagating laser beams with the laser tuned somewhat off line center. Molecules moving away from the forward propagating beam are, by geometry, moving toward the backward propagating beam. The differential shift in fluorescence from these two beams can then be related to the Doppler shift, and,

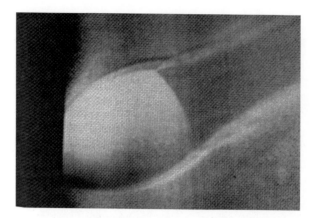

Fig. 5.2. Nitric oxide laser-induced fluorescence velocity image of a Mach 7.2 under-expanded supersonic jet (Paul, Lee & Hanson, 1989). The velocity component is 60° to the flow axis. Valid data are in the region before (to the left of) the Mach disk. White corresponds to 780 m/s, and black corresponds to -110 m/s. The vertical striations are artifacts of the laser illumination.

consequently, to the velocity (Palmer & Hanson, 1993). Fig. 5.2 is a velocity image of an underexpanded supersonic free jet (Paul *et al.*, 1989).

Recently, acetone has been used as a seed material for measurements of flow structure and flow temperature. Acetone has a very broad bandwidth absorption, extending from 0.225 to 0.320 μm (Smith & Mungal, 1998). This region is easily reached with a frequency-doubled Nd:YAG laser at 0.266 μm, and, since fluorescence occurs from the ultraviolet well into the visible region of the spectrum, acetone is much more convenient than nitric oxide for many applications. The lifetime of the upper state is dominated by intermolecular processes, and, therefore, is not affected by quenching. This means that acetone has a relatively constant fluorescence yield, and the brightness of the fluorescence can be directly related to the density of the acetone in the field-of-observation (Thurber *et al.*, 1998). Furthermore, the acetone fluorescence is a function of temperature, and by calibration this property may be used as a simultaneous temperature probe.

Laser-induced fluorescence from oxygen requires a far ultraviolet laser, most often an injection-narrowed argon-fluoride laser system (Laufer *et al.*, 1990). For room temperature or colder air, the oxygen is excited from its ground vibrational state, X, up to the B state or Schumann–Runge bands by absorbing light whose wavelength is less than 0.2 nm. Within a few picoseconds the excited oxygen molecule dissociates into atomic oxygen. As a consequence, the fluorescence signal is reduced by many orders of magnitude, but the fluorescence strength is usually unaffected by collisional quenching, which is much slower than the predissociation rate (Massey & Lemon, 1984). This has the great advantage that the oxygen laser-induced fluorescence signal is directly proportional to the density of oxygen molecules in the sample volume. By taking the ratio of the fluorescence from separate rotational lines, oxygen can be used to determine the temperature in much the same manner that nitric oxide was used (Grinstead *et al.*, 1995). The difficulty, however, is that the laser-induced fluorescence signal is so small that accurate measurements are hard to make. In addition, the absorption constant is very low because of a poor overlap between the ground state and the excited state, so the laser needs to have a high fluence (energy per unit area) in order to excite a reasonable fraction of oxygen molecules. At higher temperatures, upper vibrational states of oxygen are thermally populated. These states have a much better overlap with the excited states, so the fluorescence signal becomes significantly stronger. Fig. 5.3 shows "excitation" spectra at temperatures ranging from 100 K to 1600 K in 300 K increment (Miles *et al.*, 1988). In this case the laser is tuned across a portion of the oxygen absorption band and the total fluorescence is measured as a function of laser wavelength. The peak differential fluorescence cross section, $\partial(\sigma\eta)/\partial\Omega$, is shown on the right side, indicating the substantial increase in signal level at higher temperature. The Rayleigh signal is also shown, and is somewhat enhanced at high temperature. Note that the Rayleigh cross section is actually larger than the fluorescence below about 500 K.

The strong fluorescence from the vibrationally excited state of oxygen provides a useful way of locating and tracking vibrationally excited oxygen molecules. Since very few vibrationally excited oxygen molecules are naturally present in room temperature or colder air, this fluorescence may be used to interrogate tagged molecules for the measurement of velocity profiles, turbulent structure, etc. This is the principle associated with the RELIEF (*Raman Excitation + Laser-Induced Electronic Fluorescence*) flow tagging technique (Miles *et al.*, 1989). In this approach, oxygen molecules are vibrationally excited by a pair of high-power visible laser beams which, through a two-photon Raman process, drive the oxygen molecules into their vibrationally excited state. If these two laser beams are co-propagating, one on top of the other, and focused into the sample volume, a thin line of vibrationally excited molecules is produced in the focal region. Typically, this line has a diameter of approximately 100 μm and a length of a centimeter or so. Due to the symmetry of the oxygen molecule, the vibrational motion does not cause an electric dipole to be formed, and so the

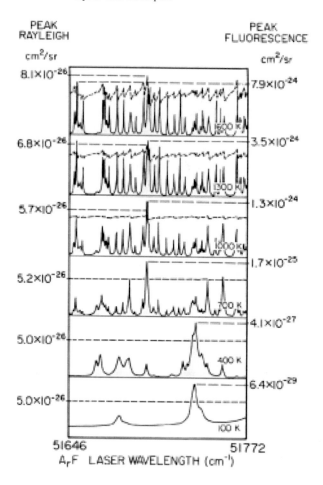

Fig. 5.3. Computer-modeled excitation scan of the argon-fluoride laser (solid line), and Rayleigh scattering (dotted line), across a portion of the oxygen Schuman–Runge absorption spectrum at temperatures from 100 K (bottom) to 1600 K (top) in 300 K increments. Values for the peak fluorescence and the peak Rayleigh cross sections are indicated to the right and left.

vibrationally excited state is metastable. That means oxygen remains in that state for a relatively long period before relaxing back to the ground state. In air, the lifetime of this vibrationally excited state is determined by collisions with water vapor, and ranges from a few microseconds in flows saturated with water vapor, to milliseconds if no water vapor is present. Since the center of the line

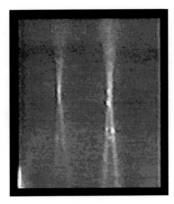

Fig. 5.4. RELIEF measurement of velocity taken in the center of the AEDC R1D one-meter diameter research facility. Flow is from right to left. The tagging cross is on the right and contains some scattering from dust particles. The time-delayed interrogation of the cross is on the left. Velocity measurement error ranged from 0.18% to 0.5% (Kohl & Grinstead, 1998).

can be found to subpixel resolution by fitting a curve to the line profile, time delays on the order of five to 10 μs are sufficient to get accurate velocity measurements. Fig. 5.4 shows an image of a displaced cross taken from experiments run at the Arnold Engineering Development Center where this technique was used to measure flow velocities under varying conditions in the R1D one-meter diameter test facility (Kohl & Grinstead, 1998). The cross on the right is the tagging location and the cross on the left has been displaced by the moving air.

The RELIEF tagging approach is particularly useful for observing turbulent structure. In that case, the time between tagging and interrogation must be short compared to the eddy roll-over time, and the line must be thin compared to the scale of the turbulent structure, which generally implies that it is thin compared to the Taylor microscale. Fig. 5.5 shows the image of a line written into a turbulent free jet and displaced for seven microseconds. Ten thousand or so such images were used to measure turbulent structure function scaling and explore the frequency of small-scale violent events (Noullez *et al.*, 1997). Other approaches to flow tagging using laser-induced chemistry are also being explored, including ozone formation (Pitz *et al.*, 1996) and water dissociation (Boedeker, 1989). Similar approaches have been demonstrated in water (Lempert *et al.*, 1995; Koochesfahani *et al.*, 1996).

Fig. 5.5. Composite image of tagged and 7 μs delayed RELIEF line in a turbulent air free jet (Noullez *et al.*, 1997).

5.3   Rayleigh Imaging from Molecules and Particles

An alternate approach to both quantitative and qualitative imaging of flow fields is to use direct light scattering from either particles or molecules in the flow. Since this process does not involve any resonant states of the molecules or particles, the scattering essentially occurs instantaneously, so light is only collected during the time the flow is illuminated. Processes such as quenching, predissociation, and intramolecular energy transfer do not play a role, and the signal is directly proportional to the number of scatterers in the volume element. For molecular scattering, this means that the signal is directly proportional to the gas density. With the use of spectral filters, velocity and temperature can also be extracted.

On a microscopic level, Rayleigh scattering occurs because the electric field of the incident laser induces an oscillating dipole moment in the molecule, which then radiates, much like a small antenna. For atomic species, this induced dipole moment is in the same direction as the electric field, so the scattered light remains well polarized. For molecules, the direction of the induced dipole moment is also affected by the nonsymmetry of the molecule itself, so there is a small amount of depolarization due to the random orientation of the molecules. In either case, scattering is minimum when the molecule is viewed along the polarization axis of the incident laser, and is a maximum orthogonal along the polarization axis. Consequently, for maximum signal strength one must insure that the laser is polarized orthogonal to the vector pointing to the camera. The "scattering plane" is defined by the three points: the light source, the scatterer, and the detector. Thus, for maximum signal intensity, the laser should be polarized orthogonal to the scattering plane.

Neglecting depolarization, the strength of the scattering is expressed in terms of the differential scattering cross section, $\partial\sigma$, into a differential solid angle element, $\partial\Omega$:

$$\frac{\partial\sigma}{\partial\Omega} = \frac{\omega^4}{c^4(4\pi)^2}\left(\frac{\alpha}{\varepsilon_0}\right)^2 \sin^2\phi \qquad (5.3)$$

where ϕ is the angle between the collection optics and the polarization vector of the illumination laser, and α is the polarizability. For a gas, the polarizability is related to the index-of-refraction, n, by the Lorentz–Lorenz relation:

$$\frac{\alpha}{\varepsilon_0} = \frac{3}{N}\left(\frac{n^2-1}{n^2+2}\right) \qquad (5.4)$$

where N is the number of molecules per unit volume. Note that since $(n^2-1)/(n^2+1)$ is proportional to gas density, α/ε_0 is a parameter associated with a particular molecular species and a particular illumination wavelength.

By integrating the differential scattering cross section over the solid angle of the collection aperture, and multiplying by the total number of molecules in the observation volume, the total light collected is then determined. From Eqn. 5.3 it is apparent that the scattering intensity can be enhanced by increasing the frequency of the illumination laser. While the scattering cross section increases as the fourth power of the frequency, most light detectors are only sensitive to the number of photons arriving at the detector surface. Since the photon flux is the intensity divided by $\hbar\omega$, this means that the actual increase in signal scales with the third power of the frequency. Nevertheless, it is of great benefit to use higher frequency laser sources for Rayleigh scattering. Some of the highest contrast images have been taken using the argon-fluoride laser at 0.193 μm. For example, Fig. 5.6 shows a Rayleigh scattering image of an underexpanded supersonic free jet, with the flow from bottom to top. The image includes a RELIEF line that was written into the flow 5 μs before the picture was taken. The Mach disk can be seen just above the center of the RELIEF line.

In many cases the Rayleigh scattering is too weak to give good image quality. This is particularly true for low density gases such as might be encountered in hypersonic flow facilities. In such cases, the Rayleigh scattering can be significantly enhanced by scattering from nanoscale particulates in the flow. Often, these particulates arise naturally from the condensation of water vapor, which occurs even in well-dried flows at temperatures below 150 K or so. Carbon-dioxide also forms small clusters and can be used for enhanced visualization. In both of these cases, the clusters are formed in the core of a supersonic flow where the temperature is low, but are usually not formed in the boundary layers where

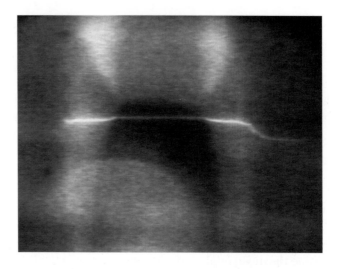

Fig. 5.6. Ultraviolet Rayleigh image of an underexpanded supersonic air jet taken with an argon-fluoride laser at 193 nm. Flow is from bottom to top.

the recovery temperature approaches the stagnation temperature. As a consequence, this condensate scattering can be used to highlight the outer portion of the boundary layer and provide a method for visualization of shock wave and boundary layer structure. By increasing the amount of carbon dioxide in the air, the CO_2 particle fog can be increased to further enhance the signal level. With CO_2 mole fractions of a percent or less, there is very little impact on the flow field itself. Measurements indicate that CO_2 condensation and sublimation occur very rapidly, so the scattering is a close indicator of the condensation temperature line and not simply a "memory" effect. Images of the interaction of a shock wave with a boundary layer taken using CO_2-enhanced are shown in Fig. 5.7. These images were taken with a 500 KHz frequency-doubled, pulse-burst Nd:YAG laser and a high-speed CCD framing camera (Wu *et al.*, 1998). The flow is from right to left, and is passing over an 14° angle wedge. The effect of the boundary layer on the shock wave can be clearly seen. It is also apparent that the boundary layer is compressed after the shock, as would be expected.[b]

---

[b]These images can be viewed as a movie on the Worldwide Web at: http://www.princeton.edu/~milesgrp/movie/pbl1.html.

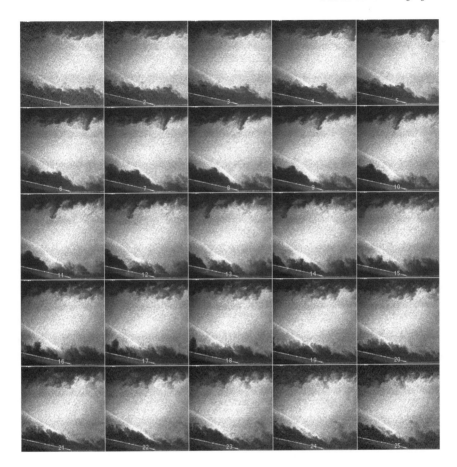

Fig. 5.7. Sequential images of a boundary layer/shock wave interaction in a Mach 2.5 flow taken at a 500 KHz framing rate.

5.4   Filtered Rayleigh Scattering

In many cases, background scattering from windows and walls either obscures or degrades Rayleigh images. This is a particularly serious problem for low density flows where the Rayleigh signal, even with augmentation from CO_2 particulates, is very small. In such cases, a sharp cut-off blocking filter can be used to eliminate the background scattering, significantly enhancing image quality (Forkey *et al.*, 1998). This is accomplished by using an injection-locked, very narrow linewidth laser which can be tuned onto the absorption line center of an atomic

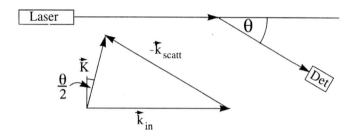

Fig. 5.8. Vector diagram for Rayleigh scattering.

or molecular vapor. The atomic or molecular vapor is then placed in a cell which, in turn, is placed in front of the camera. Light which scatters at the laser frequency is absorbed by the vapor and does not reach the camera. Light which scatters from the flow, however, is frequency-shifted due to the Doppler effect, passes through the filter, and is seen by the camera. For molecular scattering, the thermal motion of the molecules can produce a large enough frequency shift for background suppression, even when the flow is not moving. This approach works best, however, for high-speed flows where the average fluid motion produces a significant Doppler frequency shift. The Doppler shift associated with Rayleigh scattering can be represented by the vector difference between the propagation vector of the illumination light and the propagation vector of the scattered light, as shown in Fig. 5.8. The resulting vector, K, is the direction of velocity sensitivity. Molecules moving in that direction will generate a frequency shift, whereas molecules moving orthogonal to K will not. The frequency shift, $\Delta\nu$, can be represented by the expression:

$$\Delta\nu = \frac{2v}{\lambda}\sin{\tfrac{1}{2}\theta} \tag{5.5}$$

where θ is the scattering angle, as shown in Fig. 5.8, and v is the component of the velocity in the K-direction. For scattering at 90°, which is the typical configuration for imaging, the Doppler shift is 2.66 MHz per meter per second for a frequency-doubled Nd:YAG laser.

For the frequency-doubled Nd:YAG laser, molecular iodine is used in the filter. The vapor pressure is high enough and/or the cell is long enough to make the iodine absorption "optically thick," so that very little light passes through on line center. Due to a weak continuum band of iodine, the filter cannot block more than approximately five orders of magnitude before losing its out-of-band transmission. The typical filter profile is shown in Fig. 5.9 (Forkey, 1996b). Here, note that the frequency range from 90% absorption 90% transmission is approximately 300 MHz (0.01 cm^{-1}). This means that flow velocities in excess of 100 m/s or so lead to a Doppler shift large enough to move the light scattered from the flow completely into the transmission window of the filter, while the background scattering from windows and walls is simultaneously blocked. This approach has been used to observe boundary layer structure in a Mach 8 blow-down facility on flat plate and elliptical cone models (Huntley & Smits, 1999). The geometry for the elliptical cone model experiments is shown in Fig. 5.10, and a series of transitional boundary layer images captured through a molecular iodine blocking filter using this geometry is shown in Fig. 5.11. Without the filter, scattering from the model surface completely obscures the Rayleigh scattering from the flow, and the image is lost. This filter approach can also be used to highlight particular velocity components in the flow, as is shown in Fig. 5.12. These images are taken from the Mach 2.5 shock wave/boundary layer experiment discussed previously. Column 1 is without a filter. In Column 2, the high-speed components have been highlighted, and the low-speed components have been blocked. In Column 3 the high-speed components have been blocked and the low-speed components are highlighted (Wu *et al.*, 1998).

In the case of scattering from molecules, the transmission of the filter becomes a function of the thermal motion as well. In addition, acoustic waves in the gas contribute to the motion of the molecules and will change the scattering characteristics. This effect becomes particularly strong when the mean free path of a molecule is small compared to the wavelength, λ_s, of the scattering vector, K ($\lambda_s = 2\pi/|K|$). This ratio is typically described in terms of the "Y" parameter, which is the scattering vector wavelength divided by the mean free path (Tenti *et al.*, 1974). For Y parameters much larger than one, acoustic sidebands occur in the scattering spectrum, whereas for Y parameters much less than one, the scattering spectrum is Gaussian, as would be expected for molecules whose motion is dominated by thermal effects. Fig. 5.13 shows the frequency profiles of the scattered light for a variety of Y parameters. An approximate expression for the Y parameter is:

$$Y = 0.230 \left[\frac{T(K) + 111}{T^2(K)} \right] \left[\frac{p(atm)\,\lambda(nm)}{\sin \frac{1}{2}\theta} \right] \tag{5.6}$$

Here it can be seen that, for high temperatures or low pressures, the Y parameter is small and acoustic effects can be neglected. At atmospheric pressure for 90°

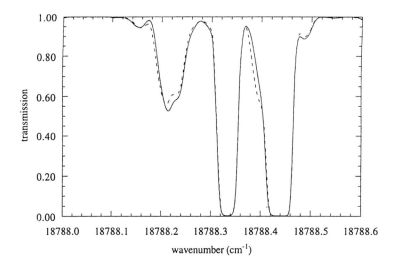

Fig. 5.9. Measured (solid line), and predicted (dashed line) transmission profile for the 9.88 cm long iodine absorption cell, with cell temperature of 353 K and cell pressure of 1.03 torr (side arm temperature = 40 C). The measured data has been normalized to unity (Forkey, 1996b).

scattering in the visible region, the Y parameter is on the order of one, so acoustic effects must be taken into account for accurate measurements of flow parameters.

Molecular Filtered Rayleigh Scattering (FRS) can be used in several ways for quantitative flow field imaging. The most straightforward approach is to frequency scan the laser and record multiple images through the atomic or molecular filter (Forkey *et al.*, 1998). The brightness of each resolvable element in the image, as a function of laser frequency, will be a convolution of the filter profile and the scattering frequency shift and lineshape. The absolute frequency of the features of that profile gives the flow velocity. A careful fitting of the profile yields temperature and pressure for each resolvable flow element. This is the approach taken to get quantitative temperature and pressure velocity from a Mach 2 free jet containing a weak crossing shock structure. An image of that free jet is shown in Fig. 5.14. A RELIEF line has been written into this jet to get an accurate measurement of the velocity profile. Filtered Rayleigh temperature, pressure, and velocity images are shown in Figs. 5.15 to 5.17. Note that the pressure and temperature data are scalar fields, and the velocity is a vector field.

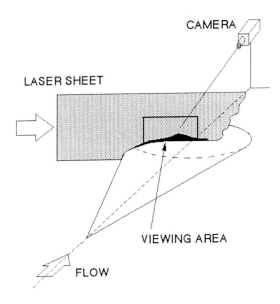

Fig. 5.10. Imaging geometry for spanwise boundary layer visualizations on 4:1 elliptic cone (Huntley & Smits, 1999).

With a single camera, this approach gives one component of the velocity vector, that lying along the vector $\vec{K}$, which is parallel to the bisector of the angle between the illumination vector and the scattering vector. Neglecting out-of-plane motion, that vector projects along the illumination direction and is responsible for the apparent nonuniformity of the velocity field in Fig. 5.17. The temperature uncertainty in these measurements is ± 5 K, the pressure uncertainty is ± 20 mbar, and the velocity uncertainty is ± 2 m/s.

The fact that the laser must be frequency scanned in order to capture the spectrum means that this approach is not suitable for imaging time-varying or turbulent flow fields. Various approaches are being examined to enable instantaneous quantitative imaging. These include simultaneous observation from many angles, and simultaneous observation through many filters. In many cases, the pressure can be assumed to be constant and the temperature field can be measured with a single pulse by using a separate monitor to measure the pressure. Since pressure equilibrium exists through out the sample volume, only one image plus a calibration image is required. This temperature measurement can be made with the laser tuned within the extinction region of the filter, so back-

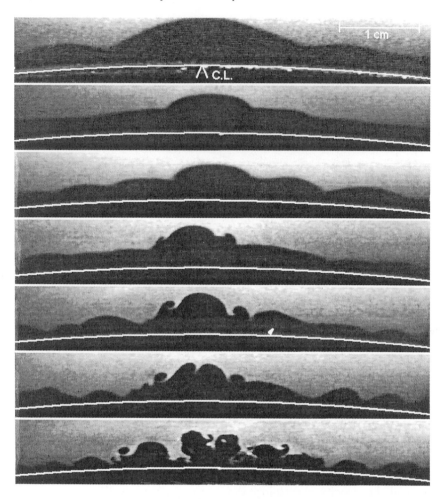

Fig. 5.11. Elliptic cone spanwise boundary layer visualizations at frestream unit Reynolds numbers from 2×10^6 (top) to 1.04×10^7 (bottom) (Huntley & Smits, 1999).

ground scattering from windows and walls, and scattering from particulates can be eliminated (Yalin *et al.*, 1999). Early LIDAR work recognized this property of atomic filters (Shimizu *et al.*, 1983), and more recent efforts are concentrating on measurements of temperatures in flames, even in the presence of small concentrations of soot (Elliott *et al.*, 1997; Hoffman *et al.*, 1996). This same approach is expected to be applicable in weakly ionized plasmas.

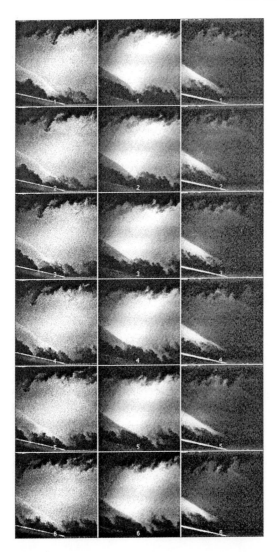

Fig. 5.12. 500 KHz images of Mach 2.5 boundary layer/shock wave interactions. Column 1: without filter. Column 2: with filter tuned to highlight high velocity components. Column 3: with filter tuned to highlight low velocity components.

5.5   Planar Doppler Velocimetry

In many cases, Rayleigh scattering from air molecules is too weak to provide adequate signal levels for flow field diagnostics. In these cases, the Rayleigh

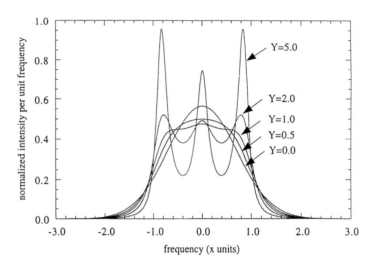

Fig. 5.13. Rayleigh-Brillouin scattering profiles for various Y values. Frequency is given in normalized units: $x = 2\pi\nu/(\sqrt{2}K v_0)$, where ν = laser frequency, and $v_0 = \sqrt{kT/m}$.

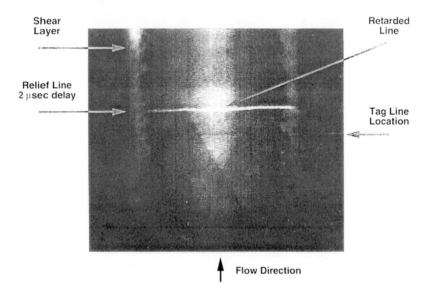

Fig. 5.14. Image of pressure-matched, Mach 2 free jet taken with an argon-fluoride laser. The velocity profile is highlighted with RELIEF flow tagging and there is a weak crossing shock structure that is apparent in filtered Rayleigh images.

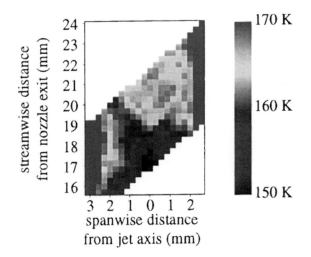

Fig. 5.15. Filtered Rayleigh temperature field of a Mach 2 pressure-matched jet, showing weak crossing shock structure (Forkey, 1996b).

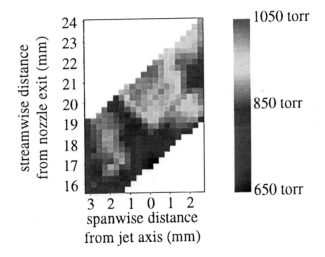

Fig. 5.16. Filtered Rayleigh pressure field of a Mach 2 pressure-matched jet, showing weak crossing shock structure (Forkey, 1996b).

signal can be significantly enhanced by seeding the flow with small particles, as has been previously mentioned. As long as the circumference of the particles is

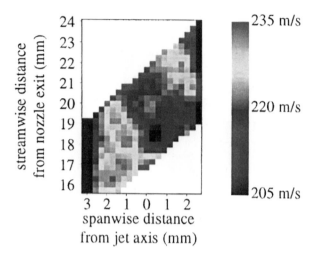

Fig. 5.17. Filtered Rayleigh velocity field of a pressure-matched Mach 2 jet showing velocity component along the illumination laser beam (Forkey, 1996b).

small compared to the wavelength of the light, then this scattering falls into the Rayleigh range. For visible light sources, this typically means that the particles must be less than 0.1 μm. In this regime, the induced polarizability for each particle scales with its volume V:

$$\frac{\alpha}{\varepsilon_0} \approx 3V \left(\frac{n^2 - 1}{n^2 + 2} \right) \qquad (5.7)$$

(Jones, 1979). This leads to a sixth power intensity dependence on the particle diameter, and the total light collected from one resolvable element in the flow will be proportional to that factor times the number of particles in the element. Thus, the Rayleigh signal is biased towards larger diameter particles, as might be expected. If the particle density is high enough to assure a large number of particles per resolvable element, but each particle is far enough apart to eliminate significant interactions between particles and multiple scattering is negligible, then the light scattered from each volume element is directly proportional to the number of particles.

The fact that these particles are heavy compared to the gas molecules means that they have very little thermal or Brownian motion associated with them compared to the gas molecules. As a consequence, the light scattered from the particles is frequency-shifted due to the average motion of the particles, but is not thermally broadened, as would be the case for Rayleigh scattering from gas molecules. This has two consequences: the first is that temperature cannot be measured using this approach, and the second is that the signal intensity is distorted by "speckle." The lack of temperature sensitivity in some ways simplifies the measurement since the Doppler shift is now only proportional to the bulk velocity of the fluid element. In the case of turbulence, there will be some broadening representing whatever distribution of velocities are present in the fluid element. If the fluid element is small compared to the structure of the turbulence, a unique velocity can be established.

The iodine absorption filter can also be used to capture images of the velocity structure in the flow by using the slope of the filter transmission as a velocity discriminator (Smith *et al.*, 1996). In this case, a reference image must be simultaneously taken so the brightness can be directly related to the filter transmission. If the laser frequency is set half-way down the slope of the transmission curve, then the corrected brightness of the image can be directly related to a value of the velocity component. For flows with non-zero average velocity, the laser frequency is selected so the average velocity falls at the center of the transmission curve and fluctuations about that average are seen as variations in transmissivity. This approach works well for scattering from particulate vapor fogs where the thermal motion of the particles is negligible. This approach to velocity measurement has come to be called *Planar Doppler Velocimetry* (PDV) and has been shown to give good quality images of velocity structure (McKenzie, 1997). In some cases, the frequency range between cut-off and full transmission is too narrow to capture the full range of velocities in the flow field. In such cases, the iodine filter absorption profile can be broadened by introducing a foreign gas such as nitrogen into the iodine cell (Elliott, 1994).

Laser speckle occurs due to the coherent interaction of light scattered from the various particles. There are two important factors that contribute to speckle: frequency broadening and collection optics. If there is no frequency broadening, that is, the scattering particles are stationary, then the speckle pattern remains constant. Nonstationary scatterers, on the other hand, cause the light to broaden in frequency, and, therefore, modulate the speckle pattern in time. The larger the frequency width, the more rapid the modulation. Particle motion arises from the random thermal motion of the scatterers, and is inversely proportional to the square root of the particle mass. For molecules, the broadening is on the order of a gigahertz or so. This means that the speckle pattern only stays constant for times on the order of a nanosecond. For particles, on the other hand, the thermal motion is small, so the speckle pattern remains constant for much longer times. Typical high-power laser systems such as the frequency-doubled Nd:YAG laser have pulse lengths on the order of 10 ns, so speckle from particle scattering is present, whereas speckle from molecular scattering is not.

The second important factor regarding speckle is the spatial frequency of the speckle pattern itself. That spatial frequency is determined by the maximum path difference between light rays. If the light is collected through a small aperture, that maximum path difference is small, and the speckle pattern features are large. If, on the other hand, the collection aperture is large, that path difference is large and the speckle pattern becomes much finer. The scale of the speckle pattern structure, λ_{sp}, is determined by the spatial Fourier transform of the collection aperture:

$$\lambda_{sp} \approx \frac{\lambda d}{D_a} \tag{5.8}$$

where D_a is the aperture diameter, d is the distance from the collection lens to the image plane, and λ is the laser wavelength. This reduces to:

$$\lambda_{sp} \approx \lambda f^{\#} \left(1 + m\right) \tag{5.9}$$

where $f^{\#}$ is the "f-number" of the collection lens (focal length/diameter), and m is the magnification ratio. If the spatial frequency of the speckle pattern is significantly smaller than the resolution scale of the detector, then the fringes are averaged and the pattern is no longer observed. This can be accomplished by choosing low f-number collection optics, low magnification, or short wavelength.

Full, three-dimensional velocity vector information can be obtained by using three cameras situated so that velocity vector components along all three orthogonal axes can be acquired. Simultaneous reference images taken without the filter are needed to calibrate the transmission and eliminate fluctuations in the illumination intensity due to laser beam nonuniformities and the reduction of light intensity due to scattering losses. Even though PDV has the capability of instantaneously capturing velocity images, data quality improves as the square root of the number of images acquired (assuming shot noise limit). The structure of speckle noise varies from image-to-image, so image averaging also significantly mitigates that problem. Fig. 5.18 is a pair of PDV images from an overexpanded supersonic jet showing both instantaneous and time-averaged velocity fields. Fig. 5.19 shows time-averaged PDV velocity data from a pair of vortices behind the trailing edge of a Mach 0.2 flow over a delta wing at 23.2 degree angle-of-attack (Mosedale *et al.*, 1998).

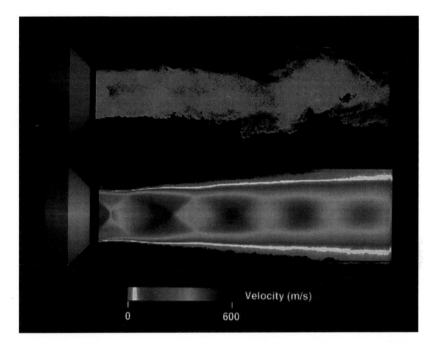

Fig. 5.18. Instantaneous and time-averaged velocity fields of an over-expanded supersonic jet (Smith & Northam, 1995). Figure also shown as Color Plate 2.

5.6 Summary

Spectrally selective laser imaging of complex flows has now come-of-age in the context of Planar Laser-Induced Fluorescence, Flow Tagging, and Rayleigh Imaging from molecules and particles. While the Planar Laser-Induced Fluorescence work has been primarily directed towards the study of combustion, properties of boundary layer structure in both supersonic and hypersonic flows have been visualized, and, in many cases, quantified using this approach. Of particular importance has been the use of nitric oxide laser-induced fluorescence for velocity and temperature measurements, sodium laser-induced fluorescence for boundary layer studies and imaging hypersonic structure, iodine laser-induced fluorescence for supersonic mixing studies, and oxygen laser-induced fluorescence for temperature measurement and as an interrogation for flow tagging. Rayleigh imaging through molecular filters has become an important technique for capturing boundary layer structure in high-speed flows, and for the imaging of temper-

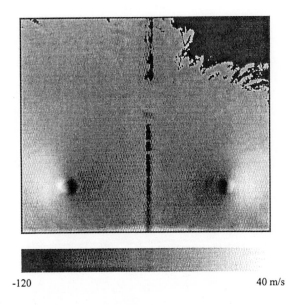

-120   40 m/s

Fig. 5.19. Time-averaged Planar Doppler Velocimetry images of a vortex pair behind the trailing edge of a Mach 0.2 flow over a delta wing at 23.2° angle-of-attack (Mosedale *et al.*, 1998).

ature, velocity, and pressure fields. Rayleigh scattering from particle fogs has been used to significantly enhance scattering intensities for imaging in lower density flows, and has also lead to Planar Doppler Velocimetry, where field velocity images can now be taken instantaneously with high resolution. With the new development of high-speed CCD imaging cameras and pulse-burst lasers, these approaches are now being used to acquire dynamic images of complex flow phenomena.

5.7   References

Boedeker, L.R. 1989. Velocity measurement by H_2O photolysis and laser-induced fluorescence of OH. *Optics Letters* **14**, 473.

Eklund, D.R., Fletcher, D.G., Hartfield, R.J., McDaniel, J.C., Northam, G.B., Dancy, C.L., and Wang, J.A. 1994. Computational experimental investigation of staged injection into a Mach 2 flow. *AIAA Journal* **32** (5), 907–916.

Elliott, G.S., Samimy, M., and Arnette, S.A. 1994. A molecular filter-based velocimetry technique for high-speed flows. *Experiments in Fluids* **18** (1–2),

107–118.

Elliott, G.S., Glumac, N., Carter, C.D., Nejad, A.S. 1997. Two-dimensional temperature field measurements using a molecular filter-based technique. *Combustion Science and Technology* **125** (1–6), 351.

Erbland, P.J., Etz, M.R., Lempert, W.R., Smits, A.J., and Miles, A.J. 1998. Optical refraction from high Mach number turbulent boundary layer structures. *Paper AIAA-98-0399*, AIAA 36th Aerospace Sciences Meeting and Exhibit, Reno, NV, Jan. 12–15, 1998.

Forkey, J. Cogne, S., Smits, A.J., Bogdonoff, S., Lempert, W.R., and Miles, R.B. 1993. Time-sequenced and spectrally filtered Rayleigh imaging of shock wave and boundary layer structure for inlet characterization. *Paper AIAA-93-2300*, AIAA/SAE/ASME/ASEE 29th Joint Propulsion Conference and Exhibit, Monterey, CA, June 28–30, 1993.

Forkey, J.N., Finkelstein, N.D., Lempert, W.R., and Miles, R.B. 1996a. Demonstration and characterization of Filtered Rayleigh Scattering for planar velocity measurements. *AIAA Journal* **34** (3), 442–448.

Forkey, J.N 1996b. *Development and Demonstration of Filtered Rayleigh Scattering — A Laser Based Flow Diagnostic for Planar Measurement of Velocity, Temperature and Pressure*. Ph.D. Thesis, Dissertation 2067-T, Department of Mechanical and Aerospace Engineering, Princeton University, Princeton, NJ.

Forkey, J.N., Lempert, W.R., and Miles, R.B. 1998. Accuracy limits for planar measurements of flow field velocity, temperature, and pressure using Filtered Rayleigh Scattering. *Experiments in Fluids* **24** (2), 151–162.

Greenblatt, G.D. and Ravishankara, A.R. 1987. Collisional quenching of NO by various gases. *Chemical Physics Letters* **136** (6), 510.

Grinstead, J.H., Laufer, G., and McDaniel, J.C. 1995. Single-pulse, two-line temperature measurement technique using KrF laser-induced O_2 fluorescence. *Applied Optics* **34** (24), 5501–5512.

Hanson, R.K. 1988. Planar Laser-Induced Fluorescence imaging. *J. Quant. Spectrosc. Radiat. Transfer* **40** (3), 343–362.

Hiller, B. and Hanson, R.K. 1990. Properties of the iodine molecule relevant to Laser-Induced Fluorescence experiments in gas flows. *Experiments in Fluids* **10** (1), 1–11.

Hoffman, D., Münch, L.-U. and Leipertz, A. 1996. Two-dimensional temperature determination in sooting flames by Filtered Rayleigh Scattering. *Optics Letters* **21** (7), 525–527.

Huntley, M. and Smits, A.J. 1999. Transition studies on elliptic cones in Mach 8 flow using Filtered Rayleigh Scattering. *European Journal of Mechanics*

B/Fluids. To appear.

Jones A.R. 1979. Scattering of electromagnetic radiation in particulate laden fluids. *J. Prog. Energy Combust. Sci.* **5**, 73–96.

Kohl, R.H. and Grinstead, J.H. 1998. RELIEF velocimetry measurements in the R1D Research Facility at AEDC. *Paper AIAA-98-2609*, 20th AIAA Advanced Measurement and Ground Testing Technology Conference, Albuquerque, NM, June 15–18, 1998.

Koochesfahani, M.M., Cohn, R.K., Gendrich, C.P., and Nocera, D.G. 1996. Molecular tagging diagnostics for the study of kinematics and mixing in liquid phase flows. Plenary Session, Eighth International Symposium on Applications of Laser Techniques to Fluid Mechanics, July 8–11, 1996.

Lachney, E.R. and Clemens, N.T. 1998. PLIF imaging of mean temperature and pressure in a supersonic bluff wake. *Experiments in Fluids* **24** (4), 354–363.

Laufer, G., McKenzie, R.L., and Fletcher, D.G. 1990. Method for measuring temperatures and densities in hypersonic wind tunnel air flows using laser-induced O_2 fluorescence. *Applied Optics* **29** (33), 4873–4883.

Lee, M.P., McMillin, B.K., and Hanson, R.K. 1993. Temperature measurements in gases by use of Planar Laser-Induced Fluorescence imaging of NO. *Applied Optics* **32** (27), 5379–5396.

Lempert, W.R., Magee, K., Gee, K.R., and Haugland, R.P. 1995. Flow tagging velocimetry in incompressible flow using Photo-Activated Nonintrusive Tracking Of Molecular Motion. *Experiments in Fluids* **18**, 249–257.

Massey, G.A. and Lemon, C.J. 1984. Feasibility of measuring temperature and density fluctuations in air using laser-induced O_2 fluorescence. *IEEE J. Quantum Elect.* **QE-20**, 454–457.

McDermid, I.S., Laudenslager, J.B. 1982. Radiative lifetimes and electronic quenching rate constants for single photon excited rotational levels of NO ($A_2\Sigma+$, $\nu' = 0$). *J. of Quantitative Spectroscopy and Radiative Transfer* **27** (5), 483–492.

McKenzie, R.L. 1996. Measurement capabilities of Planar Doppler Velocimetry using pulsed lasers. *Applied Optics* **35** (6), 948–964.

McKenzie, R.L. 1997. Planar Doppler Velocimetry performance in low-speed flows. *Paper AIAA-97-0498*, AIAA 35th Aerospace Sciences Meeting and Exhibit, Reno, NV, Jan. 6–10, 1997.

Miles, R.B., Connors, J.J., Howard, P.J. Markovitz, E.C., and Roth, G.J. 1988. Proposed single-pulse, two-dimensional temperature and density measurements of oxygen and air. *Optics Letters* **13** (3), 195–197.

Miles, R.B., Connors, J.J., Markovitz, E.C., Howard, P.J., and Roth, G.J.

1989. Instantaneous profiles and turbulence statistics of supersonic free shear layers by Raman Excitation + Laser-Induced Electronic Fluorescence (RELIEF) velocity tagging of oxygen. *Experiments in Fluids* **8** (1–2), 17–24.

Miles, R.B. and Lempert, W.R. 1997. Quantitative flow visualization in unseeded flows. *Annual Review Fluid Mechanics* **29**, 285–326.

Mosedale, A.D., Elliott, G.S., Carter, C.D., Weaver, W.L., Beutner, T.J. 1998. On the use of Planar Doppler Velocimetry. *Paper AIAA-98-2809*, AIAA 29th Fluid Dynamics Conference, Albuquerque, NM, June 15–18, 1998.

Noullez, A., Wallace, G., Lempert, W., Miles, R.B., and Frisch, U. 1997. Transverse velocity increments in turbulent flow using the RELIEF technique. *J. of Fluid Mechanics* **339**, 287–307.

Palmer, J.L. and Hanson, R.K. 1993. Planar Laser-Induced Fluorescence imaging in free jet flows with vibrational nonequilibrium. *Paper AIAA-93-0046*, AIAA 31st Aerospace Sciences Meeting and Exhibit, Jan. 11–14, 1993, Reno, NV.

Paul, P.H., Lee, M.P., and Hanson, R.K. 1989. Molecular velocity imaging of supersonic flows using pulsed Planar Laser-Induced Fluorescence of NO. *Optics Letters* **14**, 417–419.

Pitz, R.W., Brown, T.M., Nandula, S.P., Skaggs, P.A., DeBarber, P.A., Brown, M.S., and Segall, J. 1996. Unseeded velocity measurement by ozone tagging velocimetry. *Optics Letters* **21** (10), 755–757.

Seasholtz, R.G., Buggele, A.E., and Reeder, M.F. 1997. Flow measurements based on Rayleigh scattering and Fabry-Perot interferometer. *Optics and Lasers in Engineering* **27** (6), 543–570.

Shimizu, H., Lee, S.A., and She, C.Y. 1983. High spectral resolution LIDAR system with atomic blocking filters for measuring atmospheric parameters. *Applied Optics* **22** (9), 1373–1381.

Smith, M.W. and Northam, G.B. 1995. Application of absorption filter-Planar Doppler Velocimetry to sonic and supersonic jets. *Paper AIAA-95-0299*, AIAA 33rd Aerospace Sciences Meeting and Exhibit, Reno, NV, Jan. 9–12, 1995.

Smith, M.W., Northam, G.B., and Drummond, J.P. 1996. Application of absorption filter Planar Doppler Velocimetry to sonic and supersonic jets. *AIAA Journal* **34** (3), 434–441.

Smith, S.H. and Mungal, M.G. 1998. Mixing, structure and scaling of the jet in crossflow. *Journal of Fluid Mechanics* **357**, 83–122.

Tenti, G., Boley, C.D., Desai, R.C. 1974. On the kinetic model description of Rayleigh-Brillouin scattering from molecular gases. *Canadian Journal of Physics*

52, 285.

Thurber, M.C., Grissch, F., Kirby, B.J., Votsmeier, M., and Hanson, R.K. 1998. Measurements and modeling of acetone Laser-Induced Fluorescence with implications for temperature-imaging diagnostics. *Applied Optics* **37** (21), 4963–4978.

Vancruyningen, I., Lozano, A., and Hanson, R.K. 1990. Quantitative imaging of concentration by Planar Laser-Induced Fluorescence. *Experiments in Fluids* **10** (1), 41–49.

Wu, P., Lempert, W.R., and Miles, R.B. 1998. Tunable pulse-burst laser system for high-speed imaging diagnostics. *Paper AIAA-98-0310*, AIAA 36th Aerospace Sciences Meeting, Reno, NV, Jan. 12–15, 1998.

Yalin, A., Ionikh, Y., and Miles, R. 1999. Ultraviolet Filtered Rayleigh Scattering temperature measurements using a mercury filter. *Paper AIAA-99-0642*, 37th AIAA Aerospace Sciences Meeting, Reno, NV, Jan. 11–14, 1999.

Zimmermann, M. and Miles, R.B. 1980. Hypersonic-helium flow field measurements with the Resonant Doppler Velocimeter. *Applied Physics Letters* **37** (10), 885–887.

CHAPTER 6

DIGITAL PARTICLE IMAGE VELOCIMETRY

Morteza Gharib and Dana Daribi[a]

6.1 Quantitative Flow Visualization

Particle Image Velocimetry (PIV) can be considered one of the most important achievements of flow diagnostic technologies in the modern history of fluid mechanics. In this chapter, our intent is to provide a general understanding of the concepts behind this powerful global quantitative flow visualization method as well as some if its novel applications. Various technical aspects of digital particle image velocimetry have been the subject of numerous papers and books in the available literature. Throughout this paper, some key references will be introduced for the reader to consult for deeper exposure to the subject.

Quantitative flow visualization has many roots and has taken several approaches. The advent of digital image processing has made it possible to practically extract useful information from every kind of flow image. In a direct approach, the image intensity or color (wavelength or frequency) can be used as an indication of concentration, density and temperature fields or of gradients of these scalar fields in the flow (Merzkirch, 1987). These effects can be part of the inherent dynamics of the flow (for example, gradients of density are used in shadowgraph and schlieren techniques), or generated through the introduction of optically passive or active dye agents (fluorescent tracers, liquid crystals), or various molecular tagging schemes.

In general, the optical flow or the motion of intensity fields can be obtained through time sequenced images (Singh, 1991). For example, the motion of patterns generated by dye, clouds or particles can be used to obtain such a time sequence. The main problem with using a continuous-intensity pattern

[a]Center for Quantitative Visualization at the Graduate Aeronautical Laboratories, Mail Stop 205-45, California Institute of Technology, Pasadena, CA 91125, U. S. A.

generated by scalar fields (for example, dye patterns) is that it must by fully resolved (space/time), and contain variations of intensity at all scales before mean and turbulent velocity information can be obtained (Pearlstein, 1995). In this respect, the discrete nature of images generated by seeding particles has made particle tracking the method of choice for whole field velocimetry. The technique recovers the instantaneous two- and three-dimensional velocity vector fields from multiple photographic images of a particle field within a plane or volumetric slab of a seeded flow, which is illuminated by a light source. Various methods for individual tracking of particles can be used to obtain the displacement information and subsequently the velocity fields. The spatial resolution of this method depends on the number density of the particles. A major drawback in tracking individual particles has been the unacceptable degree of manual work that is required to obtain the velocity field from a large number of traces or particle images. Digital imaging techniques have helped to make particle tracking less laborious (Gharib & Willert, 1990). However, because of the errors involved in identifying particle pairs in high particle-density images, the design of an automatic particle tracking method, especially for three-dimensional flows, has been extremely challenging. Therefore, applications of the automatic particle tracking methods have been limited to low particle density images. In this respect, an alternative method which concentrates on following a pattern of particles has been implemented by various investigators in order to resolve the above mentioned issues with a particle tracking method. This method is known as particle image velocimetry.

6.2   DPIV Experimental Setup

For most fluid flow applications, experiments are performed either in air or water. Since these fluids are transparent, the flow must be quantitatively visualized through the use and motion of flow markers. Fig. 6.1 shows a standard acquisition setup for Digital Particle Image Velocimetry (DPIV) image acquisition either in a water or wind tunnel. For water, fluorescent, polystyrene, silver-coated particles, or other highly reflective particles must be used to seed the flow; while olive oil or alcohol droplets are generally used for wind tunnels. Since the fluid velocity is inferred from the particle velocity, it is important to select markers that will follow the flow to within acceptable uncertainties without affecting the fluid properties that are to be measured. This implies that the fluid marker must be small enough to minimize velocity differences across its dimensions, and to have a density as close as possible to the density of the fluid being measured.

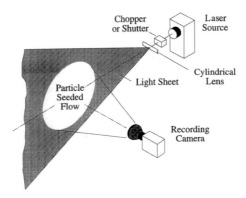

Fig. 6.1. DPIV experimental setup.

Further discussions of types of particles, and error analysis associated with particle motion, is given by Merzkirch (1987), Adrian (1986b, 1991), and Melling (1997). Upon proper selection, the particles are illuminated with a pulsed laser sheet, most typically a Nd:YAG laser. The images of the reflected particles are then acquired with a CCD camera, typically at a 30 frame per second video rate, where each image is singly exposed. Though this video rate may seem too slow, the DPIV technique has overcome this limitation through the evolution of the CCD chip, which shall be discussed in the following sections. The images are captured onto digital memory using a computerized data acquisition system. Finally, the shift of the particle images between sequential image pairs is measured using a cross-correlation technique. As the process is entirely digital, we refer to this approach as digital particle image velocimetry or DPIV.

6.3 Particle Image Velocimetry: A Visual Presentation

In contrast to the tracking of individual particles, the particle image velocimetry (PIV) technique follows a group of particles through statistical correlation of sampled numbers of the image field. This scheme removes the problem of identifying individual particles. Through the use of the statistical evaluation of PIV, it is possible to obtain the displacement field between two time-evolved patterns of particle images where the images are independently recorded by photographic or video cameras. An in-depth review of various particle imaging techniques is given by Adrian (1986b, 1991) and Keane & Adrian (1990, 1991). Here, we follow methodologies initially described by Willert & Gharib (1991) and Westerweel (1993). To demonstrate the cross-correlation concept, the reader can consult Adrian (1986b, 1991) and Hinsch (1993) for mathematical treatment of autocorrelation techniques, which are suitable for image fields with multiply exposed particle fields. Consider two instantaneous images of a particle-laden flow field taken by a video camera at two consecutive times τ and $\tau + \triangle\tau$. A sample of such images is shown in Figs. 6.2a and 6.2b. Assume that the particle field is translated by a one-dimensional parallel flow field, thereby generating another image at the time $\tau + \triangle\tau$. By combining these two time-elapsed images, one can obtain a composite, such as the one depicted in Fig. 6.2c, by carefully choosing a proper $\triangle\tau$. Through an intriguing coordination of eye and brain, linear motion of the particle field in the composite image can be sensed. In another example (Fig. 6.2d), a rotational motion can be generated by rotating image 2b with respect to 2a to generate a composite image such as 2d.

It is remarkable that the human eye-brain coordination can correlate the two time-evolved images in order to sense the motion. In the next section, the mathematical foundation for this process is laid, which makes it possible to automate this procedure. This process is known as statistical image correlation and its digital implementation as DPIV.

6.4   Image Correlation

In DPIV, two sequential digital images are subsampled at one particular area via an interrogation window (Fig. 6.3). Within these image samples an average spatial shift of particles may be observed from one sample to its counterpart in the other image, provided a flow is present in the illuminated plane. This spatial shift may be described quite simply with a linear digital signal-processing model shown in Fig. 6.4.

One of the sampled regions obtained through an interrogation window, $f(m,n)$ may be considered the input to a system whose output $g(m,n)$ corresponds to the sampled region of another image taken a time $\triangle\tau$ later. The system itself consists of two components, a spatial displacement function $d(m,n)$ (also known as the system's impulse response) and an additive noise process $N(m,n)$. This noise process is a direct result of particles moving off the sampling region, particles disappearing through three-dimensional motions in the laser sheet, the total number of particles present in the window and other components that may add to the measurement uncertainty. Of course the original sample $f(m,n)$ and $g(m,n)$ may be noisy as well. The major task in DPIV is the estimation of the spatial shifting function $d(m,n)$, but the presence of noise

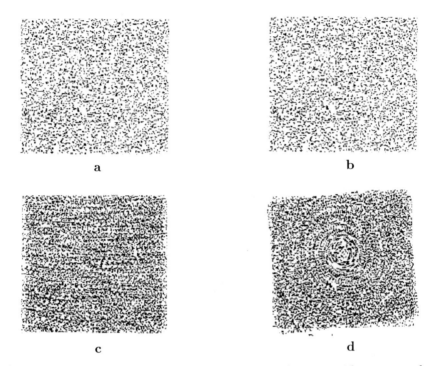

Fig. 6.2. **a** and **b** are sample particle images. By translating **a** with respect to **b** and overlaying the two, a simulated translational shift is obtained and shown in **c**. By rotating **a** with respect to **b** and overlaying the two, a rotational shift is obtained and shown in **d**.

$N(m, n)$ complicates this estimation. A description of how the output sample $g(m, n)$ relates to the input sample $f(m, n)$ can be given mathematically through the use of the discrete cross-correlation function.

The idea is to find the best match for the shifted particle pattern in the interrogation windows in a statistical sense. This can be mathematically formulated through the discrete cross correlation function,

$$C(i', j') = \sum_{i=-k}^{k} \sum_{i=-l}^{l} f(i,j)g(i+i', j+j') \qquad (6.1)$$

where f and g are intensity values or gray levels of pixels in the interrogation windows f and g shown in Fig. 6.4. The size of the interrogation window f is usually chosen to be smaller than g for the purpose of linearly shifting it

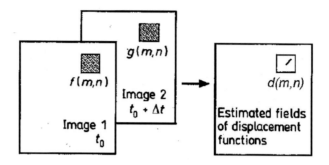

Fig. 6.3. Sequential images are subsampled with interrogation windows producing displacement vectors.

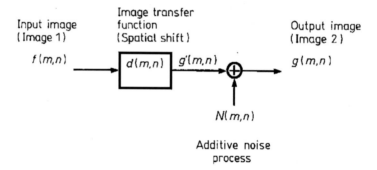

Fig. 6.4. Linear digital signal processing model describing the DPIV method

within the boundaries of interrogation window g (Keane & Adrian, 1992). For each set of correlations performed where and a correlation plane with the size of $(2M + 1) \times (2N + 1)$ will be formed. The sum of the products of pixel intensity values attain its maximum value whenever a particle pattern match occurs in a given location within the interrogation window. For a given shift value, C gives a statistical measure of matching of two particle patches. For example, Figs. 6.5a and 6.5b show sample particle images displaced by 8 pixels with respect to each other, and Fig. 6.5c shows the cross-correlation of these two images. The best estimate of the particle displacements is given by the maximum value within the cross-correlation domain.

It is noted that this correlation process only recovers a linear shift. This is due to the first order approximation of the correlation function. To ensure that the second order effects such as velocity gradient would not hinder the

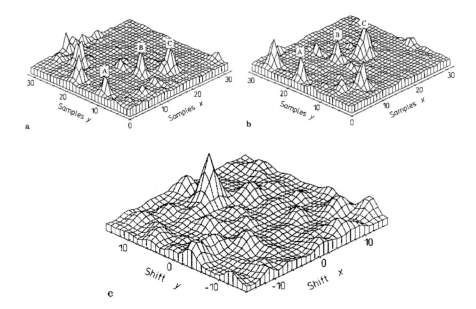

Fig. 6.5. Cross-correlation estimate between image **a** and **b**, resulting in the correlation domain **c**. Particle displacements are 8 pixels in the y-direction.

basic correlation process, the window size needs to be small enough so that the velocity gradient effect within the window can be neglected. This shall be discussed further in Section 6.7.

6.4.1 Peak finding

Perhaps the most important step in DPIV is locating the position of the correlation peak to sub-pixel accuracies. Typically, correlation results without special peak finding schemes are accurate only to within +/- half pixels. However, with peak finding schemes, it is possible to obtain accuracies as low as 0.01 pixels. Several sub-pixel peak finding schemes have been studied. Centroiding, defined as the ratio of the first order moment to the zeroth order moment (Alexander & Ng, 1991), has been used initially, and required that the correlation domain be thresholded in order to define the region containing the correlation peak. For fractional displacements, this scheme strongly biases the sub-pixel measurements towards integer values (Westerweel, 1997); an effect referred to as "peak-locking." For particle images in the 2 to 3 pixel range, more reliable methods such as parabolic and Gaussian curve fits (Westerweel, 1993; Willert & Gharib, 1991) have also been developed. Of these, the Gaussian three-point curve fit produces the least uncertainty since the cross-correlation peak itself displays a Gaussian intensity profile (Westerweel, 1993; Raffel *et al.*, 1998).

6.4.2   Computational implementation of DPIV in frequency space

The reader will appreciate that the number of multiplications per correlation value increases quadratically with sample size, which burdens heavy computational duties. To resolve this, Willert & Gharib (1991) suggested the use of a fast Fourier transform (FFT) to simplify and significantly speed up the cross-correlation process. Rather than performing a sum over all the elements of the sampled region for each element as in Eqn. 6.1, the operation can be reduced to a complex conjugate multiplication of each corresponding pair of Fourier coefficients. This reduces the number of computational operations from N^4 to $N^2 \log_2 N$ operation per sample correlations. Furthermore, Willert (1992) suggested taking advantage of the symmetric properties of the Fourier transform to allow for even further reduction of computational time. The Nyquist sampling criterion associated with the discrete Fourier transforms limits the maximum recoverable spatial displacement in any sampling direction to half the window size in that direction. In reality, even this displacement is often too large for the technique to work properly, since the signal to noise ratio in the cross-correlation decreases with increasing spatial shift.

6.5   Video Imaging

As described earlier, images are acquired with a CCD camera. An understanding of CCD cameras is therefore imperative in order to be able to take full advantage of their features. CCD cameras contain an array of photosensitive pixels that are sensitive to light. Standard CCD video cameras are capable of acquiring video at 30 frames/s. The full-frame CCD cameras read out pixel values sequentially in a row-by-row manner, requiring almost one full frame time (1/30 s) to read out completely. This presented a severe limitation, as this type of CCD necessitated the light source to be pulsed at exactly the same location within each frame. Therefore, initial applications of DPIV were limited to slow flows, since sequential images could only be pulsed synchronously at 1/30 s time difference (Fig. 6.6i). To overcome this limitation, Dabiri & Roesgen (1991) suggested exposing each frame asynchronously. To do so, they suggested using

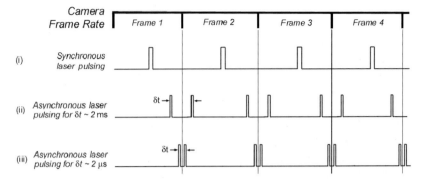

Fig. 6.6. Image acquisition timing diagram. (i) pulsed exposure of full frame CCD cameras (ii) pulsed exposure for frame transfer CCDs allowing for a minimum of 2 ms pulse separation (iii) pulsed exposure for full frame interline transfer CCDs allowing for a minimum of 2 μs pulse separation.

the frame transfer CCD. This CCD is exactly the same as a full frame CCD, except that the lower half is masked off and used only for storage. Using the frame transfer CCD marked a significant improvement as shifting the image from the exposed section to the masked-off section took about 2 ms, making it possible to reduce the time separation between the laser exposure pulses to 2 ms (Fig. 6.6ii). This increased the use of DPIV to study fluid flows that were one order of magnitude faster than what had been previously possible (Willert & Gharib, 1997; Weigand, 1996; Weigand & Gharib, 1996). Most recently, the full-frame interline transfer CCD has allowed even shorter pulse separations to be implemented for DPIV. Rather than use half of the CCD array as storage, this CCD placed the masked storage area adjacent to the pixel itself, making the total image shift time into storage approximately 1 microsecond. This vastly broadened the applications of DPIV to study even faster fluid flows as it reduced the pulse separation by 3 orders of magnitude with respect to the frame transfer CCD and 4 orders of magnitude with respect to the full frame CCD (Fig. 6.6iii). Dabiri & Gharib (1994) first implemented this technology using a 1-microsecond pulse separation to quantitatively visualize a high speed jet with exit velocity of 220 m/s (Fig. 6.7). More technical descriptions of these CCDs are explained by Raffel *et al.* (1998).

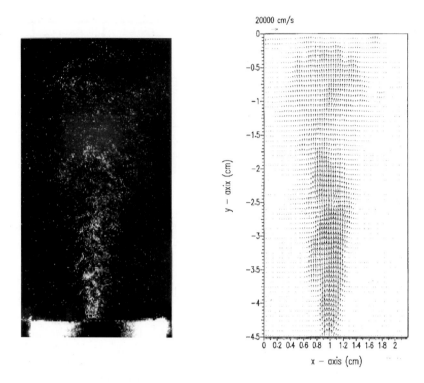

Fig. 6.7. Velocity field of a high speed jet using a pulse separation of 1 microsecond. Single image showing particulated flow is shown on the left. The resulting vector field is shown on the right.

6.6 Post Processing

6.6.1 Outlier removal

Flow seeding, though random, is not entirely uniform. Thus, it is possible to have small patches within the illuminated region that is devoid of particles. Likewise, severe three-dimensional motions can cause few and often erroneous particle-pairs. In such instances, the correlations provide incorrect data, resulting in outliers. If left untreated, these erroneous vectors will further result in erroneous differentiable and integrable quantities such as vorticity, shear and normal strains, and the streamline calculations. For example, an outlier removal scheme can be devised where each vector is compared with each of its eight surrounding neighbors. If the difference between the vector and each of its neighbors exceeds a given threshold by more than four instances, it is labeled as an outlier, and is re-interpolated from the remaining surrounding vectors using a bi-linear approach (Willert & Gharib, 1991).

6.6.2 Differentiable flow properties

Once all outliers have been removed, it is possible to post-process the velocity fields to obtain higher order properties of the flow such as the vorticity, and the strain fields. The deformation tensor is:

$$\frac{d\vec{U}}{d\vec{X}} = \begin{bmatrix} \frac{\delta u}{\delta x} & \frac{\delta \nu}{\delta x} & \frac{\delta w}{\delta x} \\ \frac{\delta u}{\delta y} & \frac{\delta \nu}{\delta y} & \frac{\delta w}{\delta y} \\ \frac{\delta u}{\delta z} & \frac{\delta \nu}{\delta z} & \frac{\delta w}{\delta z} \end{bmatrix} \tag{6.2}$$

and using strain and vorticity terms can be expressed as:

$$\frac{d\vec{U}}{d\vec{X}} = \begin{bmatrix} \varepsilon_{xx} & \frac{1}{2}\varepsilon_{xy} & \frac{1}{2}\varepsilon_{xz} \\ \frac{1}{2}\varepsilon_{yx} & \varepsilon_{yy} & \frac{1}{2}\varepsilon_{yz} \\ \frac{1}{2}\varepsilon_{zx} & \frac{1}{2}\varepsilon_{zy} & \varepsilon_{zz} \end{bmatrix} + \begin{bmatrix} 0 & \frac{1}{2}w_z & -\frac{1}{2}w_x \\ -\frac{1}{2}w_z & 0 & \frac{1}{2}w_y \\ -\frac{1}{2}w_x & \frac{1}{2}w_y & 0 \end{bmatrix} \tag{6.3}$$

where ε is the strain field, and ω is the vorticity field. Since DPIV, as a global two-dimensional technique, can only measure the u and v velocity components, the w and $\delta/\delta z$ terms in the deformation tensor are non-measureable, since velocities and gradients in a direction normal to the illumination plane cannot be determined. For measurement purposes, this reduces the deformation tensor to:

$$\frac{d\vec{U}}{d\vec{X}} = \begin{bmatrix} \varepsilon_{xx} & \frac{1}{2}\varepsilon_{xy} \\ \frac{1}{2}\varepsilon_{yx} & \varepsilon_{yy} \end{bmatrix} + \begin{bmatrix} 0 & \frac{1}{2}w_z \\ -\frac{1}{2}w_z & 0 \end{bmatrix} \tag{6.4}$$

Therefore, the only differentiable quantities that can be calculated from the velocity field are:

$$\omega_z = \frac{\delta \nu}{\delta x} - \frac{\delta u}{\delta y}, \ \varepsilon_{xy} = \frac{\delta \nu}{\delta x} + \frac{\delta u}{\delta y}, \ \eta = \varepsilon_{xx} + \varepsilon_{yy} = \frac{\delta \nu}{\delta y} + \frac{\delta u}{\delta x} \tag{6.5}$$

where η is a measure of the out-of-plane strain. In order to be able to calculate the above quantities, it is important to identify the different type of schemes available. Raffel *et al.* (1998) studied several finite-difference schemes such as forward and backward schemes, and second-order schemes, such as central differencing, the Richardson extrapolation, and the least-squares schemes. The study concluded that the least-square scheme produced the least uncertainty. Several alternative methods have also been suggested for calculating the vorticity, shear

and normal strains terms. By using Stokes theorem, the vorticity can be related
to the circulation by:

$$\Gamma = \oint \vec{u} \cdot d\vec{l} = \int \vec{\omega} \cdot d\vec{S} \tag{6.6}$$

where l is the integration path around a surface S. This can be rewritten as:

$$(\omega_z)_{avg} = \frac{1}{2}\Gamma = \frac{1}{2} \oint \vec{u} \cdot d\vec{l} \tag{6.7}$$

to give the value of the average vorticity within the enclosed area. In practice, the
following formula provides this vorticity estimate (Reuss *et al.*, 1986; Landreth
& Adrian, 1990a):

$$\begin{aligned}
[\omega_z]_{i,j} &= [u_{i-1,j-1} + 2u_{i,j-1} + u_{i+1,j-1} - u_{i+1,j+1} - 2u_{i,j+1} - u_{i-1,j+1}] \frac{1}{8\delta y} \\
&+ [v_{i+1,j-1} + 2v_{i+1,j} + v_{i+1,j+1} - v_{i-1,j+1} - 2v_{i-1,j} - v_{i-1,j-1}] \frac{1}{8\delta x} \tag{6.8}
\end{aligned}$$

Westerweel (1993) found that this method resulted in the best vorticity esti-
mator, as it provided the least measurement uncertainty when examined under
ideal and noisy conditions. In actuality, this formula is equivalent to using a
central difference scheme with a 3 by 3 smoothing kernel. Likewise, the normal
strain can be found by calculating the total entrainment along a closed path,
and dividing by the enclosed area. The resulting formula is:

$$\begin{aligned}
\left(\frac{\delta w}{\delta z}\right)_{i,j} &= \frac{\oint_A \vec{u} \cdot d\vec{l}}{A} \\
&= [u_{i-1,j+1} + 2u_{i-1,j} + u_{i-1,j-1} - u_{i+1,j-1} - 2u_{i+1,j} - u_{i+1,j+1}] \frac{1}{8\delta x} \\
&+ [v_{i-1,j-1} + 2v_{i,j-1} + v_{i+1,j-1} - v_{i+1,j+1} - 2v_{i,j+1} - v_{i-1,j+1}] \frac{1}{8\delta y} \tag{6.9}
\end{aligned}$$

A similar approach is used to calculate the shear strain even though no direct
use of the Stokes law can be made:

$$\begin{aligned}
[\varepsilon_{xy}]_{i,j} &= [u_{i+1,j+1} + 2u_{i,j+1} + u_{i-1,j+1} - u_{i+1,j-1} - 2u_{i,j-1} - u_{i-1,j-1}] \frac{1}{8\delta y} \\
&+ [v_{i+1,j-1} + 2v_{i+1,j} + v_{i+1,j+1} - v_{i-1,j-1} - 2v_{i-1,j} - v_{i-1,j+1}] \frac{1}{8\delta x} \tag{6.10}
\end{aligned}$$

6.6.3   Integrable flow properties

It is equally possible to obtain integrable quantities, such as circulation and streamlines. As mentioned above, the circulation is the line integral of the dot product of the local velocity vector and the incremental path length vector over a closed path length. The line integral shown in Eqn 6.6,

$$\Gamma = \oint \vec{u} \cdot \vec{dl} \tag{6.11}$$

can therefore be used to calculate circulation, since the velocity measurements are less noisy than the vorticity calculations. The implementation of equation is straightforward and the main question is in regards to the path length chosen for integration. For flows where the vorticity field is of interest, the ideal choice of the integration path would be a path of constant vorticity since, by definition, circulation is the integrated vorticity over a given area. Willert & Gharib (1991), for example, used sequential concentric constant vorticity-value rings around the vortex ring's core centers as the integration path length to show the distribution of circulation.

It is also possible to integrate the velocity field to obtain streamlines. This is done by assuming that the flow is two-dimensional, while making use of potential theory to relate the stream and potential functions to the velocity field:

$$\Psi = \int_y u \, dy - \int_x \nu \, dx \ , \ \Phi = \int_y \nu \, dy - \int_x u \, dx \tag{6.12}$$

Integration of the above provide reasonable results. However, it should be understood that Eqns. 6.12 reduce the Poisson equation:

$$\nabla^2 \Psi = -\omega_z \tag{6.13}$$

to the Laplacian equation:

$$\nabla^2 \Psi = 0 \tag{6.14}$$

which assumes that the flow is irrotational. Solving the full Poisson equation is difficult, as the vorticity field is only approximated by the velocity field. Moreover, the boundary conditions must be specified prior to integration (Willert, 1992).

6.7   Sources of Error

As with all measurement techniques, it is important to be aware of the sources of errors that could contribute to the measurements. There are several sources

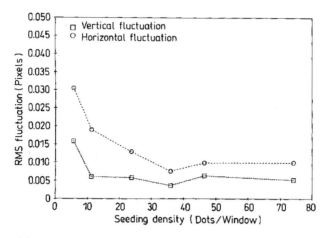

Fig. 6.8. Measurement fluctuations as a function of varying seeding densities

of error that the careful researcher must make sure to minimize in order to ensure the best measurement results. As addressing these sources of errors are difficult through direct experiments, these issues can best be addressed through simulations, since various parameters can be varied one at a time and compared with known results in order to ascertain their effects (Willert & Gharib, 1991; Raffel *et al.*, 1998; Keane & Adrian, 1990, 1991, 1992; Westerweel, 1993).

6.7.1   Uncertainty due to particle image density

Willert & Gharib (1991) were able to show that the measurement uncertainty decreases as the particle image density increases (Fig. 6.8), since there are more particles within the interrogation window that contribute to the cross-correlation peak. It is, therefore, important to maximize the particle density without changing the physical properties of the fluid or losing the particle image shape.

6.7.2   Uncertainty due to velocity gradients within the interrogation windows

Willert (1992) also showed that uncertainty of the measurements increases as the gradient within the interrogation window increases (Fig. 6.9), and further explains that the velocity is biased towards a lower velocity. This is due to the fact that there are more particles from the high-speed side of the gradient leaving the interrogation window, leaving only lower speed particles, which contribute to

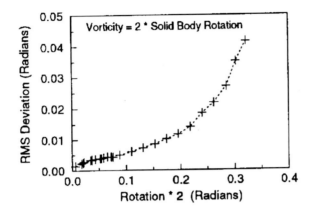

Fig. 6.9. Uncertainty due to uniform rotation for 75 particle images/32^2 pixel.

the correlation peak (Keane & Adrian, 1992). It therefore stood to reason that by reducing the window size, one could reduce the number of particles leaving the interrogation window by reducing the velocity difference within the window, and therefore obtaining more accurate results (Raffel *et al.*, 1998).

6.7.3 Uncertainty due to different particle size imaging

Westerweel, Dabiri & Gharib (1997) show that the uncertainty of using 2-pixel particle images are half the uncertainty of using 4-pixel particle images (Fig. 6.10) when using 32 by 32 pixels interrogation windows. This is confirmed by Raffel *et al.* (1998), who show that for this interrogation window size, the optimum particle image size achieves a minimum of ~2.2 pixels. This figure also shows that the error as a function of the pixel displacements increases with increasing particle image displacements. Again, this is due to the fact that for a given window size, with increasing particle shifts, there are fewer particles contributing to the correlation peak.

6.7.4 Effects of using different size interrogation windows

Raffel *et al.* (1998) show that for pure translation, the larger the window size, the smaller the uncertainty in measurements, since there are more particles that contribute to the correlation peak. It is, therefore, important to find a window size small enough to minimize gradient errors, while large enough to provide

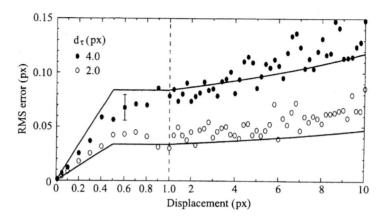

Fig. 6.10. RMS error as a function of displacement and two particle sizes. (Figure courtesy of Westerweel *et al.*, 1997).

enough particle pairs for proper correlations.

Perhaps more interesting is that the error between [0, 0.5] pixels is linear. It is therefore most beneficial to make the second image's interrogation window large enough to encompass all the particle images within the first image's interrogation window (Keane & Adrian, 1992), or to move the second image's interrogation by the mean particle shift with respect to the first image's interrogation window (Westerweel *et al.*, 1997). This procedure is at least a two step procedure. First, a non-shifting window processing is performed. The resulting particle displacements are then used to guide the window shifting processing as described above. This decrease in uncertainty resulting from this procedure is actually quite significant, since examination of the resulting velocity spectra shows that it is possible to obtain one more decade of reliable data (Fig. 6.11).

6.7.5   Mean-bias error removal

Perhaps the most difficult uncertainty is what is referred to as the mean-bias error. This occurs when the correlation peak shape does not match the shape of the fitted curve, or results due to the use of a finite window size. Westerweel (1993, 1997) suggests that this bias is due to the fact that the number of summations contributing to each correlation value (Eqn. 6.1) for finite, equally-sized image domains is not constant. This bias can be corrected by dividing the cor-

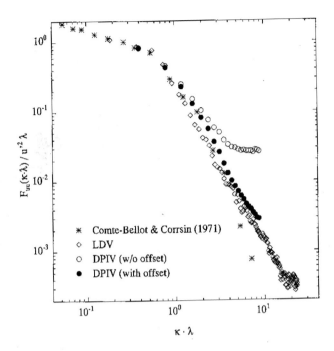

Fig. 6.11. The normalized power spectrum of the fluctuating streamwise velocity of a turbulent glow behind a grid. The open dots represent the result obtained with DPIV window without window offset; the closed dots the same image data but now with window offset. Also plotted are the result obtained with LDV in the same facility and at the same location as for the DPIV, and the result obtained with hot-wire anemometry by Comte-Bellot & Corrsin (1971). (Figure courtesy of Westerweel *et al.*, 1997).

relation domain by the convolution of the interrogation windows (Fig. 6.12a). Another method suggested by Keane & Adrian (1992) is to use interrogation windows of different sizes, so that the convolution of the interrogation windows is flat within the area of interest (Fig. 6.12b). As a result, no correction to the computed correlation is necessary.

Huang, Dabiri, & Gharib (1997) suggested a third procedure to reduce the mean-bias error. By normalizing the cross-correlation given by Eqn. 6.1 as:

$$C'(m,n) = \frac{C(m,n)}{\left[\sum_{i,j} f^2(i,j) \sum_{i,j} g^2(i,j)\right]^{\frac{1}{2}}} \qquad (6.15)$$

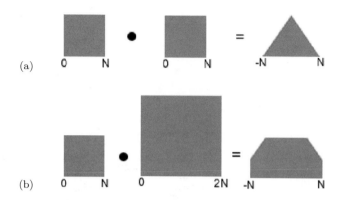

(a)

(b)

Fig. 6.12. The mean-bias can be corrected by dividing the cross-correlation by the convolutions of the window sizes of the interrogation windows of the two sequential images. (a) For interrogation windows of the same size, the resulting convolution is a triangular shape; (b) For interrogation windows of different sizes, the resulting convolution is flat in the area of interest.

it was possible to reduce the RMS error dramatically. However, this procedure still maintained a mean-bias error of 0.01 pixels, which was comparable to the RMS error, and therefore could not be neglected. This error was attributed to the asymmetry of the cross-correlation peak, and was compensated by the following procedure. First, upon calculation of the correlation domain, the neighboring correlation values about the peak value are compensated by

$$R'_n = \alpha R_n \qquad (6.16)$$

where $R_n \in R(x_0 + 1), y_0), R(x_0 - 1, y_0), R(x_0, y_0 + 1), R(x_0, y_0 - 1), R(x_0, y_0)$ is the integer peak correlation; and

$$\alpha = 1 + k\frac{R(x_0, y_0) - R_{nmax}}{R(x_0, y_0)} \qquad (6.17)$$

where R_{nmax} is the maximum valued of the integer peak neighbors Rn. The optimum value of k is shown to be 0.143, which results in a mean-bias error of the order of 0.001 pixels. Note that his error is one order of magnitude less than that reported earlier in Section 6.4.1.

6.8 DPIV Applications

6.8.1 Investigation of vortex ring formation.

DPIV is currently considered a method of choice in investigating and revisiting conventional flow problems like boundary layers, separated shear flows or unsteady flows such as vortex rings. Gharib *et al.* (1998) used DPIV to investigate the formation process of axisymmetric vortex rings.

Fig. 6.13 depicts a velocity vector field for a vortex ring generated by an impulsively started jet through a piston/cylinder arrangement. According to Gharib *et al.* (1998), for long stroke ratios (> 4), the vortex ring pinches off from its trailing jet. This phenomenon can be more clearly seen in Fig. 6.14 where a gap exists between the vorticity field of the trailing jet and that of the front vortex. The example shows the unique capability of DPIV in the process of discovery of the new phenomenon in fluid mechanics.

6.8.2 A novel application for force prediction DPIV

By having the whole flow field available by DPIV, ample applications can be thought of in terms of extracting various flow field properties. Perhaps one of the more interesting applications of DPIV is the force measurements in fluid-structure interactions. The instantaneous force on a body using a control volume approach for momentum conservation is: where ρ is the fluid density, u is the fluid velocity, and $\sum$ is the stress tensor. The material volume $V_m(t)$ is bounded by an inner surface, which corresponds to the body surface, and an outer material surface $S_m(t)$, with an outward unit normal n of an arbitrarily chosen control surface (Fig. 6.15). For example, Lin & Rockwell (1996) and Noca *et al.* (1997) showed that by using the proper expressions for the fluid velocity and the stress tensor in terms of the velocity and vorticity field, it is possible to obtain the time history of forces on an oscillating, circular cylinder. Fig. 6.16 depicts one such calculation by Noca *et al.* (1997).

$$\vec{F} = -\frac{d}{dt} \int_{V_m(t)} \rho \vec{u} dV + \oint_{S_m(t)} \vec{n} \cdot \sum dS \qquad (6.18)$$

6.8.3 DPIV and a CFD counterpart: a common ground

DPIV offers a unique opportunity for defining a common ground with computational fluid dynamic (CFD). For example, the velocity field information obtained

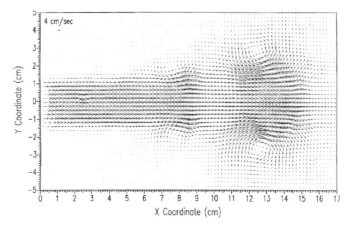

Fig. 6.13. The instantaneous velocity field of a vortex ring generated by an impulsively started jet.

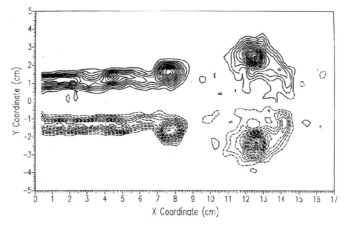

Fig. 6.14. The instantaneous vorticity field of a vortex ring depicted in Fig. 6.13.

through DPIV can be used for the purpose of straight validation or checking on the flow dimensionality, geometric definition and velocity or vorticity measurement comparisons. Fig. 6.17 shows a simulation of vortex shedding from a circular cylinder using Direct Numerical Simulation (DNS), compared with its DPIV counterpart (Henderson & Karniadakis, 1995). The circulation magnitudes agree very well between the sets of data which, by itself, is very valuable in checking the range of Reynolds numbers for proper application of DNS.

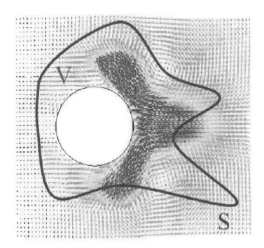

Fig. 6.15. The control surface and volume used with the instantaneous velocity field of an oscillating cylinder. Figure also shown as Color Plate 3. (Figure courtesy of F. Noca.)

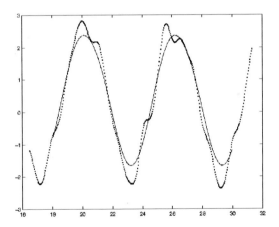

Fig. 6.16. Calculation of the drag coefficient as a function of non-dimensional time on an oscillating, circular cylinder using DPIV. The sinusoidal curve is the theoretical calculation, and the dotted line shows the experimental results. (Figure courtesy of F. Noca.)

6.9   Conclusion

The advent of DPIV has provided one of the most important milestones in our experimental observations of fluid flow. The power of this technique comes from

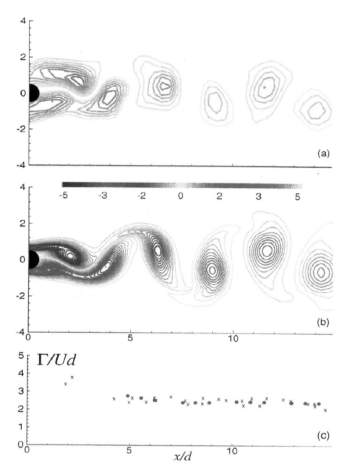

Fig. 6.17. Vorticity in the wake of a circular cylinder, Re = 100: (a) DPIV measurements (b) 2-D numerical simulations (c) computed values of the circulation of wake vorticies from experiments, ×, and simulations, ●. Figure also shown as Color Plate 4.

the fact that it can provide measurements on a scale beyond the capabilities of single-point measurement techniques such as hot wire or laser Doppler velocimetry. Furthermore, other quantities such as vorticity, deformation, and forces can also be derived which, as shown, provides a common ground with CFD theoretical modeling of fluid flow phenomena. In this respect, DPIV can heavily contribute to fluid flow research in that it can provide detailed information for various flows and at conditions that until only recently could only have been simulated computationally. However, its application to complex three-dimensional flows should be carried out with care and concern so as to prevent possible sources of error and misinterpretation. Extensions of this valuable technique to three component and volume mapping, as well as technological improvements such as faster, high-resolution CCD cameras, are the next few necessary steps that are currently undergoing research in various groups.

6.10   References

Adrian, R. J. 1986b. Multi-point optical measurements of simultaneous vectors in an unsteady flow-a review. *Int. J. Heat & Fluid Flow*, **7**, 127–145.

Adrian, R. J. 1991. Particle-imaging techniques for experimental fluid mechanics. *Annual Rev. Fluid Mech.*, **22**, 261–304.

Alexander, B. F. and Ng K. C. 1991. Elimination of systematic-error in subpixel accuracy centroid estimation. *Opt. Eng.*, **30**, 1320–1331.

Comte-Bellot, G. and Corrsin S. 1971. Simple eulerian time correlation of full and narrow-band velocity signal in grid-generated, "isotropic" turbulence. *J. Fluid Mech.*, **22**, 261–304.

Dabiri, D. and Roesgen T. 1991. Private communications.

Dabiri, D. and Gharib M. 1994. Internal GALCIT Report.

Gharib, M. Rambod, R. and Shariff K. 1998. A universal time scale for vortex ring formation. *J. Fluid Mech.*, **360**, 121–140.

Gharib, M. and Weigand A. 1996. Experimental studies of vortex disconnection and connection at a free surface. *J. Fluid Mech.*, **321**, 59–86.

Gharib, M. and Willert C., 1990. Particle tracking-revisited. In *Lecture Notes in Engineering: Advances in Fluid Mechanics Measurements*, **45**, ed. M. Gadel-Hak, Springer-Verlag, New York, 109–126.

Henderson, R. and Karniadakis G. E. 1995. Unstructured spectral element methods for simulation of turbulent flows. *Journal of Computational Physics*, **122**, 191–217.

Hinsch, K. D. 1993. Particle image velocimetry. In *Speckle Metrology*, ed. R. S. Sirohi, Marcel Dekker, New York, 235–323.

Huang, H. Dabiri, D. and Gharib M. 1997. On errors of digital particle image velocimetry. *Meas. Sci. Technol.*, **8**, 1427–1440.

Keane, R. D. and Adrian, R. J. 1990. Optimization of particle image velocimeters. Part I: Double-pulsed systems. *Meas. Sci. Technol.*, **1**, 1202–1215.

Keane, R. D. and Adrian, R. J. 1991. Optimization of particle image velocimeters. Part II: Multiple-pulsed systems. *Meas. Sci. Technol.*, **2**, 963–974.

Keane, R. D. and Adrian, R. J. 1992. Theory of cross-correlation of PIV images. *Appl. Sci. Res.*, **49**, 191–215.

Landreth, C. C. and Adrian, R. J. 1990a. Impingement of a low Reynolds number turbulent circular jet onto a flat plate at normal incidence. *Exp. Fluids*, **9**, 74–84.

Lin, J. and Rockwell, D. 1996. Force identification by vorticity fields: techniques based on flow imaging. *J. Fluids and Structures*, **10**, 663–668.

Melling, A. 1997. Tracer particles and seeding for particle image velocimetry. *Meas. Sci. Technol.*, **8**, 1406–1416.

Merzkirch, W. 1987. *Flow Visualization*. 2nd ed., Academic Press, Orlando.

Noca, F., Shiels, D. and Jeon, D. 1997. Measuring instantaneous fluid dynamic forces on bodies, using only velocity fields and their derivatives. *J. Fluids and Structures*, **11**, 345–350.

Pearlstein A. J. and Carpenter B. 1995. On the determination of solenoidal or compressible velocity fields from measurements of passive and reactive scalars. *Physics of Fluids*, **7** (4), 754–763.

Raffel, M. Willert C. and Kompenhans J. 1998. Particle image velocimetry — a practical guide. Ed. R. J. Adrian, M. Gharib, W. Merzkirch, D. Rockwell, H. Whitelaw, Springer-Verlag, Heidelberg

Reuss, D. L. Adrian, R. J. Landreth, C. C., French, D. T. and Fansler T. D. 1989. Instantaneous planar measurements of velocity and large-scale vorticity and strain rate in an engine using particle-image velocimetry. *SAE Technical Paper Series 890616*.

Singh, A. 1991. *Optic flow computation*. IEEE, Computer Society Press.

Weigand A. 1996. Simultaneous mapping of the velocity and deformation field at a free surface. *Exp. Fluids.*, **20**, 358–364.

Westerweel J. 1993. *Digital particle image velocimetry-Theory and application*. Ph.D. Thesis, Delft University Press, Delft.

Westerweel J. 1997. Fundamentals of digital particle image velocimetry. *Meas. Sci. and Technol.*, **8**, 1379–1392.

Westerweel, J. DabiriI, D. and Gharib M. 1997. The effect of a discrete window offset on the accuracy of cross-correlation analysis of PIV recordings. *Exp. Fluids*, **23**, 20–28.

Willert, C. E. and Gharib M. 1991. Digital particle image velocimetry. *Exp. Fluids*, **10**, 181–193.

Willert, C. E. 1992. *The interaction of modulated vortex pairs with a free surface*. Ph.D. Thesis, Dept. of Applied Mechanics and Engineering Sciences, University of California, San Diego, CA.

Willert, C. E. and Gharib M. 1997. The interaction of spatially modulated vortex pairs with free surfaces. *J. Fluid Mech.*, **345**, 227–250.

CHAPTER 7

SURFACE TEMPERATURE SENSING WITH THERMOCHROMIC LIQUID CRYSTALS

C. R. Smith, T. J. Praisner and D. R. Sabatino[a]

7.1 Introduction

The last decade has seen a rapid evolution in the application of thermochromic Liquid Crystals (LCs) for field-wise surface temperature measurements. Liquid crystals are unique because their optical properties are dependent on temperature in a predicable and repeatable manner. When applied in a thin layer to a black surface, LCs selectively reflect light depending on the temperature of the surface. The reflected light is within the visible color spectrum and the dominant wavelength, or hue [from hue, saturation, and brightness (HSB) color space], varies monotonically with temperature. This relationship of color to temperature has allowed researchers to quantitatively map surface and flow-field temperature distributions with high spatial resolution and, more recently, high accuracy.

Although current LC-based heat transfer research has focused on quantitative measurement techniques, LCs are also an excellent temperature field visualization tool. Fig. 7.1 shows raw LC images for two different flow geometries where LCs are applied to a constant heat flux surface. Fig. 7.1a illustrates the temperature patterns created by a cool fluid jet impinging perpendicular to the surface; Fig. 7.1b shows the endwall patterns for a linear turbine cascade (Sabatino & Praisner, 1998). Both images illustrate the high spatial resolution which is achievable, as well as the naturally brilliant colors displayed by the LCs. Images such as those in Fig. 7.1 can be converted to high-resolution heat

[a]Dept. of Mechanical Engineering and Mechanics, Lehigh University, Bethlehem, PA 18015, U. S. A.

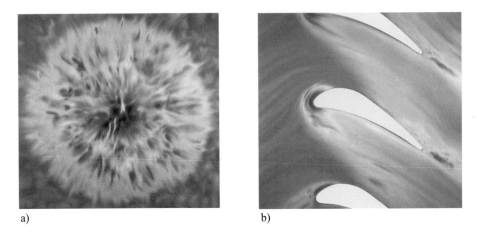

a)   b)

Fig. 7.1. Images of *instantaneous* liquid crystal surface temperature patterns generated by: (a) a jet impinging perpendicular to a heated surface; and (b) a turbulent juncture end-wall in a linear turbine blade cascade (Sabatino & Praisner, 1998). Figure also shown as Color Plate 5.

transfer distributions by applying appropriate thermal boundary conditions and employing a calibration algorithm relating color to local temperature. blade

7.1.1   Calibration techniques

To date there are two primary calibration techniques used to derive quantitative information from the color play of thermochromic LCs. The "narrow-band" technique employs LCs with a narrow activation bandwidth (typically 1°C or less) and has been used successfully by a number of researchers (Ireland & Jones, 1986; Hippensteele & Russel, 1988; Giel *et al.*, 1998). Using this technique, the temperature at which a single "event" color appears (usually yellow) is established. Yellow is generally chosen because it is displayed over the narrowest temperature range for most thermochromic LCs, thus minimizing the measurement uncertainty. Experimentally, this technique is employed to accurately establish an instantaneous isotherm on a LC-coated surface. The primary benefit of employing the narrow-band calibration technique is that only a single calibration point needs to be established.

An alternative "wide-band" calibration technique has been employed by a number of researchers (Hollingsworth *et al.*, 1989; Camci *et al.*, 1993; Farina *et al.*, 1994; Babinsky & Edwards, 1996; Wang *et al.*, 1996; Takmaz, 1996; Praisner *et al.*, 1997). For this technique, a color, or hue versus temperature calibration is established over the full range of colors displayed by LCs with a relatively wide temperature bandwidth (typically a range of several degrees or more). The primary disadvantage of this technique is that it requires a significant number of calibration points to accurately resolve the highly non-linear hue-temperature variation that is typical of thermochromic LCs. However, the particular benefit of this technique is that the entire surface heat transfer distribution can be established from a single image. Typically, wide-band calibration techniques use a single reference point on the LC surface, and apply this "single-point" calibration to the entire test surface.

7.1.2 Convective heat transfer coefficient measurement techniques

While LC thermography can be employed for simple temperature measurements, the potential of the technique is truly realized when used to determine convective heat transfer properties in fluid flows over solid boundaries. Liquid crystal thermography, when combined with known boundary conditions can yield detailed convective heat transfer information in complex flow configurations. Examples of these flows, both laminar and turbulent, include: boundary layers, transitional flows, three-dimensional separation, and bluff-body flows. Time-mean convective heat transfer coefficients are generally determined using one of two experimental approaches coupled with LC thermography. These time-mean heat transfer techniques are termed "steady-state" and "transient" techniques (where "transient" is a misnomer used to describe a technique that employs a *transient* phenomena to determine *time-mean* convective heat transfer coefficients). Additionally, a third type of LC-based technique is used to determine *instantaneous* fluctuating convective heat transfer coefficients.

Time-mean techniques

For the *steady-state* technique, a constant heat flux surface is typically created by passing an electrical current through a thin film of electrically resistive material. The material of choice is typically thin metal foil. However, vapor-deposited gold on polycarbonate sheets has also been employed (Simonich, 1982; Hippensteele & Russell, 1988). Stainless steel foil is generally chosen because of its uniform thickness and electrical resistivity. Using temperature maps obtained from LCs applied to a constant heat flux surface, the convective heat transfer equation can be used to establish distributions of the convective heat transfer coefficient. Either narrow or wide-band calibration techniques may be employed with the steady-state technique.

Incorporating the narrow-band calibration technique with a constant heat flux surface, the position of the yellow color, which generally appears as a line contour, is varied by systematically adjusting the applied power to establish distributions of the convective heat transfer coefficient. While this technique requires only one temperature/hue calibration point, it necessitates a large number of images to completely map a surface; additionally, resolution may be poor in regions of low transverse thermal gradients (Babinsky & Edwards, 1996). Note, that use of a wide-band calibration technique in conjunction with a constant heat flux surface provides greater flexibility and requires less data acquisition than the narrow-band calibration technique.

Transient techniques

The transient technique has been employed by a large number of researchers (Giel *et al.*, 1998; Wang *et al.*, 1996; Babinsky & Edwards, 1996; Camci *et al.*, 1993; Jones & Hippesteele, 1988; Ireland & Jones, 1986) because it does not require the test surface to be continuously heated, and hence is simpler to employ than the steady-state technique. The name "transient" is used because the technique relies on the transient thermal response of a test surface to a step change in free-stream temperature. In practice, a test surface is typically heated to an elevated temperature, and then a step change in the free stream temperature imposed. For the narrow-band transient technique, the temporal position of the isothermal yellow contours are optically monitored after the application of a step change in free-stream temperature. In a similar manner, a wide-band technique may be employed, where local temporal variations of the temperature (color) are monitored. For either method, the temperature-time information is used in conjunction with a semi-infinite wall model to establish the time-mean convective heat transfer coefficients (Camci *et al.*, 1993). An extension of the wide-band transient technique (Camci *et al.*, 1993) employs the use of a mixture of LCs, each with its own distinct, non-overlapping bandwidth. Using a mixture of LCs significantly extends the range of detectable temperatures, and hence the time duration over which the time-temperature history can be established.

Instantaneous technique

The primary limitation of both the narrow and wide-band techniques, when employed with either the steady-state or transient techniques, is that only time-mean surface heat transfer distributions can be established. However, for determination of true transient behavior, a technique has been developed which allows determination of *instantaneous* surface heat transfer distributions (Praisner *et al.*, 1997, 1999). The technique reported by Praisner *et al.*, developed for use in (but not limited to) water flows, is based on the point-wise, wide-band calibration technique of Sabatino *et al.* (1999). The technique employs a fast-response, thin film constant heat flux surface which allows the optical separation of heat transfer measurements from the flow field, and hence, the capability to simultaneously record both instantaneous flow-field (via particle image velocimetry) and heat transfer data (Praisner *et al.*, 1999).

A constant heat flux surface, which is the primary component of the system, is configured using stainless steel foil stretched around a Plexiglas plate with rounded streamwise ends (see Praisner *et al.*, 1999). The important feature of this constant heat flux surface is a shallow cavity machined into the Plexiglas plate below the stretched stainless steel foil (Fig. 7.2a). This air-filled cavity serves as a near perfect insulator compared to the water flow conditions on the opposing side of the foil. In the working configuration, the stainless steel foil is stretched tightly over the cavity, creating a seal with the rim of the cavity, and thereby maintaining an insulating, moisture-free environment.

A constant heat flux condition is generated by galvanically heating the stainless steel foil using an adjustable low voltage AC or DC power supply. The use of low-voltage and high amperage power is dictated by safety considerations and the low resistance characteristics of the heater foil. Calibration of the heating foil is done by concurrent monitoring of voltage and the corresponding current flow across the extent of the heating foil. Constant heat flux levels between 8,000 and 16,000 W/m^2 are generally appropriate for both laminar and turbulent flows, yielding convective heat transfer coefficients between 500 and 4000 W/m^{2}K. It should be noted that significantly lower heat flux levels are required for experiments performed in wind tunnels.

To facilitate temperature measurements, the cavity side of the foil is painted with LCs (Fig. 7.2a). Using this LC arrangement, the flow conditions are generated on the side of the foil opposite that of the LCs. Detailed analysis has shown that the total thermal response time of the stainless steel foil and LC coating is on the order of 200 Hz, which is sufficient to resolve the frequency characteris-

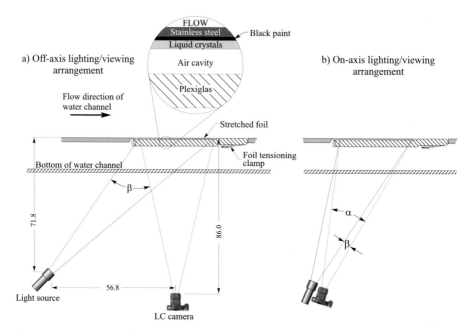

Fig. 7.2. Schematic illustrating: (a) the off-axis lighting/viewing arrangement used to illuminate the LC surface from below a water channel, and (b) the on-axis lighting/viweing arrangement. All dimensions in cm.

tics of low Reynolds number turbulent flows generated in typical water channel applications (Praisner *et al.*, 1999), and reasonably responsive for low-speed air flows. The primary benefits of this wide-band, instantaneous technique are:

(1) Instantaneous surface heat transfer distributions can be established;

(2) Heat transfer measurements are physically separated from the flow field, facilitating the possibility for the simultaneous measurements of the flow-field;

(3) LCs are isolated from potential shear effects imposed by the flow;

(4) LCs are separated from the detrimental effects of water exposure (their useful life-span, even the water resistant types, is as little as 30 minutes when exposed to water (Takmaz, 1996)).

The primary limitation of this technique is that it is restricted to test surfaces with curvature in only one direction.

7.2 Implementation

This section presents a description of an application of the instantaneous technique as employed by the authors for the study of a variety of turbulent flows. The technique described was specifically developed for the test surface described in the previous section.

7.2.1 Sensing sheet preparation

Prefabricated LC sheets, which include a black backing material and protective clear polyester layer for the LCs, may be applied to a test surface in lieu of sprayed coatings. The primary disadvantage of prefabricated sheets is that they have slower thermal response characteristics and increased thermal contact resistance compared to sprayed applications. In addition, the colors displayed by prefabricated sheets are typically muted compared to sprayed coatings. To overcome the limitations of prefabricated sheets, a micro-encapsulated form of chiral nematic thermochromic LCs has been generally employed. This micro-encapsulated form is employed by nearly all researchers because it is less sensitive to contaminants such as dust and moisture, and less affected by shear effects than unencapsulated forms. Additionally, of the many types of commercially available thermochromic LCs, the perceived color of micro-encapsulated chiral nematic LCs has the lowest sensitivity to variations in lighting/viewing angle.

The micro-encapsulation process typically results in nominally 10–15 μm capsules of LCs encased in a protective polymer coating. As for all LC applications, a black backing is required beneath the LCs to absorb impinging light not reflected by the LC layer. This is achieved by applying a thin coat of black paint to the test surface prior to the application of the LCs. Fig. 7.2a illustrates the layering of the black paint and LCs on the cavity side of a stainless steel foil. Prior to the application of the black paint and LCs, the region of application is thoroughly cleaned with a solvent. After masking the area to be sprayed, a gloss black enamel paint is applied in two thin layers to the stainless steel. The resulting thickness of the fully dried paint layer is typically 15 μm. Thermochromic liquid crystals are commercially available in a variety of forms including water resistant types. In addition, LC temperature bandwidths are available ranging from a fraction of a degree to over $20°$C.

Once the black paint is completely dry, a 40 μm thick layer of the micro-encapsulated LCs is applied to the test surface (32 cm × 33 cm for the authors' application) at approximately 2.25 ml per cm^2 of test surface. Before spraying, the LCs should be filtered through a 40 micron filter to remove extraneous aggregates, and then diluted with approximately 50% distilled water (Farina *et al.*, 1994). The LC/water mixture is applied via an air brush applicator (at about 18 psi). Micro-encapsulated chiral-nematic liquid crystals (Hallcrest Inc. type C17-10) with a bandwidth of 7°C (where red starts at 25°C and blue at 32°C) were used for the example results presented in Section 7.3.

The LCs should be applied in smooth, sweeping motions both parallel and diagonal to the edges of a sprayed region. Typically, applications are made continuously until half of the LC/water mixture is exhausted. The first coat is then allowed to dry completely before a second application of the remaining half of the mixture is done. This two-coat application procedure prevents pooling of the LC/water mixture, which can result in thickness variations.

7.2.2 Test surface illumination

Liquid crystals reflect specific wavelengths of light as a function their molecular orientation, which is dictated by the temperature of the crystals (Simonich, 1982). The LCs are illuminated with a white light source (the entire visible spectrum), to ensure all temperatures within the LC bandwidth will be visualized. Typical LC color response is red at the lower temperatures, passing through yellow, green, and then blue at the highest temperatures. Outside of the upper and lower clearing points (the extremes of the temperature bandwidth) the crystals appear translucent, and thus one sees only the black backing layer (Section 7.2.1).

The color displayed by thermochromic LCs has been established to be strongly dependent on the angle between the light source and the viewing position (β in Fig. 7.2) (Farina *et al.*, 1994). To minimize variations in the light/viewing angle, β, a collimated light source is employed. In general, high wattage (about 1000 W), omni-directional photographic light sources yield ineffectual results in comparison to a collimated source. The optics employed in a slide projector have been found to be effective because they provide an inexpensive, variable-focus colimated light source as well as an infrared filter which reduces radiative heating effects (Sabatino *et al.*, 1999; Hacker & Eaton, 1995). For the present application (where the light must pass through two layers of polycarbonate and appoximately 10 cm of water), illuminate the LCs, and then pass back out through the same materials, a 300 W halogen bulb with parabolic reflector was employed. Considering that approximately 4% of the illumination light is lost to reflection at each surface interface, significantly less power would be required, for example, in a wind tunnel application with only one view-port to pass through.

In general, the smallest lighting/viewing angle variations are achieved with coincident viewing and lighting axes. Farina *et al.* (1994) accomplished this by positioning their light source (a slide projector) immediately adjacent to their camera. They found that this was as effective as employing a ring-light mounted around the lens of the camera. As detailed in Sabatino *et al.* (1999), both on and off-axis lighting arrangements have been evaluated for uniformity of color (hue) displayed under a uniform temperature condition. However, when experiments are viewed through glass or Plexiglas viewports, as in the case of many water channel and wind tunnel studies, the camera and lighting source cannot be oriented normal to the surface because large specular reflections are produced. These reflections significantly increase the signal-to-noise ratio in the recorded images. Therefore, the light source/camera axis must be skewed *from* a viewing angle normal to the surface to eliminate specular reflections, as shown in Fig. 7.2b. This skewing imposes a parallax distortion to the recorded image, which must be digitally removed during post-processing. Although an on-axis arrangement yields a smaller variation of displayed hue (as established in Section 7.2.3) than an off-axis arrangement, spatial variations in hue across a viewed surface can still exceed 15% of the full hue range displayed by the LCs. Thus utilization of either an on-axis or off-axis technique requires a calibration method which minimizes the hue variations across a surface. In general, it has been determined (Sabatino *et al.*, 1999) that an off-axis arrangement is easier to implement, primarily because it does not introduce parallax distortion.

Normally, a continuous light source is employed primarily because of the low cost and ease of use. A pulsed light source is equally effective, and may generate a brighter surface; however, the light pulses must be synchronized with the data recording cycle.

To minimize specular reflection from the LC surface, crossed grey linear polarizers should be employed on both the light source and camera. Circular polarizers could also be employed, such as utilized in electronic liquid crystal display units. However, to the authors' knowledge these types of polarizers have not been used for thermochromic LC experiments.

7.2.3 Data reduction

Liquid crystal images may be recorded using either a CCD camera or color film. When digitizing photographic images, a scanner, which linearly converts the intensity of each red, green, and blue color component to a 24-bit digital value, should be employed. The primary advantage of a film-based system is the excellent color saturation achievable with film. However, uncertainty is introduced by the chemical developing process, because the amount and application of the chemicals during the developing process can affect the resulting color. Therefore, the calibration data must be processed *identically* to the experimental data to ensure the accuracy of the quantitative information. For studies performed in the authors' laboratory, both the calibration and the data images are developed *simultaneously* by a Kodak Q Lab (highest level of Kodak-certified computer controlled developing process).

When employing a CCD video camera to capture LC images, the highest accuracy is obtained when the camera output is linearly related to the intensity of the red, green, and blue components of the incident light (Hacker & Eaton, 1995). Modern scientific CCD cameras can linearize the output as well as employ separate CCD chips for each red, green, and blue channel. Using three chips instead of one improves the color saturation, which more closely approximates the characteristics of photographic film.

An advantage of a digital system is that it does not require chemical developing, and thus provides results more quickly and with less effort. However, a disadvantage of CCD cameras has traditionally been their limited spatial resolution and depth of color. A typical NTSC camera is capable of a maximum frame resolution of 640 x 480 pixels, which is composed of two vertically interlaced fields recorded 1/60 sec apart. New, more expensive cameras have increased the resolution to $1K \times 1K$ pixels and continue to improve (note, 400 speed color film has an equivalent resolution of over $5K \times 4K$ pixels).

Regardless of the method employed to obtain images in RGB color space, the hue component [from hue, saturation, and brightness (HSB) color space] of the color is extracted and used as the calibration parameter. The hue physically represents the dominant wavelength of the light being displayed by the LCs, and is determined by establishing the angle between the orthogonal red, green, and blue components in RGB space (Foley *et al.*, 1990). Using hue from HSB color space can reduce possible sources of uncertainty, such as variations in the brightness of the light source.

Various methods have been developed to calculate hue from the RGB components of the digitized images. Commercial software packages typically employ a conditional algorithm (Foley *et al.*, 1990), while some researchers have employed a closed-form, arc-tangent equation (Hacker & Eaton, 1995). It must be noted that commercial imaging software often stores the calculated hue values as

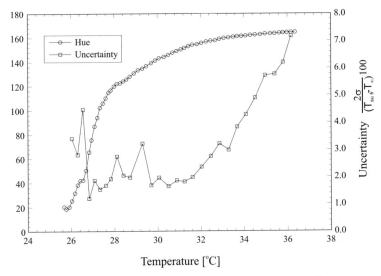

Fig. 7.3. Typical single-point calibration curve and the corresponding uncertainty for liquid crystal-based temperature measurements constructed using a point-wise calibration technique. After Sabatino *et al.* (1999).

an 8-bit grayscale image. This effectively truncates the calculated hue to whole numbers ranging from 0 to 255, limiting the resolution of the calculations.

7.2.4 Calibration

Fig. 7.3 shows a typical single-point hue vs. temperature calibration curve. This figure illustrates that the relationship is very non-linear, as well as double-valued at the low end of the temperature range. This double-valued region must be accurately identified so that the experimental conditions do not fall outside the single-valued temperature region.

Establishing the temperature reflected by a measured hue value is accomplished using one of several methods. Some researchers record approximately 10–15 data points and perform a polynomial fit of the data. However, this approach only approximates the true hue-temperature correlation, and can result in errors that are often large at the extremes of the bandwidth (Hollingsworth *et al.*, 1989). Because of this resultant error, some researchers have restricted their data to the "linear" central third of the LC bandwidth (Moffat, 1982). This restriction reduces the uncertainty in the data, but severely limits the useful temperature range.

It is to be noted that all single-point calibration methods inherently assume negligible variation of the displayed hue across a test surface *under a uniform temperature condition*. However, the use of the hue component of color does not completely eliminate variations in the LC color. Variations in the lighting/viewing angle, the illumination source, defects in the LC coating, and reflected light from viewport surfaces all can contribute to variations in the displayed hue across a test surface. Therefore, a constant and uniform surface temperature will not necessarily yield a uniform color over the entire test surface. Additionally, it has been determined by Sabatino *et al.* (1999) that the non-uniform hue distributions displayed under a uniform temperature condition vary with temperature across the bandwidth of the LCs. In fact, the non-uniform spatial distributions of hue can change dramatically in magnitude and pattern across the entire range of the LCs. Consequently, every point on the LC surface can display a unique hue vs. temperature calibration relationship.

A calibration technique which addresses these inherent spatial variations has been reported by Sabatino *et al.* (1999). This technique essentially creates a spatial point-wise calibration for the entire test surface from a sequence of "uniform temperature" images. As described in Sabatino *et al.*, the test surface, designed for use in a water channel, is surrounded by an insulated baffle, that creates a captive volume of water above the surface. By heating the fluid within the baffle, a uniform and constant surface temperature condition is generated over the test surface. Liquid crystal images are recorded while systematically modulating the bath temperature through the useful range of the LCs. Approximately 60 separate surface temperature images spanning the full bandwidth of the LCs are typically recorded. Following digitization and conversion to hue, the uniform temperature images are used to generate a calibration curve for every spatial point on the test surface.

This technique is powerful because it accounts for all sources of uncertainty in the perceived hue. For example, specular reflections from the LC test surface are systemic and constant throughout the temperature range. To remove these reflective components, previous researchers have first established baseline images for their test surfaces below the lower clearing point of the LCs. The R, G, and B components of these baseline images are then digitally removed the from subsequent data images before their conversion to hue (Farina *et al.*, 1994; Babinsky & Edwards, 1996). However, using the point-wise calibration technique, it is unnecessary to determine and isolate the reflected components, or any other specific sources of uncertainty. However, for optimal accuracy, the point-wise calibration technique requires that the uniform and constant temperature images of the LC surface are recorded in-situ and that the lighting/viewing arrangement remains invariant throughout the experiments.

It is important to note that LCs do not exhibit hysterisis effects within the active temperature bandwidth (Sabatino *et al.*, 1999). However, hysterisis behavior can become significant if the LCs are heated above the upper clearing point (the upper temperature limit at which the crystals become translucent). Therefore, regardless of the method of calibration, it is important that the temperature of the LCs does not exceed the upper clearing-point temperature at any time following calibration (Sabatino *et al.*, 1999).

7.2.5   Establishing uncertainty

An accurate assessment of the temperature measurement uncertainty is necessary because of the many previously discussed factors that can affect LC measurements. As detailed in Praisner (1998) and Sabatino *et al.* (1999), a quantitative measure of the uncertainty is established by application of the calibration routine to a series of independent images obtained at uniform temperature, and employing the spatially established *rms* of the corrected images as the measure of the uncertainty. Fig. 7.3 shows an example of the relative uncertainty for LC-measured temperature across an entire LC bandwidth, illustrating that uncertainty is a strong function of the temperature location within the bandwidth. See Sabatino *et al.* (1999) for a detailed uncertainty assessment and discussion.

7.3   Examples

7.3.1   Turbine cascade

Figs. 7.4a and 7.4b show raw LC images of a test surface employed by Hippensteele & Russell (1988) to investigate endwall heat transfer with a linear turbine cascade model in a wind tunnel. The authors employed the narrow band technique, calibrating the "yellow band" temperature by means of a constant temperature water bath. The LCs were applied to a constant heat flux surface, and the isothermal contours established by varying the applied heat flux. Combining the temperature contours determined from multiple data images (for example, Figs. 7.4a and 7.4b), Hippensteele and Russell generated the Stanton number ($St = h/(\rho U_\infty C_p)$) contour plots shown in Fig. 7.4c.

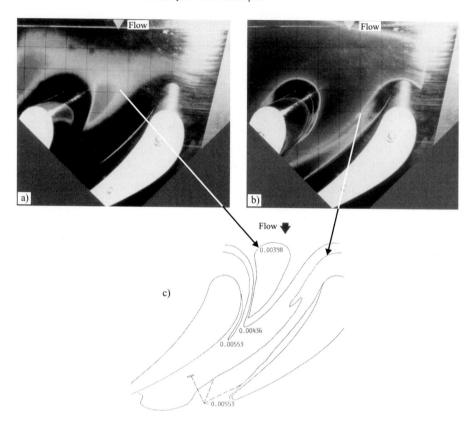

Fig. 7.4. Photographic images illustrating the employment of a constant heat flux surface in conjunction with a narrow-band calibration technique. Images in (a) and (b) were used to determine the contours indicated in (c). After Hippensteele & Russell (1998). Figures (a) and (b) also shown as Color Plate 6.

7.3.2   Turbulent spot and boundary layer

The test section described in Section 7.1.2, in conjunction with the point-wise calibration technique described in Section 7.2.4, was used to examine the surface heat transfer behavior for a turbulent spot (a finite region of developing turbulence in an otherwise laminar boundary layer). Fig. 7.5a shows the quantitative local heat transfer for a passing turbulent spot, reflected as spatial distributions of instantaneous local heat transfer coefficients in the form of the Stanton number (after Sabatino, 1997). Qualitatively, the data images are quite similar to previous free-surface and dye-injection visualizations. Note how the streamwise

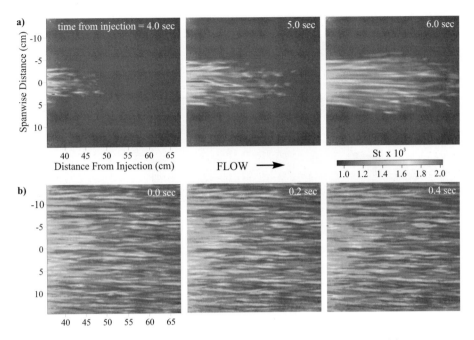

Fig. 7.5. Instantaneous surface heat transfer patterns generated by (a) a turbulent spot (produced by an upstream injection into a laminar boundary layer), passing over a constant heat flux surface at $Re_x = 2 \times 10^5$; (b) a fully turbulent boundary at $Re_\theta = 10,000$. After Sabatino (1997). Figure also shown as Color Plate 7.

oriented streak-like patterns generated by the turbulent spot are similar to those produced by the low-speed streaks characteristic of a fully turbulent boundary layer which are shown in Fig. 7.5b. Fig. 7.5b also illustrates the effect of the streamwise development of the turbulent boundary layer, which is reflected by the diminishing magnitude of the Stanton number with downstream distance (Sabatino, 1997).

7.3.3 Turbulent juncture flow

Fig. 7.6 is a projection of the instantaneous surface heat transfer at the base of a 5:1 tapered cylinder with a turbulent approach boundary layer (Praisner, 1998). Only one half of the upstream region is shown in order to illustrate the stream-wise, symmetry-plane heat transfer profile. For clarity, only one third of the data resolution was employed to construct this image. Fig. 7.6b shows a plan view image of the projected data. The distinguishing features of the symmetry-

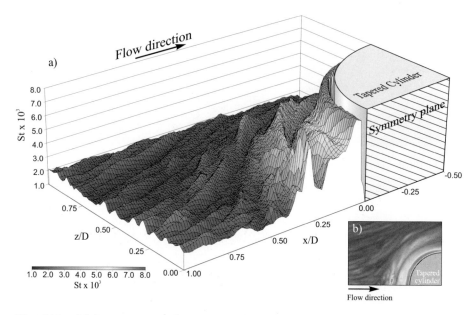

Fig. 7.6. (a) Instantaneous Stanton number projection upstream of a 5:1 tapered cylinder for $Re_D = 2.4 \times 10^4$; (b) Plan view of the same Stanton number distribution. After Praisner (1998).

plane heat transfer profile are the peaks in Stanton number near $x/D = 0.05$ and $x/D = 0.28$ (where D is the diameter of the tapered cylinder). These peaks reveal two bands of high heat transfer which wrap around the base of the cylinder. By employing simultaneous flow-field measurements, Praisner (1998) was able to illustrate that a switching of the instantaneous heat transfer behavior near $x/D = 0.28$ is a result of eruptive turbulent bursting events occurring upstream of the cylinder and impacting on a resident horseshoe vortex which circumscribes the upstream base of the cylinder. distribution.

Fig. 7.7, from Praisner (1998), is a composite time-mean image constructed from ensemble averages of instantaneous, PIV-based flow field data (in the form of vorticity distributions) superposed over the associated time-mean endwall heat transfer (in the form of Stanton number). The time-mean heat transfer data was obtained from an ensemble average of 317 instantaneous distributions. This image clearly reveals the spatial relationships between the time-mean flow structures and the resulting endwall heat transfer.

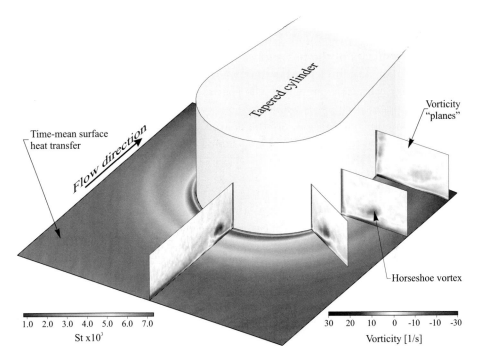

Fig. 7.7. Composite image of time-mean vorticity and end-wall heat transfer for a turbulent end-wall juncture. Image is to scale except for the height of the cylinder which was twice the diameter. After Praisner (1998). Figure also shown as Color Plate 8.

7.4   Concluding Remarks

Thermochromic Liquid Crystals can be an excellent visualization tool as well as function as a quantitative thermographic detection medium. To determine quantitative information from the color change of LCs, one needs to employ one of two types of calibration techniques which correlate color to temperature. The "narrow-band" calibration is the simplest to implement, but requires the most post-processing and yields only time-averaged information. The "wide-band" technique, while more complicated to implement, results in simpler post-processing for time-mean results. Both calibration techniques may be applied to a constant heat flux surface or in a transient technique to obtain convective heat transfer distributions. Finally, a combination of a wide-band calibration technique with a high-frequency response constant heat flux surface has proven effective for establishing instantaneous heat transfer distributions.

In qualifying any type of LC-based temperature measurement system, uniform-temperature images of the test surface should be recorded and assessed to establish the magnitude of possible hue variations across the bandwidth of the LCs. If these variations are small, the spatial deviations of the hue within the uniform-temperature images will be indicative of the calibration uncertainty using a single-point calibration. If the spatial hue variations are significant, a point-wise spatial calibration technique should be considered.

7.5   References

Babinsky, H., and Edwards, J.A. 1996. Automatic liquid crystal thermography for transient heat transfer measurements in hypersonic flow. *Experiments in Fluids*, **21**, 227–236.

Camci, C., Kim, K., Hippensteele, S.A., Poinsatte, P.E. 1993. Evaluation of a hue capturing based transient liquid crystal method for high-resolution mapping of convective heat transfer on curved surfaces. *Journal of Heat Transfer*, **115**, 311–318.

Farina, D.J., Ahcker, J.M., Moffat, R.J., and Eaton, J.K. 1994. Illuminant invariant calibration of thermochromic liquid crystals. *Experimental Thermal and Fluid Science*, **9**, 1–12.

Foley, J.D., van Dam, A., Feiner, S.K., Hughes, J.F. 1990. *Computer Graphics, Principles and Practice*. Addison-Wesley Publishing Company, 590–593.

Giel, P.W., Thurman, D.R., Van Fossen, G.J., Hippensteele, S.A., and Boyle, R.J. 1998. Endwall heat transfer measurements in a transonic turbine cascade. *Journal of Turbomachinery*, **120**, 305–313.

Hacker, J.M., and Eaton, J.K. 1995. Heat transfer measurements in a backward facing step flow with arbitrary wall temperature variations. Thermosciences Division, Department of Mechanical Engineering, Stanford University, *Report No. MD-71*.

Hippensteele, S.A., and Russell, L.M. 1988. High resolution liquid-crystal heat-transfer measurements on the end wall of a turbine passage with variations in Reynolds number. *NASA Technical Memorandum 100827*.

Hollingsworth, D.K., Boehman, A.L., Smith, E.G., and Moffat, R.J. 1989. Measurement of temperature and heat transfer coefficient distributions in a complex flow using liquid crystal thermography and true-color image processing. *Collected Papers in Heat Transfer 1989*, The American Society of Mechanical Engineers, HTD-Vol. 123.

Ireland, P.T. and Jones, T.V. 1986. Detailed measurements of heat transfer on and around a pedestal in fully developed passage flow. *Proceedings, 8th International Heat Transfer Conference*, C.L. Tien *et al.*, eds., Hemisphere Publishing Corporation, Washington DC, **3**, 975–980.

Ireland, P. T., Jones, T. V. 1987. Response time of a surface thermometer employing encapsulated thermochromic liquid crystals. *Journal of Physics. E, Scientific Instruments*, **20** (10), 1195–1199.

Jones, T.V. and Hippensteele, S.A. 1988. High-resolution heat-transfer-coefficient maps applicable to compound-curve surfaces using liquid crystals in a transient wind tunnel. *NASA Technical Memorandum 89855*.

Moffat, R.J. 1990. Some experimental methods for heat transfer studies. *Experimental Thermal and Fluid Science*, **3** (1), 14–32.

Praisner, T.J. 1998. *Investigation of Turbulent Juncture Flow Endwall Heat Transfer and Flow Field*. Ph.D. Dissertation, Lehigh University.

Praisner, T.J., Sabatino, D.R., and Smith, C.R. 1998. Simultaneously combined liquid-crystal surface heat transfer and PIV flow field measurements. *Experiments In Fluids*, in review.

Praisner, T.J., Seal, C.V., Takmaz, L., Smith, C.R. 1997. Spatial-temporal turbulent flow-field and heat transfer behavior in end-wall junctions. *International Journal of Heat and Fluid Flow*, **18** (1), 142–151.

Sabatino, D.R. 1997. *Instantaneous Properties of a Turbulent Spot in a Heated Boundary Layer*. Master's thesis, Lehigh University, Bethlehem, PA.

Sabatino, D.R., Praisner, T.J., and Smith, C.R. 1998. A high-accuracy calibration technique for thermochromic liquid crystal temperature measurements. *Experiments In Fluids*, in review.

Sabatino, D.R., Praisner, T.J. 1998. The colors of turbulence. *Physics of Fluids*, **10**, S8.

Simonich, J.C. 1982. *Local Measurements of Turbulent Boundary Layer Heat Transfer on a Concave Surface Using Liquid Crystals*. Ph.D. dissertation, Stanford University, Stanford CA.

Takmaz, L. 1996. *Spatial-Temporal End-wall Heat Transfer*. Ph.D. dissertation, Lehigh University, Bethlehem, PA.

Wang, Z., Ireland, P.T., Jones, T.V., and Davenport, R. 1996. A color image processing system for transient liquid crystal heat transfer experiments. *Journal of Turbomachinery*, **118**, 421–427.

CHAPTER 8

PRESSURE AND SHEAR SENSITIVE COATINGS

**Rabindra D. Mehta, James H. Bell, Daniel C. Reda,
Michael C. Wilder, Gregory G. Zilliac and David M. Driver**[a]

8.1   Introduction

In practical aerodynamics and experimental fluid mechanics, a major challenge
is to accurately measure the two surface force distributions, namely those associated with static pressure and with shear stress. The pressure distributions
are either integrated for load analysis or they are used to study specific flow
phenomena such as boundary layer separation. Measurements of surface shear
stress are equally important since skin friction can account for over half the drag
on a flight vehicle and drag reducing mechanisms are often investigated experimentally. With rapid improvements in computational fluid dynamics, the need
for accurate pressure and shear stress data has become even more urgent so that
new computational codes can be adequately verified before using them in design
processes.

   An additional challenge is to measure these distributions with sufficient spatial resolution. Prior to the present decade, scientists and engineers had at their
disposal only point-measurement methods, such as taps for pressure measurements and Preston tubes or floating balances for shear stress measurements.
An adequate spatial resolution with point measurements is at best tedious and
costly in terms of time and money required to instrument the model.

   With the recent availability of pressure- and shear-sensitive coatings as sensors and high-sensitivity charge-coupled device (CCD) arrays as optical imaging
devices, full-surface distributions are now becoming available. Although the
initial cost of developing and installing such systems can be quite high, the in-

[a]Experimental Physics Branch, Mail Stop 260-1, NASA-Ames Research Center, Moffett Field,
CA 94035-1000, U. S. A.

vestment can be amortized over several tests, with the only additional cost being that for the coating itself.

The present chapter describes three optical techniques, one for surface pressure measurements and two for the measurement of surface shear stress, which have been developed at NASA Ames Research Center for ultimate application in the production wind tunnels. In the Pressure Sensitive Paint (PSP) technique, the model surface is covered with an oxygen-permeable paint and excited using illumination of a specific wavelength. The luminescence of the PSP at a given point turns out to be inversely proportional to the local pressure. Thus by imaging the whole surface, the pressure distribution can be evaluated, with the spatial resolution determined by the specifications of the optical set up. When Shear-Sensitive Liquid Crystal Coating (SSLCC) is illuminated from the normal direction with white light and observed from an oblique above-plane view angle, its color changes in response to a change in shear vector magnitude and/or direction. In the Fringe-Imaging Skin Friction (FISF) technique, a drop of oil is placed on the model surface, and with the wind turned on, a wedge with a nearly linear thickness profile is formed. When illuminated with a quasi-monochromatic light source oriented perpendicular to the surface, a fringe pattern is formed, and the distance between the destructive interference bands is proportional to the skin friction magnitude.

In this chapter, these three measurement techniques are discussed in some detail, with emphasis on how to apply each technique and how to accurately reduce the data.

8.2   Pressure-Sensitive Paint

"Pressure-sensitive paint" or PSP is a technique for measuring pressure on a surface using a luminescent coating whose brightness varies with air pressure. In the most common approach (Fig. 8.1a), a wind tunnel model is painted with PSP and illuminated with light at an excitation wavelength λ_1, that excites the luminescent material in the paint, causing it to emit light at an emission wavelength λ_2. During testing, the model is imaged using a CCD camera equipped with a filter which only admits light at the emission wavelength of the luminescent material.

As shown in Fig. 8.1b, when a luminescent material is excited by absorbing a photon with energy $h\nu_1$, it can return to the ground state through one of several mechanisms, each occurring at a different rate. The predominant mechanisms are radiative decay (luminescence), in which a photon with energy $h\nu_2$ is emitted

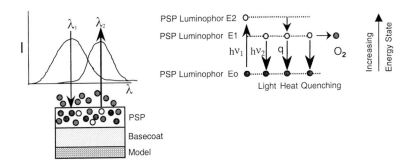

Fig. 8.1. (Left) Diagram showing physical arrangement of PSP and reaction with oxygen. (Right) Energy state diagram for PSP.

and which occurs at rate k_r, and non-radiative decay through the release of an amount of heat q, occurring at rate k_n. Some materials can also return to the ground state by colliding with an oxygen molecule, a process known as "oxygen quenching." The rate of quenching is proportional to the local oxygen partial pressure, which is in turn proportional to absolute pressure, and so can be written as $k_q p$. Thus, the intensity of light (I) emitted when the material is illuminated at λ_1 is:

$$I \propto \frac{k_r}{k_r + k_n + k_q p} \tag{8.1}$$

Eqn. 8.1 cannot be used to directly measure pressure because the constant of proportionality is unknown. Furthermore, this constant depends on both the local excitation light intensity and luminophor (dye) concentration and so it varies from point to point. This dependence can be eliminated by forming a pixel-by-pixel ratio of intensities measured at two conditions: a reference condition (I_0) where the pressure is known (for example, the no-flow or wind-off condition) and the test (wind-on) condition. Under conditions of constant excitation, this results in the Stern-Volmer equation:

$$\frac{p}{p_o} = A + B \frac{I_o}{I} \tag{8.2}$$

where A and B are constants derived from the decay rates. A and B are temperature dependent because k_n can vary with temperature, and because the luminophor is generally suspended in a "binder" with temperature-dependent oxygen permeability. Temperature effects are one of the most important sources of uncertainty in PSP measurements.

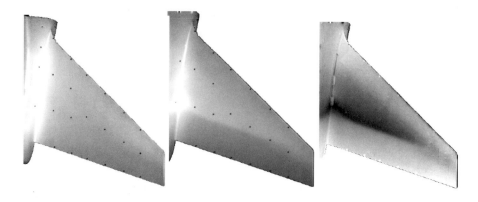

Fig. 8.2. PSP image of the upper surface of a B747-SP horizontal stabilizer. (a) Wind off (uniform pressure on model); (b) Wind on at M=0.88, α=2.5°; (c) Ratioed PSP image, with grey-scale indicating pressure.

An example of the image intensity ratio procedure defined by Eqn. 8.2 is shown in Fig. 8.2, which shows PSP applied to the upper surface of a horizontal stabilizer. In Fig. 8.2a, the wind tunnel is turned off, while in Fig. 8.2b, the wind tunnel is running at transonic speed. As air flows over the model, the paint in regions of lower pressure becomes brighter, while that in comparatively high pressure regions becomes dimmer. This effect is most striking in regions where the pressure change is abrupt, and so the feature most easily seen in Fig. 8.2b is the shock formed as the flow decelerates abruptly from supersonic to subsonic speed. Fig. 8.2c shows a pixel-by-pixel ratio of Fig. 8.2a with respect to Fig. 8.2b. The dark, low-pressure supersonic region terminated by a shock is clearly evident.

Equations. 8.1 and 8.2 demonstrate some fundamental characteristics of the PSP technique. Unlike conventional pressure sensors, PSP provides an absolute rather than a differential measurement. Thus, in order to measure small fluctuations around some mean pressure, the PSP system must be able to measure small changes in emitted light intensity. Also, a PSP's sensitivity to pressure is determined by the size of k_q compared to k_r. Paints with relatively high k_q are more sensitive, but also emit less light, since they are more highly quenched at nonzero pressure. By manipulating k_q and k_r, the sensitivity of PSP can be optimized for a given pressure range (Oglesby *et al.*, 1995). Finally, k_n should be small to maximize the paint's absolute brightness.

The effect of oxygen quenching is to increase the total decay rate of the luminophor. This is the basis of "lifetime" PSP methods, which seek to measure the decay rate directly. Since the decay rate is independent of excitation light intensity or luminophor concentration, there is no need to normalize the image by data from a reference condition. Lifetime methods require pulsed or time-varying illumination and sensors such as "gated" cameras whose sensitivity can be varied with time. Lack of space prohibits the discussion of this as well as other versions of the PSP technique. The interested reader is urged to consult one of the available review papers (McLachlan & Bell, 1995; Liu *et al.*, 1997).

PSP brightness drops slowly during use because the highly reactive singlet state oxygen produced by the quenching process can occasionally react with the luminophor in such a way as to destroy its fluorescence properties. This reaction is called photodegradation since its rate depends on the illumination level of the PSP. Photodegradation rates of 0.5 to 1%/hr are typically observed for PSPs in actual use in wind tunnels.

8.2.1 Obtaining and applying pressure-sensitive paint

As shown in Fig. 8.1, pressure sensitive paints generally consist of two layers: a base coat and a top coat, which are applied to the model surface. The base coat is a white paint used to give the model an optically smooth, low contrast surface, as well as to isolate the pressure-sensitive top coat from any chemically active regions of the model which might react with it. Commercially-available white paint may be used as a base coat for some top coats; others require a special base coat. The top coat contains the actual pressure-sensitive luminophor, suspended in an oxygen-permeable binder (usually a polymer).

Purchasing pressure-sensitive paint

As of this writing, ready-made PSP can be bought from Optrod Ltd.[b] in Russia, and Innovative Scientific Solutions Inc. (ISSI)[c] in the US. These corporations hold licenses to manufacture paints developed at the Central Aero-Hydrodynamic Institute (TsAGI), Russia and the University of Washington, respectively. The field of PSP is developing rapidly, and other paint sources are likely to emerge. An Internet search is recommended for more up-to-date

---

[b]Optrod Ltd., Dugin str. 17-31, Zhukovsky, Moscow reg., 140160 Russia, Fax: 07(095)939-2484, Email: optrod@photo.chem.msu.su
[c]Innovative Scientific Solutions, Inc., attn: Jeffrey Jordan, 2766 Indian Ripple Road, Dayton, OH 45440-3638, Tel: 937-252-2706, Fax: 937-656-4652, email: jdjordan@innssi.com, www.innssi.com

information. Commercial users are encouraged to consult the patent holders (University of Washington and TsAGI) for more detailed licensing information.

Making pressure-sensitive paint

A simple pressure-sensitive paint can be made by mixing the following ingredients:

Top coat: 1000 ml General Electric S4044 silicone polymer solution (contains xylene)[d] 1000 ml Occidental Chemical Co. Oxsol 100 (parachlorotrifluorotoluene)[e] 100 mg platinum tetraphenyl fluoro-porphine (PtTFPP)[f]

Base coat: 1000 ml SR9000, 1000 ml Oxsol 100, 100 g titanium dioxide (TiO_2). Mix in a blender (for example, Osterizer) for 30 minutes or in a ball mill for 2 days to disperse the TiO_2 in the mixture.

This paint has a peak excitation frequency $\lambda_1 = 380\text{-}400$ nm and a peak emission frequency $\lambda_2 = 630\text{-}670$ nm. The coefficients for Eqn. 8.2 are $A = -0.11$ and $B = 1.11$, where $p_0 = 1$ atm. The temperature dependence of this paint is about $0.6\%/^\circ C$ at vacuum and $0.8\%/^\circ C$ at 1 atm. The 95% response time to a pressure jump from 0 to 1 atm is 65 ms. The paint is smooth and rubbery, but buffable. The top coat can also be used with a conventional white paint base coat, although temperature sensitivity and photodegradation rate will increase.

Paint application

Pressure sensitive paints are typically applied to the test surface with either an airbrush, for small models, or with an automotive spray gun for larger models. Either air or nitrogen (clean and dry) can be used as propellants.

The first step in applying PSP is to thoroughly clean the model by wiping with detergent followed by acetone. Then apply the base coat, taking care to produce a smooth, even coating thick enough to hide any marks or high-contrast features on the model. Once the base coat is dry to the touch, black target dots can be applied to the model (their use is shown in Fig. 8.2, and discussed in Section 8.2.4). Rub-on transfers, such as those manufactured by Letra-set, Inc. or Chart-Pak, Inc. make good target dots and are available at most art supply stores. Dots should be sized so that they are 3 to 5 pixels across in the images.

---

[d]GE S4044 is available from: GE Silicones, 260 Hudson River Rd., Waterford NY 12188, Tel. 518-237-3330.

[e]Oxsol 100 available from: Oxy Chem. Co., www.oxychem.com

[f]PtTFPP is available from: Porphyrin Products, PO Box 31, Logan Utah 84323-0031, USA. Tel. 435-753-1901, Fax 435-753-6731, email: porphym@porphyrin.com, www.porphyrin.com.

PSP top coats often have a very low concentration of binder to solvent, compared with commercial paints. They must be applied very "dry," allowing ample time for the solvent to evaporate. One technique is to hold the airbrush about 30 to 60 cm from the model, make one or two spray passes, wait 2 to 3 s for the solvent to evaporate, and then repeat the process. Top coats are usually nearly transparent, and it can be hard to discern how much paint has been applied to a given area. This problem can be alleviated by lighting the model at the paint excitation wavelength, so that the paint luminesces as it is being applied. Hard PSP coatings can often be buffed to reduce surface roughness. The authors have had effective results with 9 μm sanding disks from 3M Inc.

After the model has been painted, it should be touched as little as possible. Gloves should be worn, since PSP is often vulnerable to damage by human skin oil. To minimize the need for handling, it is best to paint the final model configuration in the test section. If configuration changes will require the replacement of model parts, it is best to paint the model and any separate parts in one session. Several paint coupons should also be painted at this time for later use in a calibration chamber to determine the coefficients A and B in Eqn. 8.2.

8.2.2   Lamps

A lamp for illuminating PSP must have high output at the PSP excitation wavelength and essentially zero output at the PSP emission wavelength. In addition, the lamp should be very stable, as any change in brightness between the taking of the wind-off and wind-on images will be sensed as a change in pressure.

Different types of lamps used for PSP

For UV-excited paints, UV lamps sold for non-destructive evaluation and for UV curing of plastics are a good choice. These lamps are robust, bright and have fairly stable output, but may not be suitable for time-resolved measurements due to 60Hz ripple in the light output. The authors have had fairly good results with the Electro Lite ELC-251.[g] Stock filters used on these lamps often allow some undesirable transmission in the deep red, which is near the emission frequency of some paints. This can be reduced by adding an extra blue glass filter. Blue-excited PSPs can be illuminated with any good conventional light source which

---

[g]Electro-lite Co., 43 Miry Brook Rd., Danbury CT 06810-7414, Tel 203-743-4059 Fax 203-792-2275, eluv@electro-lite.com, www.pcnet.com/ eluv/home.htm

has been properly blue-filtered. One popular choice is a quartz halogen overhead projector bulb paired with a 450 nm bandpass interference filter (Morris *et al.*, 1993). These lights are very stable when the bulb is cooled with a fan and when driven by a stabilized power source. Another choice that has received much interest recently is blue LEDs. These have high light output efficiency compared with conventional lights, and can be assembled in arrays to produce a bright light source. When driven by a stabilized power supply, they too produce stable light output.

Lamp placement

Within the constraints of wind tunnel optical access, lamps should be mounted so as to evenly illuminate the model. "Hot spots" in the image should be avoided, since the requirement to avoid saturation of the CCD will result in other parts of the image being very dim. Also, the effect of registration errors (see Section 8.2.4) is higher in regions of high brightness gradients. Large changes in image brightness can also occur as the model traverses through its full range of motion. Images should be evaluated at the extreme model positions in the test section in order to look for hot spots and to determine if exposure time should vary as the model moves. Lamps should also be placed so that illumination is as nearly normal to the model surface as possible. When the model is lit obliquely, small changes in model position under airloads can result in large changes in illumination. In general, any illumination difference between wind-on and wind-off conditions can be misinterpreted as a change in pressure.

8.2.3   Cameras

Desired camera characteristics

The most important characteristic of a PSP camera is a high signal-to-noise ratio (SNR), especially in low-speed testing where small changes in light intensity must be measured to accurately determine pressure. In a typical low-speed wind tunnel test at a mean flow speed of 50 m/s and a stagnation pressure of one atmosphere (1.031×10^5 Pa), the dynamic pressure is $q = 1500$ Pa. Assume that the desired measurement precision is 150 Pa (that is, $\triangle p = 0.0015 p_0$), and that the paint being used has coefficients $A = -0.25$, $B = 1.25$ in Eqn. 8.2. Using Eqn. 8.2, if p/p_0 must be known within 0.15%, I_0/I must be known to within 0.12%. Since I and I_0 are measured independently, their individual errors sum in a root mean square fashion, and so light intensity must be measured to a precision of 1:1200. This is significantly beyond the 1:256 precision associated with standard video cameras. While the SNR requirement is not so stringent for transonic and supersonic applications, it remains the fundamental limitation on PSP measurement precision. Within this constraint, it is also desirable to have a camera with high light sensitivity and a fast framing rate to reduce data acquisition times. Time exposure capability is useful for cases where the PSP brightness is low. Color capability is not necessary, since the camera will be filtered to see only the wavelength emitted by the paint.

The requirement for high SNR can be met by using a scientific grade digital CCD camera. For PSP, these cameras are typically used in such a way that photon shot noise is the dominant noise source. Readout and digitization noise are negligible, except for very dim images. Noise from dark current and pixel-to-pixel sensitivity variation are corrected for using the methods described in Section 8.2.4.

"Shot noise" is an inherent physical limitation on measurement accuracy. Quantum mechanics requires that the SNR of a light intensity measurement cannot be greater than the square root of the number of photons counted to produce that intensity measurement. A CCD pixel can accept only a finite number of photons, referred to as its "full well capacity," before saturating. Thus, the full well capacity defines the limiting SNR of an image from a camera. For example, a typical mid-range CCD camera might have a full well capacity of 50,000 photons. If the exposure time is chosen to nearly saturate the brightest pixels, a typical pixel might be illuminated to within 75% of saturation, for a SNR of $\sqrt{50,000 \times 0.75} = 193$. To get the high SNR needed for measurements at low speeds, either a CCD with a much higher full well capacity must be chosen (capacities of up to 700,000 photons are available), or multiple successive images must be summed to achieve the effect of a higher full well capacity. Although time consuming, the latter process allows SNR to approach the limit set by readout and digitization noise.

Analog video cameras

Many robust and inexpensive analog video cameras are readily available. The output of these cameras can be digitized with a framegrabber in order to make PSP measurements. Since these cameras rarely have SNR's higher than 300 or so, they are (as a rule of thumb) unsuitable for most testing below Mach 0.5.

When using a standard format camera, features which cause the camera to mimic the nonlinear brightness response of the human eye (such as automatic gain control and non-unity gamma) should be disabled. Care should be taken if the camera is placed a long distance away from the framegrabber. Long analog lines are vulnerable to electromagnetic interference, especially in the high EMI environment typical of wind tunnels, which significantly degrades the quality of the resulting images. Finally, any storage of data on videotape, even using professional high quality formats, significantly reduces SNR. Image data should be digitized "live" whenever possible.

8.2.4   Data reduction

The data reduction procedure described here starts by calibrating the camera to remove any non-uniform response to light. A pixel by pixel ratio of the wind-off over the wind-on image is then obtained, and the intensity ratio image is converted into a pressure image by applying the known coefficients A and B from Eqn. 8.2. Finally, image data are mapped to model surface coordinates.

Camera calibration

While the voltage output of a CCD pixel for a given light input is highly linear, exact gain and offset values can vary several percent from pixel to pixel. To remove this variation, the intensity at each pixel (I_{xy}) must be corrected by use of dark (D) and flat field (F) images. A dark image is simply one taken with the camera shutter closed, which determines the level reported by the CCD in the absence of light. To obtain a flat field image, the CCD is illuminated so that all pixels receive the same amount of light.[h] The corrected intensity at each pixel, C_{xy}, is obtained as follows:

$$C_{xy} = \frac{I_{xy} - D_{xy}}{F_{xy} - D_{xy}} \tag{8.3}$$

Wind-on/wind-off image registration

The wind-off over wind-on image ratio will be in error if the model is not in exactly the same position in both images. Since in general the model moves due to differing airloads between the wind-on and wind-off conditions, the wind-on image must be transformed to match the model position in the wind-off image.

---

[h]While the most accurate way to get a flat field image is to use an integrating sphere, usable flat fields can be produced by pointing the camera at an evenly illuminated ground glass screen. Note that most lenses put more light on the center of the CCD than the edges, and that this nonuniformity too, is corrected by calibration with a flat field image.

The necessary transformation is determined by measuring the locations of small marker dots, or targets (applied on the model as described in Section 8.2.1) in both images. Targets should be located with at least 0.05 pixel accuracy, which typically requires the use of centroid finding techniques. One convenient registration transform from wind-off (x, y) to wind-on (x', y') coordinates is given by a third order polynomial:

$$
\begin{aligned}
x' &= a_0 + a_1 x + a_2 y + a_3 x^2 + a_4 xy + a_5 y^2 + a_6 x^3 + a_7 x^2 y + a_8 xy^2 + a_9 y^3 \\
y' &= b_0 + b_1 x + b_2 y + b_3 x^2 + b_4 xy + b_5 y^2 + b_6 x^3 + b_7 x^2 y + b_8 xy^2 + b_9 y^3
\end{aligned}
$$

$$(8.4)$$

Coefficients a_i, b_i are determined by matching the known target locations in both images. Typically, there will be more targets in the images than free parameters in Eqn. 8.4 and so the overdetermined system is solved in the least-squares sense. Even sub-pixel motion introduces a large enough error that registration will noticeably improve data quality. Eqn. 8.4 is capable of correcting model motion to about 0.1 pixel accuracy. Model motion also causes a change in illumination between the wind-off and wind-on positions. This introduces a spurious intensity signal, which in some cases can result in significant pressure error (Bell & McLachlan, 1995). Two-luminophor paints, where the second luminophor is pressure-insensitive and so measures incident light, have been developed to solve this problem. Their use requires the normalization of each data image with a reference image acquired at the emission wavelength of the second luminophor.

Calibration

Pressure-sensitive paint calibration methods can be divided into two categories. *A priori* calibration techniques rely on a sample coupon, painted at the same time as the model, which is then placed in a pressure- and temperature-controlled chamber. The sample is illuminated and its output brightness measured at different pressures over the temperature range encountered in the wind tunnel test. The resulting intensity versus pressure curve is fit to determine A and B in Eqn. 8.2. The *in situ* calibration uses pressure taps to calibrate the paint. The intensity versus pressure relationship is determined by correlating the paint intensity ratio at the locations of several pressure taps with pressure data from the taps. *In situ* techniques automatically correct for variations in mean illumination level and temperature between wind-on and wind-off conditions, as well as any paint photodegradation. These error sources affect the paint brightness much more than its pressure sensitivity. Thus a hybrid technique has been developed which begins with an *a priori* calibration to obtain coefficient B in Eqn. 8.2. An *in situ* calibration with one or two pressure taps is then used to adjust the coefficient A.

A major part of calibration is correction for the temperature sensitivity of the paint. PSP temperature sensitivity has varied from more than 2%/°C for the original paint formulations to 0.2 to 1.2%/°C today. While a general temperature correction is complex (Woodmansee & Dutton, 1995), two approximations are commonly made to simplify it. First, point-to-point temperature variations are assumed to be small compared with mean variation between wind-on and wind-off conditions. Second, the paint is assumed to be "ideal" in that its pressure sensitivity is assumed to be unaffected by temperature. Modern paints approach this condition very well. These assumptions underlie the hybrid calibration technique described above.

Mapping to model coordinates

A final data reduction step, not required in all cases, is to associate all points on the calibrated PSP image with the corresponding points on the model. This is conveniently done using the Direct Linear Transform method of photogrammetry. In this method, model coordinates X, Y, Z are related to their corresponding image coordinates x, y through the equations:

$$x = \frac{L_1 X + L_2 Y + L_3 Z + L_4}{L_9 X + L_{10} Y + L_{11} Z + 1} \quad y = \frac{L_5 X + L_6 Y + L_7 Z + L_8}{L_9 X + L_{10} Y + L_{11} Z + 1} \qquad (8.5)$$

The coefficients $L_1 \ldots L_{11}$ are determined by inserting the known x, y and X, Y, Z coordinates of each target applied to the model into Eqn. 8.5. The resulting system of equations is linear in the unknown coefficients $L_1 \ldots L_{11}$, and can be solved provided that at least six targets are known. This technique is discussed more fully by Bell & McLachlan (1996).

8.3   Shear-Sensitive Liquid Crystal Coating Method

The shear-sensitive liquid crystal coating (SSLCC) method is an image-based technique for both visualizing dynamic surface-flow phenomena, such as transition and separation, and for measuring the continuous shear-stress vector distribution acting on an aerodynamic surface. Shear-sensitive liquid crystals belong to the cholesteric mesophase of liquid crystals (Fergason, 1964). Under aerodynamic shear, this optically active material selectively scatters incident white light as unique colors of visible light at unique orientations, that is, as a three-dimensional spectrum. This color-change response is continuous and reversible, with a response time of milliseconds.

Klein (1968) introduced the liquid crystal method to aerodynamics. Based on this early work, liquid crystal coatings were used to obtain qualitative areal visualizations of shear stress magnitudes acting on aerodynamic surfaces in both wind tunnel (Hall *et al.*, 1991) and flight-test (Holmes *et al.*, 1986) applications. More recently, Reda & Muratore (1994) showed that SSLCC color-change response to shear depends on both shear stress magnitude and the direction of the applied shear vector relative to the observer's in-plane line of sight.

8.3.1   Color-change responses to shear

When illuminated by white light directed normal to a coated surface and observed from an oblique angle above the plane of that surface, any point exposed to a shear vector with a component directed away from the observer exhibits a color-change response (Reda & Muratore, 1994) (see Fig. 8.3a). This color-change response is characterized by a shift from the no-shear color (red or orange) toward the blue end of the visible spectrum. The extent of the color change is a function of both shear magnitude and shear direction relative to the observer. Conversely, any point exposed to a shear vector with a component directed toward the observer exhibits no color change (Fig. 8.3b). The specifics of the physics and optics of liquid crystals are beyond the scope of this text. Further details are given by Chandrasekhar (1992) and De Gennes & Prost (1995).

The SSLCC color-change responses shown in Fig. 8.3 were quantified by measuring the scattered-light spectra from a point on the wall-jet centerline using a fiber-optic probe and a spectrophotometer (Reda & Muratore, 1994). Results are shown in Fig. 8.4a and Fig. 8.4b. At any fixed shear stress magnitude (τ/τ_r), the maximum color change (that is, the change in dominant wavelength, λ_D) is always measured when the shear vector is aligned with and directed away from the observer ($\beta = 0°$ in Fig. 8.4a). The color change decreases with changes in the relative in-plane view angle (β) to either side of this vector/observer aligned orientation. The color change is a Gaussian function of β, as shown in Fig. 8.4a. For any fixed in-plane view angle $|\beta| < 90°$, the color change is a monotonically increasing function of the shear stress magnitude, $|\tau/\tau_r|$. This is shown in Fig. 8.4b for $\beta = 0°$. For $|\beta| > 90°$ there is no color change, that

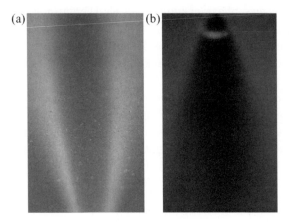

Fig. 8.3. Color-change response of liquid crystal coating to tangential jet flow, $\alpha_L = 90°$, $\alpha_C = 35°$. (a) Flow away from, and (b) flow toward observer. Figure also shown as Color Plate 9.

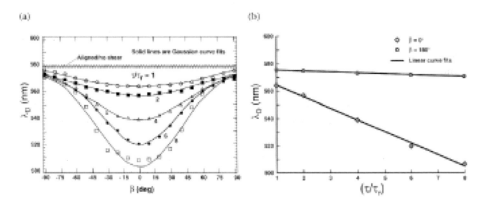

Fig. 8.4. SSLCC color-change responses. (a) Dominant wavelength vs. relative in-plane view angle between observer and shear vector, with relative shear magnitude as the parameter; (b) Dominant wavelength vs. relative shear magnitude for relative in-plane view angles of 0° and 180°.

is, λ_D is essentially independent of shear magnitude, as shown in Fig. 8.4b for $\beta = 180°$.

All of the above results were obtained on a planar surface illuminated from the normal direction. However, experiments have shown that off-normal illumination by as much as $\pm 15°$ does not influence the measured color change for $\beta = 0°$ (Wilder & Reda, 1998).

Based on these observations and measurements, full-surface shear stress vector visualization (Reda, 1995a; Reda *et al.*, 1997a) and measurement (Reda, 1995b; Reda *et al.*, 1997b) methodologies were formulated and demonstrated. Examples illustrating each method will be given.

8.3.2   Coating application

Liquid crystal materials are commercially available from the Liquid Crystal Division of Hallcrest, Inc. in Glenview, Illinois. A wide variety of SSLCC materials exists, covering a wide range of viscosities. The "correct" compound for any experiment is the one that yields a full-range (red-to-blue) color change response under the range of shear magnitudes experienced in the experiment, yet is viscous enough not to flow over the surface. Color measurements of the light scattered by the SSLCC are valid only in the absence of such macroscopic migrations.

Compounds used in aerodynamic applications include BCN/192, BCN/195, CN/R1 and CN/R3. The usable shear range for these materials is approximately 5 to 50 Pa (0.1 to 1 psf). Hydrodynamic applications should utilize more-viscous compounds, for example, CN/R7 and CN/R8. All such compounds are broadband insensitive to temperature, freezing at $0°C$ and melting at $50°C$. The shelf life is quoted as one year.

A mixture of one part liquid crystal to nine parts, by volume, of a solvent such as Freon or a Freon replacement (for example, DuPont Vertrel XF-9571) is sprayed onto the test surface. For small test areas, an artist airbrush is the preferred spray device and it should be flushed out with pure solvent after every use. The solvent evaporates at atmospheric conditions, leaving a uniform liquid crystal coating. Unlike paint, the coating does not dry, but remains viscous and should not be touched. A smooth, flat-black surface (for example, anodized aluminum) is essential for color contrast: the color response is a low-intensity scattered field that is easily overpowered by light scattered from a bright surface finish. Reference marks are required on the surface for image registration when quantitative results are desired. The solvent should be used to clean the test surface prior to the coating application, as well as to remove the coating after testing.

Recommended coating thicknesses are in the range 25 to 75 μm (0.001 to 0.003 in). Assuming a 50% spray loss, this requires a sprayed volume of the nine-to-one mixture equal to 0.15 cc/cm^2 (1 cc/in^2) of test surface. Cost to coat a test surface is less than $200/m^2$ ($20/ft^2$).

The molecules within a newly sprayed coating are generally not aligned in the molecular orientation required to disperse white light into a spectrum of colors. The optically active arrangement can be achieved by shearing the coating, either by passing a pressurized air jet over the coated surface prior to the experiment, or by the flow under investigation itself. For this second alternative, it is important that the initial flow provide the maximum shear condition that the coating will experience during the experiment, otherwise lightly sheared regions of the coating may not achieve the proper molecular arrangement.

8.3.3   Lighting and imaging

The SSLCC must be illuminated with white light (color temperature near 5600 K) in order for the color-change response to produce the full visible spectrum. The 1200-W Sylvania PAR64 BriteBeam is an example, and should be operated with a flicker-free ballast.

For qualitative flow visualizations, the color-change response can be imaged using standard color video cameras. When quantitative measurements are required, a 3-CCD, co-site sampling, RGB (Red, Green, and Blue) video camera is preferred. This type of camera, which is typically used in medical imaging applications, records the light intensity of the NTSC (National Television Systems Committee) standard RGB components of the scattered light using three CCD chips. Features that compensate for the nonlinear brightness response of the human eye, such as automatic gain and non-unity gamma, should be deactivated. Each color component is digitized using a frame grabber. An 8-bit per channel (24-bit color) digitizer provides better than 1 nm color resolution, while coating-to-coating repeatability is typically 2 to 3 nm.

The color measurements are rendered illuminant-invariant through a linear color correction (Farina *et al.*, 1994) that references all measurements to the CIE (Commission Internationale d'Eclairage) Illuminant C. Color is determined from the measured RGB intensities by calculating hue (Hay & Hollingsworth, 1996), an intensity-invariant measure of color which can be directly related to dominant wavelength through the CIE colorimetric system (Wyszecki & Stiles, 1967).

Two imaging concerns need to be addressed: reflected glare and potential saturation of one or more of the color signals. Reflected glare can be minimized by adjusting the relative angular orientation of two linear polarizing filters: one placed on the camera lens, and the second placed between the light and the

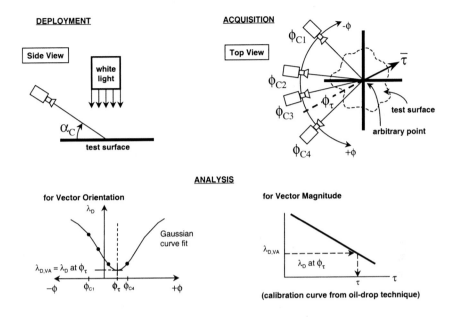

Fig. 8.5. Schematic of full-surface shear stress vector measurement methodology.

coated test surface. The circularly polarized light scattered from the SSLCC is unaltered by this approach. To overcome the second concern, images can be recorded at two or more exposure settings, and a composite image can be formed using only the correctly exposed pixels from each image. This technique is possible since color (dominant wavelength) is independent of intensity.

8.3.4 Data acquisition and analysis

The data presented in Fig. 8.4 were used to formulate the full-surface shear stress vector measurement method shown schematically in Fig. 8.5. The coated test surface is illuminated from the normal direction with white light and the camera is positioned at an above-plane view angle (α_C) of approximately 30°. For quantitative measurements, images of the SSLCC color-change response to the shear field are recorded from multiple in-plane view angles encompassing all shear vector directions to be measured (shown in Fig. 8.5 as ϕ_{C1} to ϕ_{C4}). As shown in Fig. 8.4, the color-change response to a constant shear stress vector is a Gaussian function of the relative in-plane view angle between the observer and the vector orientation. Therefore, the shear vector orientation can be determined at each physical point on the test surface by fitting a Gaussian curve to the variation in measured color (λ_D) with changing in-plane view angle (ϕ_C) at that point on the surface. In theory, a minimum of four images is required to obtain the Gaussian curve fit, but in practice this number is generally increased consistent with optical access. The in-plane angle corresponding to the maximum color-change value of the curve-fit determines the vector orientation (ϕ_τ), and the value of the vector-aligned color ($\lambda_{D,VA}$) is then related to the shear magnitude (τ) via a calibration curve acquired using conventional point-measurement techniques (for example, the fringe-imaging skin friction or "oil-drop" technique discussed in Section 8.4).

In-situ calibration of the SSLCC is optimally achieved after the unscaled data set, comprised of a vector-aligned color (proportional to shear magnitude) and a vector orientation at every grid point on the surface, has been acquired. In this manner, the point-measurement method of choice (for example, the oil-drop technique) can be employed at precise locations to encompass the complete range of vector-aligned colors (shear magnitudes) encountered in the flow under study.

Calibration of SSLCC materials in a mechanical shearing apparatus such as a rotating-disc or rotating-shaft device is not recommended. The no-slip boundary condition coupled with the relative motion between the moving and fixed surfaces forces a velocity distribution to occur within the liquid crystal material. This flow situation alters the liquid crystal molecular arrangement and thus its color-change response as compared to the application of aerodynamic shear to the exposed surface of a non-flowing SSLCC.

The signal-to-noise ratio of images can be improved by frame-averaging several images (for steady-flow applications) and/or by spatially filtering the images. Spatial filtering involves replacing the RGB values of each pixel with the average of its neighboring pixels, and sacrifices spatial resolution in favor of increased signal-to-noise ratio. Typically a 3×3 or 5×5 pixel neighborhood is used.

The color (λ_D) measurements used in the Gaussian curve-fit portion of the analysis (see again Fig. 8.5) must be obtained at the same physical location on the surface for each in-plane view angle. This requires mapping the color images onto a common grid on the physical surface through the principles of photogrammetry (Reda *et al.*, 1997b; Stacy *et al.*, 1994). See also Section 8.2.4.

SSLCC vector measurement resolution and accuracy issues were discussed in detail by Wilder & Reda (1998) and Reda *et al.* (1997b). Uncertainties of 2 to 4% in shear vector magnitude and less than 1° in shear vector orientation have been attained for absolute magnitudes in the range of 5 to 50 Pa (0.1 to 1 psf).

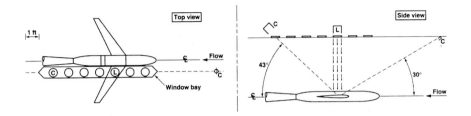

Fig. 8.6. Schematic of experimental arrangement for visualizations of transition and separation.

These uncertainties were attained using color images acquired at spacings of 15 to 25° for the in-plane view angles. Uncertainties in both shear vector magnitude and orientation increase with increasing image spacing, and are about double these values for image spacings greater than 40° (Wilder & Reda, 1998). The SSLCC method has been validated against independent oil-drop skin-friction measurements not used in the calibration process (Reda *et al.*, 1997b, 1998).

8.3.5   Example: Visualization of transition and separation

An example of an experiment that capitalized on the unique shear-direction-indicating capabilities of SSLCCs to visualize transition and separation on a model aircraft wing (Reda *et al.*, 1997a) is schematically illustrated in Fig. 8.6. The model, a generic commercial-transport aircraft with a tip-to-tip wing span of 1.7 m (67 in.), was positioned on the centerline of the Boeing 2.4 × 3.6 m (8 × 12 ft) transonic wind tunnel. The SSLCC was applied to the upper surface of the starboard wing and the inboard portion of this wing was uniformly illuminated by a white light (L) from above. Two synchronized color video cameras (C) were positioned as shown in Fig. 8.6.

In this arrangement, transition to turbulence on the wing upper surface, characterized by an abrupt increase in surface shear stress magnitude in the principal flow direction, was made visible by the SSLCC color-change response recorded with the downstream-facing camera. Conversely, regions of reverse flow enveloped by upstream-directed shear vectors were made visible by the SSLCC color-change response recorded with the upstream-facing camera. Regions of the coated test surface exposed to shear vectors directed toward either camera yielded no color-change response, appearing as either dark or reddish-brown zones, depending on the absolute light levels reaching the camera. Any regions

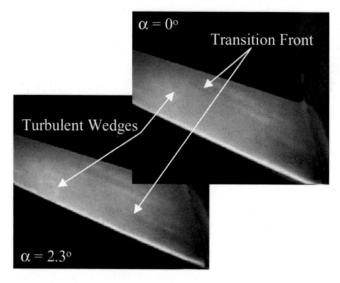

Fig. 8.7. Transition-front visualizations recorded by downstream-facing camera at $M = 0.4$ and $Re = 8.2 \times 10^6$/m. Figure also shown as Color Plate 10.

of extreme transverse flow, enveloped by shear vectors directed either inboard or outboard and approximately perpendicular to the principal flow direction, would have appeared (if present) as a yellow color-change response simultaneously to both cameras (Reda & Muratore, 1994).

Because the SSLCC color-change response is both dynamic and reversible, it is possible to visualize phenomena such as transition-front movement on a maneuvering surface. This is illustrated in Fig. 8.7, which shows two frames from a video recorded while the model was slowly pitched through an angle-of-attack (α) range at freestream Mach number $M = 0.4$ and Reynolds number $Re = 8.2 \times 10^6$/m (2.5×10^6/ft.). Regions of low shear magnitude were delineated by a red or yellow color, whereas regions of high shear magnitude appeared as green or blue. The chordwise transition front was seen to move forward with increasing angle of attack. The discrete turbulent wedges originating from the wing leading-edge region were a result of isolated roughness elements caused by freestream contaminants impacting the surface.

Fig. 8.8a, which shows images of the color-change response recorded simultaneously by both the upstream and downstream facing cameras, captures a leading-edge separation zone on the outboard upper surface of the wing at

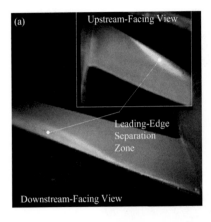
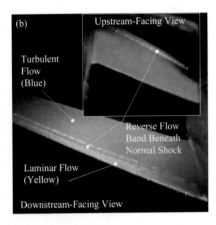

Fig. 8.8. Color-change response as recorded by opposing-view cameras: (a) leading-edge separation, $\alpha = 8°$, $M = 0.4$, $Re = 8.2 \times 10^6$/m; (b) normal-shock/boundary-layer interaction, $\alpha = 5°$, $M = 0.8$, $Re = 11.2 \times 10^6$/m. Figure also shown as Color Plate 11.

$\alpha = 8°$. The red zone in the downstream-facing view and the corresponding yellow zone in the upstream-facing view indicate a low-shear region with flow directed upstream (reverse flow).

At a higher freestream Mach number, $M = 0.8$ and $Re = 11.2 \times 10^6$/m (3.4×10^6/ft.), a normal shock wave/laminar boundary-layer interaction occurred slightly downstream of the wing leading edge for $\alpha = 5°$. This is illustrated in Fig 8.8b, which shows the SSLCC color-change responses recorded by the synchronized opposing-view cameras. Here, the yellow zone along the wing leading edge, recorded by the downstream-facing camera, indicates a low-shear (laminar) region upstream of the interaction. A narrow band of reverse flow formed beneath the interaction region and is indicated by the reddish-brown band in the downstream-facing view and, simultaneously, by the yellow band in the upstream-facing view. This reverse-flow region was breached by numerous turbulent wedges emanating from aforementioned roughness elements along the leading edge; passage of these locally attached turbulent wedges through the interaction region are illustrated by the dark breaks in the yellow band recorded by the upstream-facing camera.

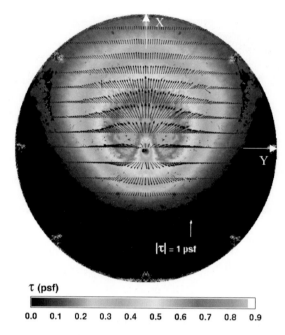

Fig. 8.9. Measured surface shear stress vector field beneath inclined, impinging jet: colors show shear magnitudes and vector profiles every $\triangle X/D = 1$ show shear orientations. Figure also shown as Color Plate 12.

8.3.6   Example: Application of shear vector method

The measurement methodology was applied to measure the shear stress vector distribution on a planar surface beneath a tangential wall jet (Reda *et al.*, 1997b) similar to that shown in Fig. 8.3, and beneath an inclined, impinging jet (Reda *et al.*, 1998). Results of the impinging-jet experiment are reviewed below.

The jet diameter, D, was 0.84 cm (0.33 in) and the jet initial velocity profile corresponded to that of a fully-developed turbulent pipe flow with a centerline exit Mach number of 0.66. The Reynolds number, based on exit centerline conditions and D, was 1.36×10^5. The jet exhausted into atmospheric pressure and the jet total temperature matched the ambient level. The jet exit plane was 13D from the geometric stagnation point (GSP), the center of a 15.24 cm (6 in) diameter test surface. The jet impingement angle was 57° relative to the plane of the surface.

Frame-averaged images of the complete test surface were recorded from each

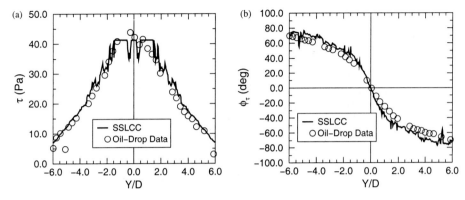

Fig. 8.10. Cross-stream profiles of shear vector field beneath inclined, impinging jet at $X/D = 2$. (a) Magnitude, (b) orientation.

of fifteen ϕ_c orientations over the arc $0 \leq \phi_c \leq 180°$. Taking advantage of the symmetry of the flow field, the images were mirrored across the plane of symmetry (the X-axis) to form a complete $0 \leq \phi_c \leq 360°$ image set. The images were analyzed according to the procedure outlined in Fig. 8.5, and the resulting surface shear stress vector distribution is shown in Fig. 8.9. Color in this figure represents shear stress magnitude, while select shear stress vector orientations are illustrated by the vector cross-cut profiles drawn every $\triangle X/D = \pm 1$ starting at the Y-axis. For clarity, only every fifth vector, spaced at $\triangle Y/D = 0.15$, is shown in each profile.

A local minimum in shear magnitude was seen to occur in the immediate vicinity of the GSP. Shear magnitude increased rapidly in all directions emanating from the stagnation zone as the inclined-jet flow turned to align itself with the plate surface, then accelerated outwards.

Fig. 8.10 shows continuous measurements from the SSLCC method versus point measurements from the oil-drop technique as acquired on a transverse cross-cut at $X/D = 2$. None of the oil-drop data shown here were used in calibration. Very good overall agreement was noted between shear vector magnitudes and shear vector orientations measured by these two methodologies. Calibration data for $\tau > 41.7$ Pa (0.87 psf) were not available, hence the SSLCC magnitude data are clipped in Fig. 8.10a.

8.4   Fringe Imaging Skin Friction Interferometry

Oil film interferometry has been used to measure skin friction since 1976 when Tanner & Blows first developed the technique. In 1993, Monson & Mateer developed a simplified form of the oil-flow equation where skin friction coefficient can be determined knowing the final thickness distribution of the oil film and a few other readily available quantities. They also demonstrated the use of standard room lighting to visualize the oil film interferometric patterns and showed that skin friction measurements can be made using a minimum of equipment and set-up time. Recently, the technique has been extended to three-dimensional flows (Zilliac, 1996) and to the use in large wind tunnels (Driver, 1997).

8.4.1   Physical principles

The oil film technique is based on the principle that oil on a surface, when subjected to shear, will thin at a rate related to the magnitude of the shear. The measurement of skin friction involves measuring the oil thickness distribution, recording a history of the tunnel run conditions, and knowing the properties of the oil.

The oil film's thickness distribution is determined from the interference patterns that can be seen in the oil as a result of interference between reflected light from the model surface and the reflected light from the air-oil interface (see Fig. 8.11). The spacing of the dark bands (or fringes) is a measure of the slope of the oil front and are contours of constant oil thickness. Fig. 8.11 shows a plan view along with a cross-sectional view a typical oil flow.

The spacing between the fringes, $\triangle$s, is proportional to the skin-friction as seen in the equation derived by Monson & Mateer (1993) from one-dimensional lubrication theory.

$$C_f = \frac{\tau_w}{q_\infty} = \left[2n_o \cos{(\theta_r)} \frac{\triangle s}{\lambda} \right] \left[\int \frac{q_\infty}{\mu_o} dt \right]^{-1}. \qquad (8.6)$$

The numerator on the right hand side of the equation is an interferometric measure of the reciprocal of the oil film slope where n_o is the oil's index of refraction, θ_r is the refracted light angle through the oil, and λ is the wavelength of the light source illuminating the oil. The integral, containing the oil viscosity, μ_o, and the tunnel dynamic pressure, q_∞, is integrated over time, t, of the tunnel run. The light refraction angle through the air-oil interface is related to the light incidence angle on the oil by $\theta_r = \sin^{-1}(\sin \theta_i / n_o)$.

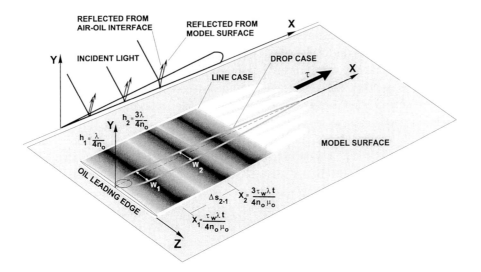

Fig. 8.11. Oil film and cross section.

The skin friction vector direction is that of the pathlines in the oil flow. If the fringe spacing Δs is measured along the direction of the oil pathline then the C_f given by Eqn. 8.6 is the magnitude of the skin friction coefficient vector.

8.4.2 Surface preparation

The model surface used in oil film interferometry (also known as Fringe Imaging Skin Friction interferometry or FISF) measurements must be smooth and also have optical properties that enable fringes to be visible. Obtaining high fringe visibility when using the FISF technique under less than ideal conditions can be difficult. Theoretically, the maximum fringe visibility is achieved on a surface with an index of refraction of 2.0 (for fluids with $n_o = 1.4$ such as silicone fluid).

Many different surfaces have been tried with dense flint glass being the best. Test surfaces made from acrylic, glass or polished stainless steel provide good fringe visibility (polished surfaces need to be 2 μ-in or better polish). Nickel plating and some high gloss black paints applied over a smooth model surface also provide good fringe visibility. Aluminum has proven to be a relatively poor optical surface (its light absorption is too low). A good rule of thumb for surface materials is that you should be able to see your reflection in the surface.

When working with existing wind tunnel models, often it is not possible (or allowable) to tamper with the surface finish. Under these circumstances, the easiest approach is to apply thin sheets of MonoKote Trim Sheets (glossy Mylar with black pigmented backing and adhesive coating) to the test surface. The combination of Mylar (index of refraction 1.67) and black pigmented backing (sandwiched between the Mylar and the adhesive compound) provides a partially reflective surface that reflects light with about the same intensity as does the air-oil interface (slightly less than 4% reflected). MonoKote Trim Sheets are sold by Top Flite Models, Inc. Champaign, IL and are often stocked in hobby shops.

Just prior to a wind tunnel run, small patches (line segments, drops, smudges, etc.) of silicon oil are applied to a series of locations on the freshly cleaned model surface. Dow Corning provides silicone fluids[i] in range of viscosities of 5 to 100,000 cs. The choice of viscosity is based on tunnel run time, dynamic pressure, temperature and desired fringe spacing. The oil patches should be spaced far enough apart so that one oil-flow patch does not obliterate another patch.

Dow Corning silicone fluid 200 is not the only type of oil that will work, but it has been the favorite choice so far due to it's low surface tension, high degree of transparency and the "relative" insensitivity of it's viscosity to temperature in comparison with other oils.

8.4.3   Lighting

After passing air over the test surface, the next step is to record the oil-flow patterns. The oil flow pattern will remain in its final thinned state for a period of time (dependent on humidity level, temperature and surface roughness). Gravity is a negligible force on a film that is only microns thick.

The oil should be illuminated with a light source that is fairly monochromatic and imaged with fiducial marks or rulers in place. A narrow notch filter, placed in front of the camera, is often used to remove undesired portions of the light-source spectrum. The light source needs to be spatially coherent over a few microns (which eliminates tungsten bulbs) and standard gas-filled bulbs usually have this characteristic. Light sources such as fluorescent bulbs, black lights, or other forms of Mercury discharge tubes provide very coherent light with a strong emission at specific wavelengths ($\lambda = 546$ nm) easily isolated with notch filters. Standard Xenon studio flashes provide good coherent light sources. Another

[i] *Information About Dow Corning Silicone Fluids*, Dow Corning Corp. Midland, MI., 1994.

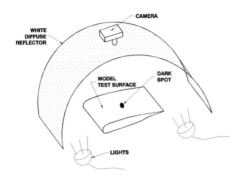

Fig. 8.12. Schematic showing a front light diffuse reflector as a light source.

excellent light source is a low-pressure sodium bulb. Typically, these bulbs emit at a single wavelength, (actually a closely spaced doublet), $\lambda = 589$ nm and $\lambda = 589.6$ nm, making filters unnecessary. Lasers are highly monochromatic and highly coherent but present difficulties due to speckle.

Many light source configurations have been used, ranging from light boxes to elaborate reflective umbrellas which surround the model (see Fig. 8.12). The light source choice and illumination techniques are discussed in greater detail by Zilliac (1996) and Driver (1997).

The most desirable way to light the surface is by normal illumination, however, this is difficult since the lights and camera are competing for the same vantage point. Half-silvered mirrors make it possible to obtain normal illumination over a small region of the model but are not very practical. The basic requirement is that light from the light source is specularly reflected from the model surface directly into the camera. The bigger (more expensive) the light source, the greater the area of the model that will specularly reflect. Alternately, the wind tunnel walls can be painted white and used as light reflectors to bounce light from a point source onto the surface of a model. The camera can be positioned in one of the windows (viewing through a small non-painted portion of window). An example of an image acquired while looking through a window with tunnel walls painted white is shown in Fig. 8.13. The black spot in the image is due to the lack of light from the reflection of the camera lens. The hole is usually a small percentage of the total area. Details of the various

Fig. 8.13. Model illuminated with test section walls.

lighting approaches are described by Driver (1997).

8.4.4 Imaging

The interference patterns can be photographed with most any camera, however, for best results it is useful to use a black and white digital camera with as high a spatial resolution as you can afford. Black and white film-based cameras are fine for visualization purposes but suffer somewhat lower contrast during the negative scanning process. Color digital cameras will also work but contain artifacts due to the doping of the individual pixels in a red-green-blue checkerboard pattern. The red and blue pixels do not contain as much signal as the green for mercury-based light sources (emission in the green portion of the spectrum).

High spatial resolution is desirable so that details of closely spaced fringes

$\nu_{o,nom}$ (cs)	10	50	100	200	500	1,000	10,000
$\rho_{o,T=25°C}$ (kg/m^3)	931	957	961	964	966	967	–
n_o	–	1.4022	1.4030	1.4032	1.4034	1.4035	1.4036
α $(cc/cc/°C)$	0.00108	0.00104	0.00096	0.00096	0.00096	0.00096	–

Table 8.1. Properties of Dow Corning 200 Fluid at 25°C (Dow Corning Corp. Midland, MI.)

can be seen within a large field of view. The accuracy of determining fringe spacing is a function of the number of pixels defining the fringe as well as the contrast of the fringe pattern (Zilliac, 1996). High-quality lenses are also desirable from the standpoint of increased resolution and also decreased image distortion.

8.4.5 Calibration

Silicone 200 Fluid manufactured by Dow Corning is typically used by most researchers in this field. This fluid is a polydimethyl-siloxane polymer and is available with the physical properties listed in Table 8.1. The manufacturer quotes a viscosity of the fluid at 25°C with a ±5% uncertainty; consequently the fluid should be independently calibrated. Furthermore, the viscosity changes by 2% per °C. A one-point viscosity calibration using a Cannon-Fenske viscometer, (Fisher Scientific, Pittsburgh, PA) at a temperature near the anticipated tunnel run temperature, is usually sufficient. The fluid's kinematic viscosity, $\nu_{o,T}$, as a function of temperature can be reliably determined using:

$$\log_{10}(\nu_{o,T}) = [C_1/(T+C_2) - C_1/(T_{cal}+C_2) + \log_{10}\nu_{o,cal}] \qquad (8.7)$$

where T is in °K, $C_1 = 774.8622$ and $C_2 = 2.6486$. This equation is most accurate for $255 < T < 310°K$ and for $100 < \nu_{o,cal} < 1000$ cs. Furthermore, the density of the oil is also slightly dependent on temperature. The oil's specific gravity is a function of temperature T, given by $\rho_{o,T} = \rho_{o,T=25°C}/[1+\alpha(T-25)]$ where the coefficient of expansion α is given in Table 8.1 and T is in °C. Hence $\mu_{o,T} = \rho_{o,T}\nu_{o,T}$.

8.4.6 Data reduction

Measuring the fringe spacing on a digital image can be done on a personal computer using various software packages such as PhotoShop (by Adobe) or customized software packages (Zilliac, 1999). It can also be done crudely with vernier calipers directly on a photograph or the test model itself. Various algorithms have been developed for determining the fringe spacing $\triangle s$ using the intensity distribution seen in each interferogram, ranging from FFT's (Monson & Mateer, 1993), to Hilbert Transforms (Naughton & Brown, 1996), to physics-based nonlinear regressions (Zilliac, 1996).

One approach developed recently (Zilliac, 1996) involves analyzing the intensity record along a line following the flow direction which is commonly deduced from the oil streak direction (see Fig. 8.11). The direction of this line, resolved in a surface coordinate system, is assumed to be the direction of the skin friction vector. A nine-parameter model can be fit to the fringe-intensity distribution using a nonlinear regression algorithm. The model is given by:

$$I = B_1 + B_2 s + B_3 s^2 + (E_1 + E_2 s + E_3 s^2) \cos{(P_1 + P_2 s + P_3 s^2)}, \qquad (8.8)$$

where I is the intensity, s is the distance along the centerline of the oil streak (in pixels), and B, E and P are the regression coefficients (Zilliac, 1996). Typically, the first two fringes in the intensity record are sufficient to obtain an accurate measure of the fringe spacing $(\triangle s)$. The advantage of this approach is that the model is derived from the physics of interferometry. It makes allowances for such effects as surface curvature, noise, small optical imperfections and nonuniform lighting. In addition, the whole intensity record is used by the regression analysis to identify the fringe spacing as opposed to schemes that simply fit the peaks of the intensity distribution.

An example of a typical oil dot streak and its associated fringe intensity distribution is shown in Fig. 8.14. The skin friction coefficient magnitude can be determined by using the fringe spacing found from the regression followed by a conversion of pixel-based fringe spacing to physical coordinates (often achieved via photogrammetry) and application of Eqn. 8.6. The direction of the skin friction vector is found by determining the orientation of the oil pathline measured in the vicinity of the leading edge of the oil.

Other elaborate approaches involve backing out the oil film thickness distribution using a Hilbert transform and then solving the partial differential equations governing the oil flow numerically to determine what skin-friction distribution must have been necessary to cause that particular oil film thickness distribution (Naughton & Brown, 1996). The use of 2D Hilbert transforms and the 2D oil-film equations allows one to solve for the skin-friction distribution as a function of location in the oil film patch.

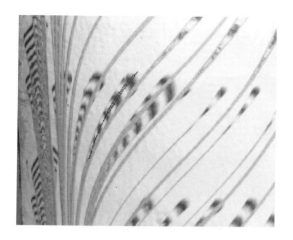
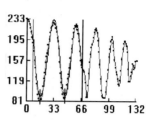

Fig. 8.14. Oil flow fringe pattern and corresponding intensity record (measured along the black line drawn normal to the fringe fronts).

Error source	Uncertainty range	Remarks
Non parallel streamlines	0% to 5% of C_f	a positive bias error
Oil viscosity	±0.2% to ±5% of ν_o	calibrate the oil
Temperature	±0.05% to ±1% of T_o	measure T_o
Pressure gradient effects	±0% to ±0.14% of C_f	estimated by $h(\delta C_p/\delta s)$
Shear gradient effects	±0% to ±20% of C_f	estimated by $\frac{1}{4} \triangle s(\delta C_f/\delta s)$
Freestream dynamic	±0.25% to ±1.0% of q_∞	use accurate transducer
Regression and imaging	±0.5% to ±5% of $\triangle s$	use a calibrated camera
Startup and shutdown	±0% to ?% of C_f	minimize tunnel startup t

Table 8.2. FISF error sources

8.4.7 Uncertainty

The uncertainty of FISF measurements is dependent on many factors the most important of which are listed in Table 8.2. Using calibrated oil, an accurate measurement of q_∞ and surface temperature along with a good measure of the incident light angle, it is possible to achieve C_f accuracy to better than ±5% in magnitude and ±1 deg. in the vector direction. The lowest limit currently achievable is estimated to be ±3% C_f magnitude and ±0.2° C_f vector direction. As Table 8.2 shows, the uncertainty caused by non-parallel oil streamlines,

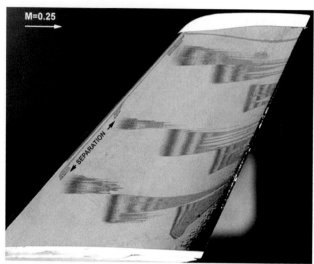

Fig. 8.15. Oil flow fringe pattern seen on the winglet of modern transport aircraft model at low Reynolds number.

and pressure and shear gradient effects can be appreciable under certain circumstances (near shocks, transition and in separated flows) when using the simplified 1D equation. Simple corrections can be applied to the C_f given by Eqn. 8.6 to minimize the effects of these errors. For instance, C_f can be multiplied by $(1 - (h/C_f)\delta C_p/\delta s)^{-1}$ to correct for pressure gradient and by $(1 + \frac{1}{4}(\triangle s/C_f)\delta C_f/\delta s)$ to correct for shear gradient effects, where h is the thickness of the oil given by $h = (\lambda N_f)/(2n_o \cos\theta_r)$, and N_f is the number of fringes from the oil's leading edge (that is, first dark fringe $N_f = \frac{1}{2}$, second dark fringe $N_f = \frac{3}{2}$). An alternate approach for FISF measurements in regions of high shear gradients is to use an equation in place of Eqn. 8.6 that is specifically tailored for high shear gradients (Zilliac, 1996).

8.4.8   Examples

l as Color

Oil film interferometry has been applied on models in numerous flow fields ranging from wind tunnel models in high and low speed production tunnels to flight experiments and rotor craft blades in hover. Fig. 8.15 shows an oil flow pattern on the winglet of a modern transport aircraft model in the 12 ft Pressure Wind Tunnel at NASA Ames at 3 atm total pressure. The interference patterns

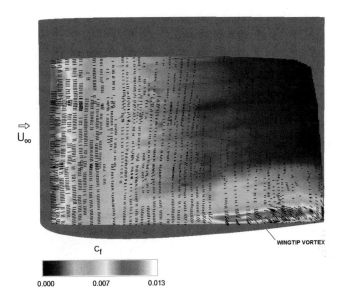

U_∞

C_f

WINGTIP VORTEX

0.000   0.007   0.013

Fig. 8.16. Measured wingtip skin friction distribution. Figure also shown as Color Plate 13.

show a laminar separation bubble at the leading edge of the winglet followed by turbulent reattachment.

Presented in Fig. 8.16 is a skin friction distribution on the suction side of a low aspect ratio wing. The 2500 FISF data points were obtained in the 32 × 48 in. wind tunnel in the Fluid Mechanics Laboratory at NASA Ames Research Center.

8.5   References

Bell, J.H. and McLachlan, B.G. 1996. Image Registration for pressure-sensitive paint applications. *Experiments in Fluids*, **22** (11), 78–86.

Chandrasekhar, S. 1992. *Liquid Crystals.* Cambridge University Press, Cambridge.

De Gennes, P. G. and Prost, J. 1995 *The Physics of Liquid Crystals.* Oxford University Press, New York.

Driver, D.M. 1997. Application of oil film interferometry skin-friction to large wind tunnels. AGARD CP-601, Paper no. 25.

Farina, D.J., Hacker, J.M., Moffat, R.J. and Eaton, J.K. 1994. Illuminant invariant calibration of thermochromic liquid crystals. *Experimental Thermal and Fluid Science*, **9** (1), 1–12.

Fergason, J.L. 1964. Liquid crystals. *Scientific American*, **211**, 76–85.

Hall, R.M., Obara, C.J., Carraway, D.L., Johnson, C.B., Wright, E.J., Covell, P.F. and Azzazy, M. 1991. Comparisons of boundary-layer transition measurement techniques at supersonic Mach numbers. *AIAA Journal*, **29** (6), 865–871.

Hay, J.L. and Hollingsworth, D.K. 1996. A comparison of trichromic systems for use in the calibration of polymer-dispersed thermochromic liquid crystals. *Experimental Thermal and Fluid Science*, **12**, 1–12.

Holmes, B.J., Gall, P.D., Croom, C.C., Manuel, G.S. and Kelliher, W.C. 1986. A new method for laminar boundary-layer transition visualization in flight: Color changes in liquid crystal coatings. *NASA TM-87666*.

Klein, E.J. 1968. Liquid crystals in aerodynamic testing. *Astronautics and Aeronautics*, **6**, 70–73.

Liu, T., Campbell, B.T., Burns, C.P. and Sullivan, J.P. 1997. Temperature- and pressure-sensitive luminescent paints in aerodynamics. *Appl. Mech. Rev.*, **50** (4), 227–246.

McLachlan, B.M. and Bell, J.H. 1995. Pressure-sensitive paint in aerodynamic testing. *Experimental Thermal Fluid Science*, **10**, 470–485.

Monson, D.J. and Mateer, G.G. 1993. Boundary-layer transition and global skin friction measurements with an oil-fringe imaging technique. SAE 932550, Aerotech '93, Costa Mesa, Ca., Sept. 1993, 27–30.

Morris, M.J., Benne, M.E., Crites, R.C. and Donovan, J.F. 1993. Aerodynamic measurements based on photoluminescence. AIAA 31st Aerospace Sciences Meeting, Jan 11–14, Reno, NV, *AIAA Paper 93-0175*.

Naughton, J.W. and Brown, J.L. 1996. Surface interferometric skin-friction measurement technique. *AIAA Paper 96-2183*.

Oglesby, D.M., Puram, C.K. and Upchurch, W.T. 1995. Optimization of measurements with pressure sensitive paints. *NASA Technical Memorandum 4695*.

Reda, D.C. and Muratore, J.J., Jr. 1994. Measurement of surface shear stress vectors using liquid crystal coatings. *AIAA Journal*, **32** (8), 1576–1582.

Reda, D.C. 1995a. Method for determining shear direction using liquid crystal coatings. *U.S. Patent #5,394,752*.

Reda, D.C. 1995b. Method for measuring surface shear stress magnitude and direction using liquid crystal coatings. *U.S. Patent #5,438,879*.

Reda, D.C., Wilder, M. C. and Crowder, J.P. 1997a. Simultaneous, full-surface visualizations of transition and separation using liquid crystal coatings. *AIAA Journal*, **35** (4), 615–616.

Reda, D.C., Wilder, M. C., Farina, D.J. and Zilliac, G. 1997b. New methodology for the measurement of surface shear stress vector distributions. *AIAA Journal*, **35** (4), 608–614.

Reda, D.C., Wilder, M. C., Mehta, R. and Zilliac, G. 1998. Measurement of continuous pressure and surface shear stress vector distributions using coating and imaging techniques. *AIAA Journal*, **36** (6), 895–899.

Stacy, K., Severance, K. and Childers, B.A. 1994. Computer-aided light sheet flow visualization using photogrammetry. *NASA TP 3416*.

Tanner, L.H. and Blows, L.G. 1976. A study of the motion of oil films on surfaces in air flow, with application to the measurement of skin friction. *J. of Physics E*, **9**, 194–202.

Wilder, M.C. and Reda, D.C. 1998. Uncertainty analysis of the liquid crystal coating shear vector measurement technique. *AIAA Paper 98-2717*.

Woodmansee, M.A. and Dutton, J.C. 1998. Treating temperature-sensitivity effects of pressure-sensitive paints. *Experiments in Fluids*, **24** (2), 163–174.

Wyszecki, G. and Stiles, W.S. 1967. *Color Science.* John Wiley & Sons, Inc., New York, 228–370.

Zilliac, G.G. 1996. Further developments of the fringe-imaging skin friction technique. *NASA TM 110425*.

Zilliac, G.G. 1999. The fringe-imaging skin friction technique. PC application users manual. *NASA TM*.

CHAPTER 9

METHODS FOR COMPRESSIBLE FLOWS

W. D. Bachalo[a]

9.1 Introduction

In this chapter, visualization methods that depend upon changes in the fluid index of refraction of the flow are described. The methods are capable of providing useful qualitative and quantitative information on the spatial variations in the fluid density, temperature, static pressure, and with some assumptions, information on the fluid flow speed and Mach contours. These methods for visualization do not require the introduction of additives into the fluid. However, in subsonic flows or flows of liquids, index of refraction changes can be facilitated with the use of local heating, introduction of additional gases with different refractive index, or with the stratification of liquid such as in the case of saline flows. The methods to be described include the shadowgraph, schlieren, and interferometry techniques. Although these methods are rather old dating back to the 1800's in some cases, they remain of value for efficiently investigating flow behavior.

In the case of compressible flows, the optical index of refraction of the gas is a function of the gas density, so the flow will produce an optical disturbance to light rays passing through the flow field. To understand these phenomena, the relationships describing variation of the gas index refraction with density will be outlined. A brief review of the basic optical systems and their functions will be given so that the optical methods can be easily understood. The flow visualization methods will then be described and information on their capabilities, sensitivity, and various applications will be given. The more recent holographic techniques will be described in some detail as these methods extend the application of the flow visualization techniques, and simplify their implementation. Holographic methods allow the researcher to record the optical wave front in-

---

[a]Artium Technologies, Inc., 14660 Saltamontes Way, Los Altos Hills, CA 94022, U. S. A.

205

formation including the amplitude and phase disturbances produced by the flow field and to later interrogate this information in the absence of the flow field. This allows greater flexibility in implementing the shadowgraph, schlieren, and interferometry techniques when analyzing the flow field.

9.2   Basic Optical Concepts

To understand the optical flow visualization methods, a basic understanding of the nature of light and of the optical components is necessary. Light is a form of electromagnetic radiation, which may be characterized by its wavelength or frequency, amplitude, phase, polarization, and speed, and direction of propagation. As light is transmitted through a transparent medium any or all of its characteristics may be altered due to the interaction with the medium. Either the geometrical or the physical optics theories may describe the behavior of light. When the wavelength of light is small compared to the size of the apparatus or optical components under consideration, the geometrical optics techniques may be used as a first approximation. If the dimensions of the apparatus are small relative to the light wavelength or if treating light interference, the physical optics treatment is required. With the physical optics theory the dominant property of the light is its wave nature. In describing the flow visualization methods, it will be convenient to use both the physical and geometrical optics representations. For the case of geometrical optics, a concept of light rays is introduced to describe the effects of the non-homogeneity on the propagation of the light. The light ray is defined as a curve or line in space that is normal to the light wave fronts and thus, corresponds to the direction of flow of radiant energy. Hence the light ray and the wave optics are necessarily interrelated. The interested reader may find a very useful and well-illustrated description of these concepts in Hecht & Zajac (1976).

The first important concept to remember is Snell's law. Recall that the absolute index of refraction, n is simply the ratio of the light speed in a vacuum relative to the light speed in the medium, so that $n = c/v$.

When a light wave interacts with a transparent medium having a different index of refraction, the direction of propagation of light wave changed at the surface in the medium. The deflection of the wave front is easily described using the Huygens principle. This as well as Fermat's principles are described in detail in Hecht & Zajac (1976) so only a brief discussion along with a simple diagram, Fig. 9.1 is provided to explain these important concepts. The diagram on the left shows light rays incident upon a transparent medium of higher refractive

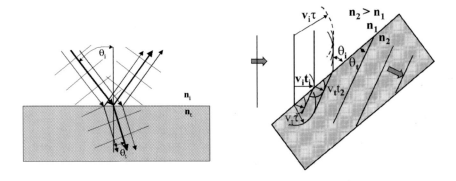

Fig. 9.1. Wave and ray diagram representations of light incident, reflected, and transmitted at a surface of a medium with different index of refraction.

index. Some of the light is transmitted whereas as a portion of the light determined by the Fresnel reflection coefficients (see Hecht & Zajac, 1976) is reflected. Snell described the deflection in the direction of light propagation in 1621. He recognized that the light incident at angle θ_i is deflected at the interface of the refracting media and changes direction according to the relationship given as:

$$n_i \sin \theta_i = n_t \sin \theta_t,$$

where n_i and n_t are the index of refraction of the incident and transmitted media, respectively and θ_i and θ_t are the angles of the incident and transmitted rays measured from the normal to the local surface (Fig. 9.1). The diagram on the right describes the phenomena from the physical optics approach. The change in the light speed changes the phase and hence, the direction of propagation. A similar result may also be derived using Fermat's *principle of least time*. This deceptively simple concept is the basis for lens design, understanding the interaction of light with various transparent media, as well as light scattering by transparent spheres (Bachalo, 1980; Bachalo & Houser, 1984).

In order to explain a broad range of optical phenomena, and primarily, those phenomena central to interferometry, light must be treated with classical wave theory. This treatment is referred to as "physical optics" and comprises those phenomena bearing on the nature of light. Typically, large-scale ($\gg$ light wavelength) effects can be explained by the treatment of light as rays and the use of geometrical optics. However, the phenomena of interference and diffraction which causes deviation from rectilinear motion of light rays that is not explainable by refraction or reflection must be treated by the wave theory.

Light waves are typical of a broad range of waves known as electromagnetic waves. On the electromagnetic spectrum, light waves are intermediate between long radio waves, which can be thought of as oscillating and propagating electric fields and short x-rays, which can be considered as energetic particles. The wavelength of light is sufficiently short so that it can be observed as rectilinear motion, as though the light is propagated as a stream of particles, and yet long enough for us to observe many interference and diffraction effects.

Compared to some other wave phenomena, light waves do not transport any material properties as, for example, sound waves, which propagate air pressure and velocity, or water waves that propagate water level and velocity. Electromagnetic waves are waves both of electric and magnetic fields. The two are, however, closely linked in free space, the ratio of the field strengths is fixed, and the field directions are mutually perpendicular while each is perpendicular to the line of propagation of the wave. The electric field is generally easier to detect, and thus, the field **E** is the variable used to describe the phenomena. The instantaneous direction of the **E** vector is referred to as the polarization direction of the light wave.

To describe a wave, four quantities are required — its wavelength, frequency, velocity, and amplitude. The wavelength is the distance between two successive crests (or other corresponding parts of the wave profile). The most general example of the solution to the one-dimensional wave equation is the form

$$\psi(x, t) = f(x - vt),$$

and for a sinusoidal wave it takes the form

$$\psi(x, t) = A \sin k(x - vt),$$

where A is the wave amplitude and $k = 2\pi/\lambda$ is the wave or propagation number. The spatial period of the wave is the wavelength. The temporal period τ is defined as the amount of time it takes for one complete wave to pass a stationary observer. Thus, the repetitive nature of the wave may be expressed as

$$\sin k(x - vt) + \sin k[x - v(t \pm \tau)] = \sin k[x - vt \pm 2\pi)].$$

It follows that

$$|kvt| = 2\pi,$$

so

$$\frac{2\pi}{\lambda}vt = 2\pi,$$

and

$$\tau = \frac{\lambda}{v}.$$

The period τ is the number of units of time per wave and the inverse is the frequency given as $\nu = 1/\tau$ (Hz), and the angular frequency is $\omega = 2\pi/\tau$ (rads/s).

The visible portion of the spectrum extends from wavelength of 0.4 μm (approaching the ultraviolet) to about 0.75 μm (approaching the infrared). The frequency is the number of waves passing a given point per second and is expressed in cycles per second (light frequencies are on the order of 10^{14} Hz). Wave velocity is the velocity at which the wave profile moves forward; it is equal to the frequency multiplied by the wavelength. Amplitude is a measure of the magnitude of the vibrations and is defined as the height of a wave crest. Light has the added requirement of specifying the polarization since it is a vector quantity.

Although the descriptions given here apply to all electromagnetic waves, the primary concern is with light which corresponds to radiation in the narrow band of frequencies from about 3.84×10^{14} to 7.69×10^{14} Hz. It is important to note that light waves are exclusively a transverse wave motion. That is, the vibrations are always perpendicular to the direction of motion of the waves. Maxwell's theory describing the dynamic electromagnetic fields required that the vibrations of light be strictly transverse and gave a definite connection between light and electricity. The theory need not be covered here but is only mentioned to connect the electric and magnetic field theory to electromagnetic radiation. That is to say, the concepts developed to describe electric and magnetic fields appear to describe many of the phenomena observed in electromagnetic radiation.

9.3 Index of Refraction for a Gas

The visualizing light beam passing through a fluid represents an electric field with strength $\mathbf{E}$ that induces a dipole moment $\mathbf{p}$ as a result of the distortion to the charge configuration of the molecules. The dipole moment is given as

$$\mathbf{p} = \alpha\mathbf{E},$$

where α is called the induced electronic polarizability. Because $\mathbf{E}$ is an oscillating field, the electric field distortion is frequency dependent. If it is assumed that the resonant frequencies of the gas molecules are significantly different from the frequency of the incident light, then the expression can be safely reduced to the form (Merzkirch, 1987)

$$n^2 - 1 = \frac{N}{\pi} \frac{e^2}{m_e} \sum_i \frac{f_i}{(\nu_i^2 - \nu^2)},$$

where N is the molecular number density, e is the charge and m_e is the mass of an electron, ν is the frequency of $\mathbf{E}$ with ν_i the resonant frequency, f_i is the oscillator strength which is a number between 0 and 1. N can be reduced to the more useful value ρ using $\rho = Nm/L$ where m is the molecular weight and L is the Loschmidt number. With the approximation $n^2 - 1 \approx 2(n-1)$,, the Gladstone–Dale (G–D) formula results as

$$n - 1 = K\rho = \frac{\rho}{2\pi} \frac{L}{m} \frac{e^2}{m_e} \sum_i \frac{f_i}{(\nu_i^2 - \nu^2)},$$

where the Gladstone–Dale constant K depends upon the gas under observation and it has the dimensions of $1/\rho$. Conversion of the light frequencies to wavelengths results in the expression for K as

$$K = \frac{e^2}{2\pi c^2 m_e} \frac{L}{m} \sum_i \frac{f_i \lambda^2 \lambda_i^2}{(\lambda^2 - \lambda_i^2)}.$$

For gas such as air, which is a mixture of several components, the value of the refractive index n of is given by

$$n - 1 = \sum_i K_i \rho_i,$$

where K_i and ρ_i are the G–D constants and the partial densities of the individual components. The G–D constant for the mixture is

$$K = \sum_i K_i \frac{\rho_i}{\rho},$$

and

$$n - 1 = K\rho.$$

For reference, Table 9.1 provides G–D constants for air at 288°K, Table 9.2 provides constants for various gases, and Table 9.3 for some representative liquids.

K cm^3/gm	wavelength λ (μm)
0.2239	0.9125
0.2255	0.6440
0.2264	0.5677
0.2281	0.4801
0.2304	0.4079

Table 9.1. Gladstone–Dale constants for air.

Gas	K (cm^3/gm)
O_2	0.190
N_2	0.238
He	0.196
CO_2	0.229

Table 9.2. Gladstone–Dale constants of various gases.

9.4   Light Ray Deflection and Retardation in a Refractive Field

With the information as to how a change in the flow density, temperature, or fluid composition can affect the refractive index, the response of light rays passing through a field with inhomogeneous refractive index can be investigated. The refractive index in a flow field is generally a function of three space coordinates and possibly, time t, and is expressed as

$$n = n(x, y, z, t)$$

Liquid	$-dn/dT$ ($^\circ K^{-1}$) for $\lambda = 0.546\,\mu$m
water	1.00×10^{-4}
ethyl alcohol	4.05×10^{-4}
n-hexane	5.43×10^{-4}
carbon tetrachloride	7.96×10^{-4}

Table 9.3. Gladstone–Dale constants for various liquids.

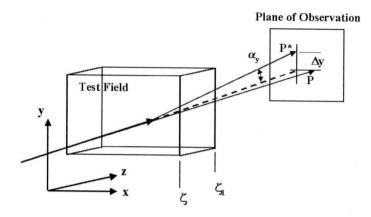

Fig. 9.2. Geometrical optics description of the deflection of light rays when passing through an inhomogeneous flow field.

Generally, only the space variables apply. Light rays passing through the refractive flow field will be deflected unless all changes in the refractive index are normal to the propagation of the light. Fig. 9.2 shows a parallel incident beam interacting with the flow. The light will be deflected by an amount predicted by Snell's law and will arrive at a point P* in the observation plane. The position and optical path length traversed by the ray will differ from an undisturbed ray and this deflection can be measured. The optical path length (OPL) is defined by the integral

$$\text{OPL} = \int_S^P n(x, y, z)\, ds,$$

where s is the arc length along the light path. The following quantities can be measured:

1. The angular deflection of the disturbed ray emerging from the flow field with respect to a fictitious undisturbed ray.

2. The displacement of the impact point of the ray on the plane of observation.

3. The phase shift between disturbed and undisturbed rays (rays passing in the absence of the flow field) due to the difference in the optical path lengths.

The optical visualization methods use one or a combination of these quantities to make observations of the flow behavior. The methods respond to either the absolute change in refractive index, the refractive index gradient or first derivative, or to the second derivative of the refractive index. In the literature, the density is often cited as the primary condition affecting the passage of the light rays. However, it is possible to have inhomogeneous mixtures of different gases with similar densities, which can deflect the light rays. Thus, referring to the refractive index provides a more general description.

When referring to light refraction at surfaces where the refractive index changes discontinuously, Snell's law applies. However, when the refractive index changes continuously as in a fluid flow field, Fermat's principle applies. Fermat's principle may be stated as: *A light ray in going from point S to point P must traverse an optical path length which is stationary with respect to variations of that path* (Hecht & Zajac, 1976). This implies that the path taken by the light traversing between the two points is one which the transit time is minimized. This principle with the variational calculus may be used to derive expressions for the inclination of the light rays leaving the inhomogeneous refractive field. The transit times for the light to pass the disturbed and the undisturbed ray required in passing the inhomogeneous refractive index field, t and t^* are given as

$$\Delta t = t^* - t = \frac{1}{c} \int_{\zeta}^{\zeta_1} [n(x,y,z) - n_0] \, dz,$$

where n_0 is the refractive index in a vacuum. This latter expression is useful in assessing the relative phase shifts induced by the flow density fluctuations when using interferometry. In general, the disturbances induced by the flow field are assumed to be small and the light rays passing the disturbance are curved whereas the geometrical rays are straight.

9.5   Shadowgraph

Perhaps the simplest method for visualizing flow fields with varying refractive index is the shadowgraph, attributed to Dvorak (1880) who was a co-worker of Ernst Mach. Fig. 9.3 shows simple optical configurations for the shadowgraph system. Either spherical lenses or mirrors may be used. In wind tunnel applications, spherical mirrors are most often used since it is possible to fabricate high quality mirrors with good optical quality that have diameters as large as one meter or greater. The objective is to produce parallel light that is incident on the transparent disturbance created by the flow field. How sharp the image will

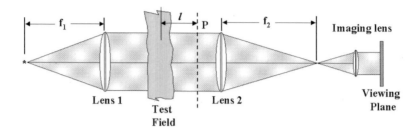

Fig. 9.3. Schematic showing typical shadowgraph systems.

be depends upon the size of the light source. The blur in the image is given by $\ell d/f_1$, where f_1 is the lens or mirror focal length, d is the size of the source, and ℓ is the distance of the observation plane (ground glass, photographic film, CCD, etc.) from the optical disturbance. The light source must be small but not so small as to suffer a lack of sharpness due to diffraction. When using large-scale optics, a second spherical lens or mirror is used to reduce the size of the image. The camera lens is placed at the focal point of the second spherical lens and focuses the reference plane P located a distance ℓ from the center of the optical disturbance field (for example, wind tunnel test section).

Light rays passing through the field are bent by refraction and have an angle of inclination with respect to their original path. If the second derivative of the density is not constant, the shadowgraph will visualize the density variations. This may be understood using simple components to simulate the local changes in density, Fig. 9.4. In the case of a rectangular transparent block with refractive index different from the surroundings, the light is not deflected but the phase of the wave fronts is delayed. If the density has a linear variation for which the gradient of the refractive index, $\partial n/\partial y$, is constant, the deflection angle of the rays is the same for all rays passing that region of the flow. The plane of observation will show a uniform illumination for this region. When the density gradient is represented by the block with a constant curvature, this corresponds to a density field with $\partial^2 n/\partial y^2$ constant. The density field with constant second derivative will also lead to a uniformly illuminated region, albeit of lower exposure since the rays are diverging approximately uniformly. Thus, the shadowgraph serves to visualize only regions of the flow that have nonuniform $\partial^2 n/\partial y^2$ which means that $\partial^3 n/\partial y^3 \neq 0$ everywhere. Strictly, the gradient in each of the three coordinates needs to be considered when analyzing the visualization of flows but the

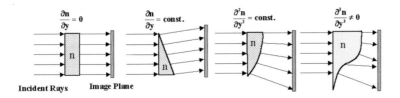

Fig. 9.4. Schematic showing the deflection of light rays by density fields that are constant, have constant index gradient, have constant second derivative of refractive index and with non-zero third derivative of the refractive index which is visualized with the shadowgraph.

one-dimensional description is easily extended for the more general case.

If a flow field with a disturbance for which $\partial^3 n/\partial y^3 \neq 0$ is considered, the light rays may be drawn to show the light intensity distribution in the image plane, Fig. 9.5. The central ray 'b' passes through a stronger disturbance and hence, is deflected more than rays 'a' and 'c.' The rays reach the plane of observation at a^*, b^* and c^*. The relative light intensity on the plane is proportional to the spacing between the rays with a darker region produced between b^* and c^* with a brighter region between a^* and b^* where the rays converge. The relative changes in the light intensity are approximately proportional to the second derivative of the index of refraction, which is proportional to the variations in the gas density.

The shadowgraph method has been used extensively in the study of supersonic and transonic flows and is of particular value because of its simplicity and the ability to easily observe such structures as shocks, Prandtl–Meyer expansions, and boundary layers in compressible flows. As an example, the case of a bow shock produced by a sphere in a supersonic flow is described. The incident light is collimated so the incident rays are parallel and perpendicular to the flow direction which is from left to right in diagram shown in Fig. 9.6. Light passing upstream of the shock passes through the test section undeflected since there is no flow disturbance upstream of the shock. As the light waves traverse the curved bow shock, they curve toward the more dense flow region downstream of the shock wave. Because the light rays passing the shock are deflected, a dark band appears on the plane of observation, Fig. 9.7. The deflected rays converge to form a caustic or region of high intensity. The leading edge of the shadow will represent an accurate position of the leading edge of the shock. In some cases, the deflected rays may cause light to appear on the shadow of the model. Clearly, the viewing screen can be moved closer to or farther from the test section to decrease or increase the width of the shadow image on the screen. Making such adjustments is useful when viewing different flow features. Note that this is also a method for changing the sensitivity of the system.

In the process of visualizing strong shocks in two-dimensional flows, the diffraction of light by the highly discontinuous density gradient will limit the sharpness of the shock image. This problem is especially evident when using laser light sources which are coherent. Prandtl–Meyer expansion fans act as essentially negative or concave lenses and produce an intensity distribution that has the bright band at the leading part of the fan followed by a less intense region.

Compressible boundary layers may also be visualized with the shadowgraph method. As a result of the lower gas densities near the wall (assuming an adiabatic wall condition) the collimated light rays entering parallel to the wall will be deflected away from the wall. Because of the density profile of the boundary layer, the light near the wall will be deflected to a greater extent than the rays entering the outer region of the boundary layer, Fig. 9.8. The result is a caustic or bright band at the outer part of the boundary layer as shown in the intensity profile for laminar flows. The band tends to disappear when the flow transitions to turbulent. There are a couple of points to remember when using the shadowgraph to visualize boundary layers. If the viewing screen is not close to the test section wall on the side opposite the light source or the imaging system is not focused on a plane approximately $\frac{2}{3}L$ where L is the width of the flow measured from the side of the light source, the boundary layer will appear

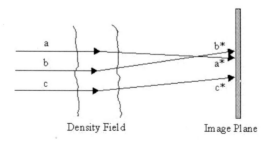

Fig. 9.5. Ray diagram showing rays passing an inhomogeneous refracting medium with variable $\partial^2 n/\partial y^2$.

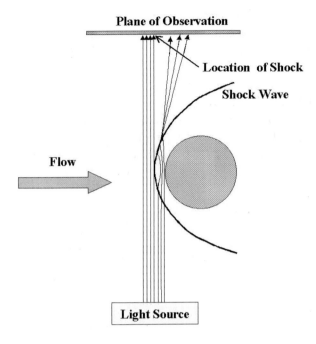

Plane of Observation

Location of Shock

Shock Wave

Flow

Light Source

Fig. 9.6. Shadow formation by a cylindrical bow shock produced by a sphere in a supersonic flow.

thicker than the true value. This value is used as a first approximation because the light rays show the strongest curvature in the latter part of the path as they are deflected through the boundary layer. It is also important to realize that the light rays are continuously deflected so that the ray entering the lower part of the layer deflects upward and passes through a continuously changing gradient before exiting.

As the boundary layer transitions to turbulent, the density fluctuations act as a range of small weak concave and convex lenses. The light rays passing through these flows will be deflected in a random pattern. Nonetheless, the mean density gradient will generally produce a shadowgraph visualization of the flow field. In fact, researchers (Uberoi & Kovasznay, 1955) have proposed methods such as the autocorrelation of the shadowgraph image to recover the statistical properties of the turbulent flow. A mistake that is often made is to expect that the shadowgraph, schlieren, or interferogram images reveal the na-

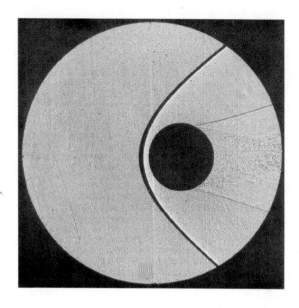

Fig. 9.7. Shadowgraph image of supersonic flow, $M = 1.7$ past a sphere (from Merzkirch, 1987).

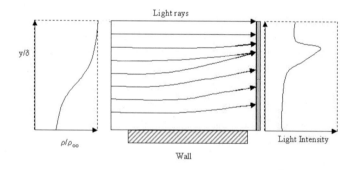

Fig. 9.8. Response of the shadowgraph to boundary layer in a compressible flow.

ture of the turbulence. In fact, the visualization methods using short duration light sources allow the observation of the density fluctuations that are induced by the turbulence. However, it is not just the density fluctuations that characterize the turbulent flow but also the velocity field, which is not visualized.

An additional point that needs to be emphasized is that turbulenc
three-dimensional. The optical methods described here integrate t
fraction information along the optical path, which results in the los
along the optical path.

9.6   Schlieren Method

The schlieren method developed by Foucault in 1859 and Toepler in 1864 is
also commonly used to visualize local optical inhomogeneities in transparent
media. This method is used to visualize flows with density gradients that are not
constant (that is, $\partial^2 n/\partial y^2 \neq 0$). Toepler used the method for the visualization
of compressible fluids. Like the shadowgraph, the schlieren method utilizes a
parallel or collimated light beam that is passed through the flow field, Fig. 9.9.
Essentially a point light source which may be a mercury vapor source or a
spark, with a circular or slit aperture, or a laser is located at the focal point
of the transmitter spherical mirror, Fig. 9.9a or lens, Fig. 9.9b. The collimated
light passes through the test region a second spherical mirror or lens focuses
the light to form an image of the light source. A knife edge is located at the
focal plane of the second mirror or lens. A camera lens is positioned beyond
the knife edge and located to form an image of the test region on the viewing
screen or on the film plane when recording the image. The knife edge is carefully
adjusted to cut off part of the light at the image of the light source. Without
any disturbances in the optical path, the original light source will have uniform
reduction in intensity due to the light cutoff by the knife edge, Fig. 9.10.

When there is a disturbance in the optical path, the light rays will be deflected by an angle α. Although any disturbance in the path will cause the
deflection of the transmitted light, it will be assumed that the only disturbances
are at the test region. These light rays will be shifted at the plane of focus by
an amount Δs given as

$$\Delta s = f_2 \tan \alpha_i,$$

where f_2 is the focal length of the mirror or lens and Δs is in a direction perpendicular to the knife edge (Fig. 9.10). The relative change in intensity at the
image plane is given as (Merzkirch, 1987)

$$\frac{\Delta I}{I} = \frac{K f_2}{s} \int_{\gamma_1}^{\gamma_2} \frac{1}{n} \frac{\partial n}{\partial z} \, dy,$$

where y is the coordinate along the optical axis through the test region. For
gaseous media, the index of refraction $n \approx 1$, and the Gladstone–Dale relation

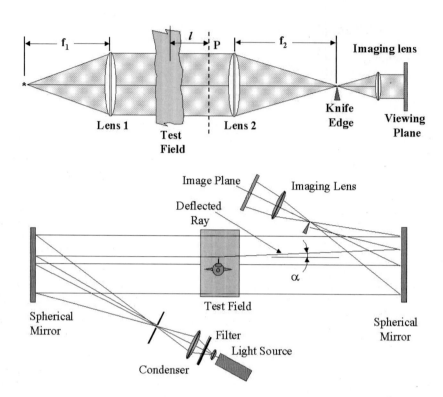

Fig. 9.9. Schematics of typical schlieren optical systems: (a) Lens system design; (b) Spherical mirror system.

can be used to reduce the relationship to

$$\frac{\Delta I}{I} = \frac{K f_2}{s} \int_{\gamma_1}^{\gamma_2} \frac{\partial \rho}{\partial z} \, dy.$$

This relationship holds for whatever orientation of the knife edge is used with the refractive index gradient being detected that is essentially normal to the knife edge. In most systems, the knife edge can be rotated so that the sensitivity to any gradient in the plane normal to the beam is achievable. In the above relationship, it can be seen that for small ratio s/f_2, the contrast on the viewing screen will be greater. By detecting changes of relative intensity as small as 0.1, the corresponding smallest deflection angles that can be detected are $\alpha_{min} =$

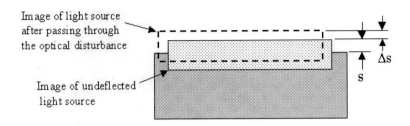

Image of light source
after passing through
the optical disturbance

Image of undeflected
light source

Δs

s

Fig. 9.10. Schematic of the knife edge and the displacement of the image due to the flow field disturbance.

$0.1(s/f_2)$.

An example of the schlieren visualization method is given in Fig. 9.11. Note that the density gradients are visible in this figure as compared to the change in the gradients that were visualized in the shadowgraph, Fig. 9.7. The Mach number is higher in Fig. 9.11 as noted from the shock angles. Note also that there is a reverse in intensity levels as gradients change sign from the top to the bottom of the figure.

There have been a number of modifications to the Toepler approach (see Merzkirch, 1987). Modifications have been made to the knife edge geometry by using circular and double cutoff knife edges. Others have used strips with gradual optical density variations, two color, and color strip filters instead of the conventional knife edge. Using color strips has the advantage that the eye is more sensitive to changes in color than to shades of gray. Colored strips are often made from commercially available gelatin filter material and the strips are cut to a width equal to the width of the image of the light source slit. Obviously, the color strips will only work well for white or broad band light sources. The color strips can also be used in a circular cutoff system to achieve sensitivity in all directions (Settles, 1970, 1982).

9.7   Interferometry

Whereas the shadowgraph responds to the second space derivative of the index of refraction and the schlieren to the first derivative or index of refraction gradient, the interferometer responds directly to the index of refraction (density in a compressible flow field). Light waves passing through the media of different index of refraction experience a change in the light speed, which results in a

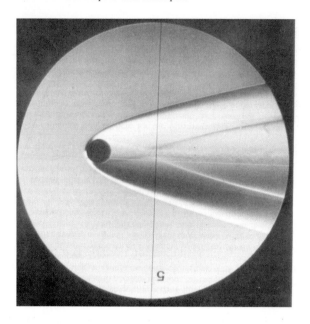

Fig. 9.11. Schlieren image of a sphere in a supersonic flow (from Merzkirch, 1987).

phase shift of the light. This phase shift may be measured using interferometry techniques.

The electric field, **E**, is a vector quantity. In the discussions of diffraction and interference the idealized description of the electric field will be used. That is, it will be assumed that the wave is monochromatic (consists of a single wavelength) and is linearly polarized. Laser light approaches these ideal conditions, which is why it is such an important light source. For these idealized conditions, the field of the linearly polarized light wave traveling in the direction of **k** can be written

$$\mathbf{E}(\mathbf{r}, t) = \mathbf{E}_0 \left(\mathbf{k} \cdot \mathbf{r} - \omega t + \varepsilon \right).$$

The intensity or irradiance is then

$$I = \varepsilon_0 c \left\langle \mathbf{E}^2 \right\rangle$$

where ε_0 is the electric permittivity of the surrounding medium, c is the speed of light ($= 3 \times 10^8$ m/s). In this case, **E** can be considered as the optical field and the brackets $\langle \ \rangle$ indicate that the **E** field is generally assumed stationary in classical optics considerations. If we are only concerned with the relative irradiance within the same medium, then the expression is simply

$$I = \langle \mathbf{E}^2 \rangle .$$

Recall that $\mathbf{E}$ is a complex quantity so

$$\mathbf{E}^2 = \left(\mathbf{E}_0 e^{i\alpha} \right) \left(\mathbf{E}_0 e^{i\alpha} \right)^* ,$$

where * implies the complex conjugate.

In speaking of light as wave phenomena, expressions are written for the waves as if the light source was ideally monochromatic and the waves were perfectly plane or spherical. However, even laser light only approaches these conditions. The frequency and amplitude of the wave varies slowly (relative to the oscillation, 10^{14} Hz). The time over which the wave train maintains its average frequency is the coherence time and is given by the inverse of the light source frequency bandwidth.

If the light source was ideally monochromatic, $\Delta \nu$ would be zero and the coherence time or length given by $c \Delta t$ would be infinite. Over an interval of time much shorter than Δt, an actual wave behaves essentially as if it was monochromatic. Coherence time is the temporal interval over which the phase of the light wave at a given point in space can be reasonably predicted. When referring to temporal coherence, this refers to the time Δt of the light source over which the wave remains at approximately a constant frequency. The length $c \Delta t$ is often referred to as the coherence length of the source and this length can be from μm for mercury lamps to several meters for some lasers. In interferometry, the path lengths of the interfering beams are matched so that the difference in the path length is less than the coherence length. This will ensure the formation of interference fringes of sufficient visibility.

In general, the time dependence of a light wave varies over a given wave front. The degree of this wave front variation is referred to as the spatial coherence. It arises because of the finite extent of light sources. Suppose that a classical broad band source of monochromatic light is used and consider two point radiators on it separated by a lateral distance, which is large, compared to λ. These two point sources will presumably behave independently and there will be a lack of correlation existing between the phases of the two emitted disturbances. Spatial coherence is closely related to the concept of the wave front. If the two laterally displaced points reside on the same wave front at a given time, the fields at those points are said to be spatially coherent.

9.8   Interference

The wave theory of the electromagnetic nature of light provides a basis from which to proceed to the phenomenon of optical interference. Because the relationship describing the optical disturbances are linear, the principle of superposition applies so that the resultant electric field (or optical field) intensity at a point in space where two or more light waves overlap, is equal to the vector sum of the individual constituent disturbances. That is, for fields $\mathbf{E}_1$, $\mathbf{E}_2$, ..., the resultant field is given by

$$\mathbf{E} = \mathbf{E}_1 + \mathbf{E}_2 + \ldots$$

Considering the case of two point sources S1 and S2 emitting monochromatic waves of the same frequency at a separation, $\mathbf{a}$, which is much greater than the wavelength, λ, the interference of the fields can be evaluated. A point P on the observation plane is taken far from the two sources so that the waves at P may be considered plane. Assuming linearly polarized light, the expressions for the two waves are given as

$$\begin{aligned}
\mathbf{E}_1(\mathbf{r}, t) &= \mathbf{E}_{01} \cos\left(\mathbf{k}_1 \cdot \mathbf{r} - \omega t + \varepsilon_1\right) \\
\mathbf{E}_2(\mathbf{r}, t) &= \mathbf{E}_{02} \cos\left(\mathbf{k}_2 \cdot \mathbf{r} - \omega t + \varepsilon_2\right).
\end{aligned}$$

The irradiance is given as

$$I = \left\langle \mathbf{E}^2 \right\rangle,$$

with

$$\begin{aligned}
\mathbf{E}^2 &= \mathbf{E} \cdot \mathbf{E} = (\mathbf{E}_1 + \mathbf{E}_2) \cdot (\mathbf{E}_1 + \mathbf{E}_2) \\
&= \mathbf{E}_1^2 + \mathbf{E}_2^2 + 2\mathbf{E}_1 \cdot \mathbf{E}_2.
\end{aligned}$$

After taking the time averages over a period much longer than the period of the optical wave $(T \gg \tau)$, the irradiance becomes

$$I = I_1 + I_2 + I_{12},$$

where $I_1 = \mathbf{E}_1^2$, $I_2 = \mathbf{E}_2^2$, and $I_3 = 2\mathbf{E}_1 \cdot \mathbf{E}_2$. The term $\mathbf{E}_1 \cdot \mathbf{E}_2$ is the interference term and in the present case is evaluated as

$$\mathbf{E}_1 \cdot \mathbf{E}_2 = \mathbf{E}_{01} \cdot \mathbf{E}_{02} \cos\left(\mathbf{k}_1 \cdot \mathbf{r} - \omega t + \varepsilon_1\right) \cos\left(\mathbf{k}_2 \cdot \mathbf{r} - \omega t + \varepsilon_2\right).$$

Taking the time average leads to

$$\langle \mathbf{E}_1 \cdot \mathbf{E}_2 \rangle = \tfrac{1}{2}\mathbf{E}_{01} \cdot \mathbf{E}_{02} \cos\left(\mathbf{k}_1 \cdot \mathbf{r} + \varepsilon_1 - \mathbf{k}_2 \cdot \mathbf{r} + \varepsilon_2\right)$$

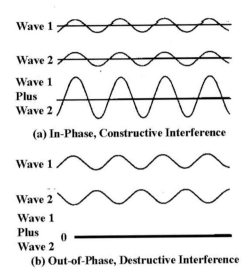

Wave 1

Wave 2

Wave 1
Plus
Wave 2

(a) In-Phase, Constructive Interference

Wave 1

Wave 2

Wave 1
Plus   0
Wave 2

(b) Out-of-Phase, Destructive Interference

Fig. 9.12. Superposition of two coherent optical waves showing interference for the various relative phase shifts between them.

which reduces to

$$I_{12} = \mathbf{E}_{01} \cdot \mathbf{E}_{02} \cos \delta,$$

where δ is the phase difference arising from the combined path length and epoch angle difference. The most common practical case is where the polarization vectors are parallel in which case the interference term reduces to a scalar as

$$I_{12} = 2\sqrt{|I_1 I_2|} \cos \delta,$$

and the total irradiance is

$$I = I_1 + I_2 + 2\sqrt{|I_1 I_2|} \cos \delta.$$

At various points in space, the resultant irradiance will be greater than, less than, or equal to $I_1 + I_2$ depending on the value of δ, Fig. 9.12. A maximum in the irradiance occurs for $\delta = 0,\ \pm 2\pi,\ \pm 4\pi, \ldots$, which is called constructive interference, and a minimum where $\delta = \pm\pi,\ \pm 3\pi,\ \pm 5\pi, \ldots$, which is referred to as total destructive interference.

For the interference pattern to be observable, the phase difference $(\varepsilon_1 - \varepsilon_2)$ between the two sources must remain fairly constant in time. This implies that the light source must be coherent. If the two beams are to interfere to produce a stable interference fringe pattern, they must have very nearly the same frequency, usually originating from the same emitter. A significant frequency difference would result in a rapidly varying time dependent, phase difference, which in turn would cause I_{12} to average to zero during the detection interval. The clearest interference fringe patterns (highest fringe visibility) will be formed when the interfering light waves have equal or nearly equal amplitudes. The central regions of the dark and light fringes will then correspond to complete constructive and destructive interference yielding maximum contrast.

The interferometers most commonly used in flow field studies make use of two beams, a reference beam passing around the flow (or representing an undisturbed light wave in holographic interferometry) and an object beam. The object beam is passed through the flow field under inspection and has undergone changes in phase as a result of the changes in the index of refraction of the flow field. The visualization and quantitative information is obtained from the interference of a wave that has passed through the field under test and a reference wave that has reached the plane of observation on a different optical path that does not produce unknown disturbances to the wave.

9.9   Mach–Zehnder Interferometer

The earliest use of the interferometry technique was by Ernst Mach in 1856 and was reduced to a practical instrument by his son Ludwig Mach (1892) and independently by Zehnder (1891). The system uses a test beam large enough to cover the field under test with a relatively wide separation between the reference and object beams, Fig. 9.13. The basic components of the Mach–Zehnder interferometer (MZI) are the coherent light source, beam splitters, first surface mirrors, and the imaging system for viewing the test region and interference fringes. The system is arranged so that the optical path lengths on both legs of the system are as near to equal as possible. All optical components must be optically flat (typically $1/10$ λ per cm) to produce reliable interferograms. The mirrors are adjustable to allow very precise control of the light beam direction. The compensating windows are used to account for the relatively large optical path length through the thick test section windows. In principle, the length of the optical path on the reference beam leg could be increased by $(n-1) \times 2T$ where T is the window thickness to avoid using the additional optical compensating windows. The lenses in the observation system are used to focus onto a plane within the test region.

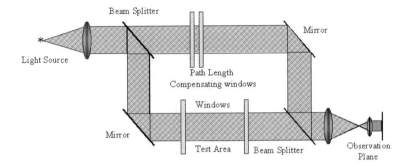

Fig. 9.13. Schematic of a Mach–Zehnder interferometer

When very high quality optics are used, there is no phase disturbance in the test region, and the system is exactly aligned so that the reference beam and object beams are collimated and parallel, the light waves reaching the image plane will be parallel and no interference fringes will be formed. This is known as the "infinite fringe" case since the interference fringes may be assumed to have infinite spacing. By adjusting the interferometer so there is a small angle of intersection between the beams as shown in Fig. 9.14, parallel fringes will appear. This is known as the "finite fringe" case. The spacing of the fringes may be adjusted by changing the angle of beam intersection, γ with the fringe spacing δ given as

$$\delta = \frac{\lambda}{2\sin\frac{1}{2}\gamma}$$

where λ is the light wavelength.

When a disturbance is present in the test region such as a compressible flow over a model, the density variations produce changes in the local index of refraction. The wave fronts of the object beam are deformed due to the phase shift caused by the variations of the light speed as the beam passes through the test section. In the infinite fringe mode, the fringes will form to map out the density variations in the flow field. This is shown schematically in Fig. 9.15. In this case, the interference fringe spacing depends upon the gradient in the index of refraction. The phase shift in the object beam is related to the index of refraction through the following expressions

$$\frac{\Delta\phi}{2\pi} = \frac{1}{\lambda}\int_{\zeta}^{\zeta_1}[n(x,y)-n_0]\,dz$$

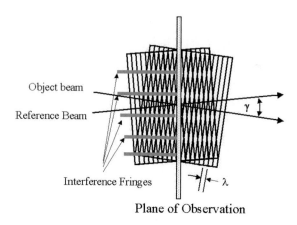

Object beam

Reference Beam

γ

Interference Fringes

λ

Plane of Observation

Fig. 9.14. Formation of fringes when the beams intersect at a finite angle.

where n_0 is known at some location in the field. Applying the Gladstone–Dale constant relating the index to the density yields the integrated relationship

$$\rho(x, y) = \rho_0 + \frac{N\lambda}{KL}$$

where N is the number of fringes from the reference density, ρ_0, which is known, and L is the optical path length through the flow. The reference value may be obtained by knowing the conditions at a undistorbed region in the flow or by making pressure measurement at a known location and using the perfect gas assumption with the total temperature, T_0, and total pressure, p_0, to convert it to density. The number of interference fringes is counted from the reference point to establish the density at each point in the flow field. The sign of the change in the phase shift is ambiguous so it is necessary to know something about the flow field when interpreting the results. However, information to determine whether the density is increasing or decreasing from the reference point is often available from a basic understanding of the fluid mechanics.

There are a number of variants to the basic Mach–Zehnder interferometer (see Merzkirch, 1987). Although the approach is able to generate very high quality interferograms, the method requires high quality optics and an extremely stable platform. Examples of infinite fringe and finite fringe interferograms are shown for the transonic flow tested by Delery *et al.* (1977), Fig. 9.16. In the infinite fringe case, each fringe occurs due to one wavelength shift in the optical

Disturbed Wave

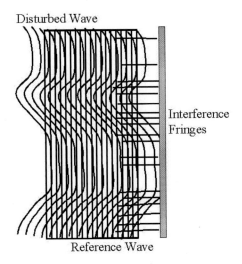

Interference
Fringes

Reference Wave

Fig. 9.15. Diagram showing the deformed object wave interfering with the plane reference wave.

path length or equivalently, 2π shift in phase as a result of changes in the flow field. In the finite fringe case, a finite angle of intersection is set by adjusting the interferometer mirrors. The fringes are displaced from their undisturbed location as a result of changes in the optical path produced by the flow field. The interferograms show the details of the shock locations, the shock-boundary layer interaction, the reflected shocks and the boundary layer thickness.

It is possible to measure fringe shifts that are as small as $\lambda/100$ to obtain high sensitivity to the flow field variations. Techniques involving phase shifting of the reference wave may be used to reach this level of sensitivity. Of course, the optical systems must be of the highest quality since the weakest component sets the level of resolution and accuracy that may be obtained. In flow visualization systems, the large mirrors and windows used usually allow $\lambda/10$ in accuracy. Other sources of error such as a lack of two-dimensionality in the flow and the fact that observations are often made through windows that have turbulent boundary layers on the inside all serve to limit the resolution and accuracy that may be achieved. Some practical results in the next sections provide an indication of the accuracy that may be achieved in realistic wind tunnel studies.

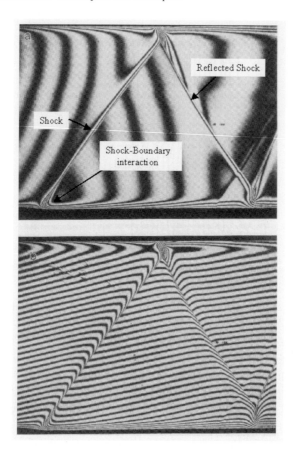

Fig. 9.16. Examples of Mach–Zehnder interferograms obtained for a transonic flow. Upper figure is an infinite fringe case and the lower figure is a finite fringe interferogram (from Delery *et al.*, 1977).

9.10   Holography

Holography offers an excellent means for recording the information on the light wave including the amplitude and phase information (see Vest, 1978, for a more detailed description). The information at an instant in time can be stored and later reconstructed for comparison to waves formed at other conditions including the case of no flow or optical disturbance in the test section. Using holographic techniques to record interferometric information greatly relaxes the stringent optical requirements that have limited the application of the interferometric techniques. In addition to the interferometry technique, the shadowgraph and schlieren methods are also available. This information is available from the same holographic recording of the flow field. The ability to reconstruct the flow field from the hologram for a particular instant of time outside of the flow facility allows much greater flexibility in the spatial filtering and photographing of the images. The recording is generally made using a pulse laser with an exposure time on the order of 10 ns and the reconstruction is accomplished with a continuous wave (CW) laser of equal or similar wavelength.

Since photographic film or any other media only responds to the irradiance, the distribution of the phase information of the light wave will be lost. Fortunately, interferometry can be used to record the phase information as an irradiance pattern. The interference patterns are obtained in much the same way as the MZI wherein a object wave is recorded using a reference wave. The resulting interference pattern is recorded on a very high-resolution photographic film or other media, Fig. 9.17. The film is developed and illuminated by a replica of the original reference beam. The interference pattern diffracts the light to recover the complex amplitude and phase of the object wave. The original method used by Gabor (1951) was an in-line system with the light deflected by the object under observation interfering with the undeflected light forming the reference wave. However, for interferometry, off-axis holography developed by Leith & Upatnieks (1962) is used. In this case, the reference beam used to record the light wave is brought to the photographic plate at an angle to the object beam. This results in a finite fringe pattern that acts as a spatial carrier frequency to record phase and amplitude information that describes the object wave.

In Fig. 9.17, the intersection angle γ between the reference and object beams is selected so that the spatial carrier frequency does not exceed the resolution limits of the photographic film. The film used must have high-resolution capability designed for scientific purposes including holography. As with the MZI, the spatial frequency, δ is given by

$$\delta = \frac{\lambda}{2 \sin \frac{1}{2}\gamma}.$$

The beam intersection angle can be set so that the spatial carrier frequency plus the frequency modulation by the phase information on the object beam do not exceed the resolution of the film. To avoid confusion, note that the first step in holographic interferometry is the recording of the light wave passing through the flow field. This is accomplished using interferometry but additional steps

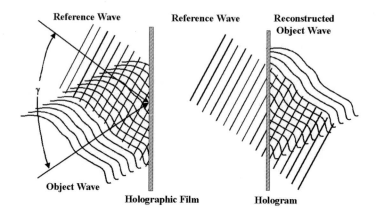

Fig. 9.17. Schematic showing the formation of an off-axis Hologram, and its reconstruction after Leith & Upatnieks (1962).

are required to obtain flow field information.

9.11   Holographic Interferometry

Holographic interferometry is possible because the object wave can be recorded in phase and amplitude and accurately reconstructed with high precision. The reconstructed wave can then be compared interferometrically with another wave recorded at a different time but that has passed on the same optical path since more than one light wave can be recorded on a single holographic (high-resolution photographic) plate. Light waves can also be recorded on separate plates and then reconstructed and superimposed using two holographic plates properly positioned in the replica of the reference beam. As a result of these capabilities, several types of interferometry are made possible. Although there are a number of possible techniques available (Trolinger 1969, 1974, 1975), generally, there are three types that are useful to flow field studies. To aid in the description, the reference wave will be designated as $U_R(t_i)$, the object wave as $U_0(t_i)$, and the reconstructed wave as $U_i(t_i)$ (Fig. 9.18).

Method 1 — Double Exposure: Using this approach, two holographic exposures are made on the same plate at times t_1 and t_2. The interval between exposures may vary from 10^{-8} to 10^{-3} second or longer, depending upon the vibration stability of the system and the characteristic time scales in the flow

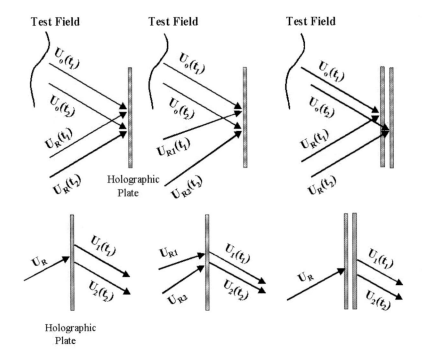

Fig. 9.18. Methods for producing Holographic Interferograms. Top: Recording Approach. Bottom: Reconstruction Approach: (a) Double exposure (b) Double Exposure, Double Reference beam, (c) Double Plate.

under investigation. After processing the hologram (photographic plate) and reconstructing it with a replica of the reference beam, the two waves $U_1(t_1)$ and $U_2(t_2)$ are reproduced simultaneously. The two reconstructed waves will interfere and form interference fringes due to time-varying density differences in the flow field between the two exposures. This technique is characteristically the easiest to perform and produces the highest quality interferograms. However, it is typically used to evaluate non-stationary phenomena in the flow field. For example, acoustic waves in the flow, vortical shedding, and turbulence-induced density fluctuations may be visualized in compressible flows.

Method 2 — Double Reference Beam: There are some advantages that can be gained by using one reference beam to object beam angle $U_{R1}(t_1)$ to record the reference test condition and a second reference beam angle, $U_{R2}(t_2)$

(second spatial carrier frequency) to record the test conditions. An electro-optical modulator (EOM) can be used to frequency shift one of the reference beams during reconstruction. This will cause the interference fringes to appear to move at the shift frequency. With the use of two-phase detectors, one at the reference location and the other to scan the reconstructed image, it may be possible to rapidly digitize the interferometric data. Where high sensitivity is required, a phase shift of $\pi/2$ or π can be introduced into one of the reference beams to produce the phase contrast mode of observation. There is also the potential for changing one reference beam angle to the other during reconstruction to introduce the finite fringe mode.

Method 3 — Double Plate: The author has found this method to be most useful and versatile for studies of compressible flows. Using this method, a reference beam $U_0(t_1)$ is recorded at one flow conditions on one plate (for example, no flow disturbance, wind tunnel off). The reference plate is then removed and subsequent test waves $U_0(t_1)$ are recorded at the various flow condition on other photographic plates. The plates are then processed and the light waves reconstructed using a replica of the reference beam. The reconstructed waves are superimposed to cause interference by properly positioning the plates with respect to each other and the reconstructing reference beam. With specially designed micrometer-controlled plate positioning allowing six degrees of movement (3 translation, 3 rotation) the plates can be positioned in their precise relative positions when the recording was made. Thus, the plates can be adjusted to produce the infinite fringe and finite fringe interferograms with any desired fringe orientation. One reference wave may be compared with several different object waves by simply changing plates which increases the efficiency of the data acquisition.

The optical components of a practical holographic interferometer are shown in Fig. 9.19. A pulsed laser which may be a Ruby or a frequency doubled Nd:YAG laser is used as the light source. A high quality dielectric beam splitter that can handle the high energy laser pulses is used to separate the beam into the reference and object beams. The reference and object beam path lengths are kept as close as possible but at least within the coherence length of the laser. The object beam is expanded to fill a spherical or parabolic mirror and produce a collimated beam that is passed through the test region. A second mirror is used to redirect the light to the holographic plate. The "Z" configuration is needed with angles that are as small as possible to minimize the astigmatism in the beam. The reference beam is directed around the flow field and is expanded onto the holographic plate to form the hologram. If a ruby laser is used, the

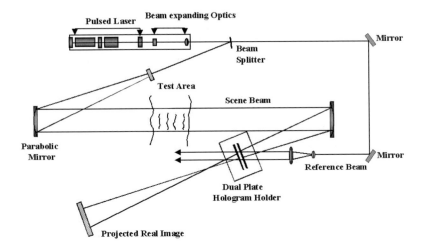

Fig. 9.19. Schematic of a holographic interferometer, which shows the optical components.

reconstruction system shown in Fig. 9.20 would use a Helium-Neon CW laser to produce a replica of the original reference beam. Although the wavelengths are slightly different (0.6943 and 0.6328 μm), this will only change the size of the image slightly. The same is true for the frequency doubled Nd:YAG laser where an argon ion laser is used for the reconstruction.

It is instructive to compare the holographic interferometer for which a typ-

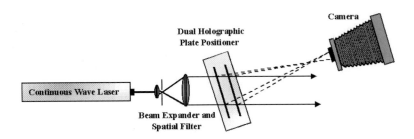

Fig. 9.20. Schematic showing the dual plate reconstruction system for the Holographic interferometer.

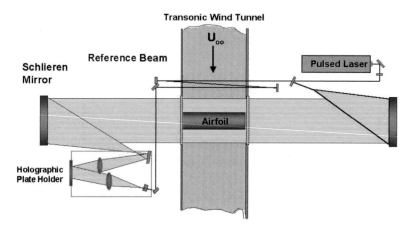

Fig. 9.21. Holographic Interferometer system using the existing wind tunnel Schlieren optics and a pulsed laser.

ical example is shown in Fig. 9.21 to the Mach–Zehnder system to understand why the mechanical and optical constraints can be significantly relaxed for the former. With the Mach–Zehnder interferometer, a coherent beam is split into two paths, one of which passes through the test region. Hence the interference is between two light waves that followed different paths at the same time. On the other hand, with holographic interferometry the interference fringes that contain flow field information is between two reconstructed light waves that were recorded at different times but that followed the same optical paths, with the exception that the optical disturbance was present in one of the recordings. It should be clear that any imperfections in the optical system would tend to cancel. Vibrations, which are ever present in large-scale wind tunnel facilities, do not present a difficulty since the pulsed lasers used to record the holograms have pulse duration on the order of 10 ns. The reconstruction and analyses of the interferograms using CW lasers is sensitive to vibrations but this part of the procedure is conducted in the laboratory. When using the dual plate approach, misalignment between the recordings can be compensated during the reconstruction of the light waves.

9.12   Applications

The accuracy of the interferometry visualization and quantitative results have been confirmed by comparisons to other data including surface pressure, pitot-static probe, Laser Doppler Velocimeter (LDV) measurements. There were reasons to question the accuracy of the method including the uncertainties in the alignment of the interferometer and three-dimensionality in a flow assumed to be two-dimensional. Because the light wave integrates the density-related phase modulations along the optical path, any three-dimensionality in the flow would produce uncertainties in the observations. For example, in wind tunnel studies using two-dimensional models, the wind tunnel boundary layers and their interaction with pressure gradients in the flow will produce undesirable three-dimensionality effects into the flow field. Extensive experiments have been carried out to test the accuracy of the method as well as to study the details of transonic flows by Bachalo & Johnson (1978), Bachalo & Spaid (1981), and others.

Work was conducted in the NASA Ames Research Center 2×2 foot Transonic tunnel to study the flow fields on various two-dimensional airfoils including a NACA 64A010 and supercritical airfoils. The existing schlieren system was easily converted into a holographic interferometer. The system had high quality parabolic mirrors with sufficiently large diameter, which allowed a collimated beam diameter of 46 cm to be passed through the test section. A 50 mJ Q-switched laser was used as the light source. Reference holograms were recorded with the wind tunnel off and the test section at atmospheric conditions. The recordings with the flow on were then made on consecutive plates positioned in the holder with a space behind the original location of the reference plates so they could be reconstructed together in the same relative positions. After processing, the reconstruction system shown in Fig. 9.20 was used to view the interferograms. The holograms were then adjusted relative to each other to recover the finite fringe and infinite fringe interferograms. Knowledge of the aerodynamics is useful at this point as clues to the proper alignment, especially if the entire field of view is disturbed by the flow.

Fig. 9.22 shows a high concentration of interference fringes in the high flow acceleration supersonic area on the leading upper surface of the airfoil terminated by a normal shock. Weak compression waves can be seen in the flow downstream of the shock, indicated by waves running vertically in the interference fringe patterns. The wake is also clearly visible. Fig. 9.23 is an enlargement of the trailing edge region showing the boundary layer, the local pressure maximum in

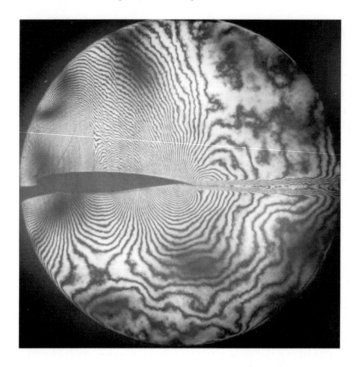

Fig. 9.22. Infinite fringe holographic interferogram of a McDonnell Douglas super-critical airfoil at $M_\infty = 0.73$, angle of attack, $\alpha = 4.32°$ (from Spaid & Bachalo, 1981).

the lower trailing edge region, and the separation in the lower cusp region of the trailing edge.

The flow visualization and the quantitative information available from the interferograms were evaluated by comparison to the surface pressure measurements. With the assumption of isentropic flow, the densities measured from the interferograms were reduced to the surface pressure coefficient, C_p, using the following relationships:

$$\frac{p}{p_0} = \left(\frac{\rho}{\rho_0} \right)^\gamma$$

and

$$C_p = \frac{2}{\gamma M_\infty^2} \left[\left(\frac{p}{p_0} \right) \left(\frac{p_0}{p_\infty} \right) - 1 \right],$$

where p is the surface or static pressure, p_0 is the total pressure, p_∞ is the free-

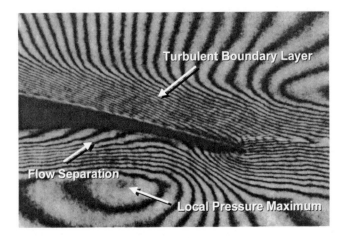

Fig. 9.23. Details of the flow in the trailing edge region of the supercritical airfoil.

stream static pressure, $\gamma = 1.4$ for air, and M_∞ is the free-stream Mach number. Since the field of view does not extend upstream to the undisturbed flow, a fringe remote from the model could be identified using the wind tunnel surface pressure, total pressure, and total temperature conditions. Interference fringes are then counted from the reference point to obtain the density at each point in the flow field. As an example, Fig. 9.24 shows the comparison for a NACA 64A010 airfoil at $M_\infty = 0.8$, and $\alpha = 3.5°$. The corresponding interferogram is shown with the plot. Once the accuracy of the infinite fringe interferograms were confirmed, the fringe patterns can be assumed to be accurate representations of the density contours for that flow condition. From the relationship given below as

$$\frac{\rho}{\rho_0} = \left(1 + \frac{\gamma - 1}{2} M^2\right)^{1/(\gamma-1)}$$

to show that the interference fringes are also lines of constant Mach number.

Finally, the boundary layer and wake density profiles obtained from the interferometric data were reduced to flow speed (velocity magnitude only) profiles using the Crocco relationship given as

$$\frac{T}{T_e} = 1 + \frac{T_w - T_{ad}}{T_e}\left(1 - \frac{U}{U_e}\right) + r\frac{\gamma - 1}{2} M_e^2\left[1 - \left(\frac{U}{U_e}\right)^2\right]$$

where the subscript "e" represents the edge condition of the boundary layer,

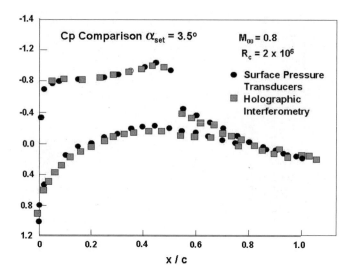

Fig. 9.24. Comparisons of the measured surface pressure coefficients obtained from the holographic interferogram and the surface pressure taps (Bachalo & Johnson, 1978).

r is the recovery factor, T_w is the temperature at the model surface, and U is the local flow speed (Bachalo & Johnson, 1978). This is used along with the ideal gas law to obtain the flow speed profiles. Good agreement with the laser Doppler velocimeter measurements shown in Fig. 9.25 shows that the boundary layers are accurately visualized.

9.13   Summary

Methods have been described that utilize the change in the refractive index of the fluid to visualize the flow and to obtain quantitative information. The shadowgraph and schlieren methods, which are over a century old, remain as very useful methods for studying compressible flows or flows with thermally induced index of refraction gradients. It is also possible to visualize incompressible flows by using gases with different refractive index. The Mach–Zehnder interferometer development in the late 1800's provides even greater detail in the flow visualization and also extends the capability to provide quantitative information. Unfortunately, the interferometer requires very high quality optics and high stability of the optical system if reliable data are to be obtained. This presented a serious limitation for large-scale applications such as in transonic

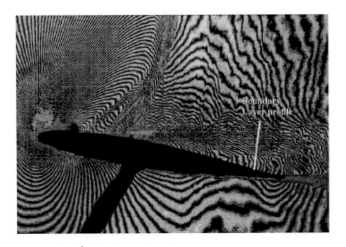

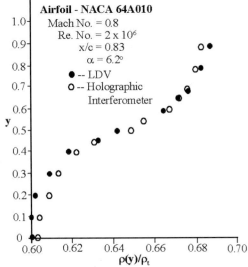

Fig. 9.25. Holographic Interferogram of the NACA 64A010 airfoil showing the mean density profiles of the boundary layer compared with that deduced from LDV data.

and supersonic wind tunnels. Holographic techniques have been introduced in the last 20 years to significantly relax the optical and mechanical requirements of the system. Although the quality of the interferograms are not as high when using holography to record the light wave phase and amplitude information, the method is applicable to large scale systems. Conversion of schlieren systems is relatively straightforward, especially with the development of pulsed lasers.

Examples of the data that may be obtained using these methods were presented. The shadowgraph and schlieren system information is useful in identifying such features as the shock locations, Prandtl–Meyer expansion fans, and the extent of the boundary layers. The interferometric techniques clearly provide greater detail in the flow visualization and have the additional capability of producing quantitative information on the flow density field. Examples of holographic information obtained in the NASA Ames Research Center wind tunnels were presented. Extensive evaluations of the interferometric results were conducted through comparisons to more basic measurements such as the surface pressure and LDV data. The good agreement of these results served to confirm the reliability of the visualization information as well as the quantitative data that could be obtained. The reader who is interested in more details and in-depth coverage of the subjects covered should turn to Merzkirch (1987) and Vest (1978).

9.14 References

Bachalo, W.D. and Houser, M.J. 1984. Phase Doppler spray analyzer for simultaneous measurements of drop size and velocity distributions. *Optical Engineering*, **23** (5), September–October.

Bachalo, W.D. 1980. A method for measuring the size and velocity of spheres by dual beam scatter interferometry. *Applied Optics*, **19** (3), February 1.

Bachalo, W.D. and Houser, M.J. 1985. Optical interferometry in fluid dynamics research. *Optical Engineering*, September.

Bachalo, W.D. 1983. An experimental investigation of supercritical and circulation control airfoils at transonic speeds using holographic interferometry. *AIAA Paper 83-1793*, AIAA Applied Aerodynamics Conference, Danvers, Massachusetts, July.

Bachalo, W.D. and Johnson, D.A. 1978. Laser velocimetry and holographic interferometry measurements in transonic flows. *Third International Workshop on Laser Velocimetry*, Purdue University, July.

Spaid, F. W. and Bachalo, W.D. 1981. Experiments on the flow about a supercritical airfoil including holographic interferometry. *Journal of Aircraft*, **18** (4), April.

Delery, J., Surget, J., and Lacharme, J-P. 1977. Interferometrie holographique quantitative en écoulement transsonique bidimensional.   *Rech. Aerosp.* **12**, 89–101.

Dvorak, V. 1880. Uber eine neue einfache Art der Schlierenbeobachtung. *Ann. Phys. Chem*, **9**, 502–512.

Gabor, D. 1951. Microscopy by reconstructed wavefronts II. *Proc. Phys. Soc.*, **64**, 449–469.

Hecht, E. and Zajac, A. 1976. *Optics*. Addison-Wesley Publishing Company, Menlo Park, CA.

Leith, E. N. and Upatnieks, J. 1962. Reconstructed wavefronts and communication theory. *J. Opt. Soc. Am.*, **52**, 1123–1130.

Merzkirch, W. 1987. *Flow Visualization*. 2nd ed., Academic Press, Inc., New York.

Settles, G.S. 1970. A direction-indicating color schlieren system. *AIAA Journal* **8**, 2282–2284.

Settles, G.S. 1982. Color schlieren optics — A review of techniques and applications. In *Flow Visualization II*, W. Merzkirch, ed., 749–759, Hemisphere, Wash, D.C.

Trolinger, J.D. 1969. Conversion of large scale schlieren systems to holographic visualization systems. 15th National Aerospace Instrumentation Symposium, Las Vegas, Nevada, May.

Trolinger, J.D. 1974. Laser instrumentation for flow field diagnostics. *AGARDograph No. 186*, March.

Trolinger, J.D. 1975. Flow visualization holography. *Optical Engineering*, **14** (5), September/October.

Uberoi, M.S. and Kovaszny, L.S.G. 1955. Analysis of turbulent density fluctuations by the shadowgraph method. *J. Appl. Phys.* **26**, 19–24.

Vest, C. M. 1978. *Holographic Interferometry*. John Wiley & Sons, New York.

CHAPTER 10

THREE-DIMENSIONAL IMAGING

R. M. Kelso[a]and C. Delo[b]

10.1   Introduction

This chapter is concerned with three-dimensional imaging of fluid flows. Although relatively young, this field of research has already yielded an enormous range of techniques. These vary widely in cost and complexity, with the cheapest light sheet systems being within the budgets of most laboratories, and the most expensive Magnetic Resonance Imaging systems available to a select few. Taking the view that the most likely systems to be developed are those using light sheets, the authors will relate their knowledge and experience of such systems. Other systems will be described briefly and references provided.

Flows are inherently three-dimensional in structure; even those generated around nominally 2-D surface geometry. It is becoming increasingly apparent to scientists and engineers that the three-dimensionalities, both large and small scale, are important in terms of overall flow structure and species, momentum, and energy transport. Furthermore, we are accustomed to seeing the world in three dimensions, so it is natural that we should wish to view, measure and interpret flows in three-dimensions. Unfortunately, 3-D images do not lend themselves to convenient presentation on the printed page, and this task is one of the challenges facing us.

10.2   3-D Imaging Techniques

Three-dimensional imaging of complex fluid flows has been attempted by many workers over many years. To suggest its originator would surely be meaningless,

[a]Department of Mechanical Engineering, University of Adelaide, Adelaide, SA 5005, Australia
[b]Dept. of Applied Physics, Columbia University, NASA Goddard Institute for Space Studies, 2880 Broadway, New York, NY 10025, U. S. A.

as the technique can take a wide variety of forms and involve an enormous range in technology. The advent of recording media such as photographic film and, later, video technologies, led to a number of imaging techniques such as stereoscopic imaging, holography and sectional imaging using light-sheets. The availability of lasers as powerful, collimated light sources, has improved the viability and accessibility of sectional imaging techniques, and allowed the use of photosensitive dye markers.

Early examples of 3-D visualization using stationary light sheets took the form of multiple cross-sections, either in orthogonal planes or as a "stack" of sheets. One such example is the visualization of a forced co-flowing jet by Garcia & Hesselink (1986), who reconstructed the flow volume to generate a 3-D image of the jet. Another example is provided by Perry & Lim (1978), who investigated the development of forced co-flowing jets and wakes, visualized by seeding the jet or wake fluid with smoke. Perry & Lim used stroboscopic light and a sweeping laser beam at several locations to image the flow, and used this information to construct wire-mesh models of the flow which provided clear visual insights into the flow physics. These stationary laser sheet techniques relied upon the highly periodic nature of the flow. However, for flows that exhibit no such periodicity, the structure can only be faithfully imaged using simultaneous or quasi-simultaneous images of multiple sectional planes at different spatial locations. Several techniques have been used to achieve this, most using one or more oscillating mirrors to sweep the beam through space. Other techniques, as will be described later, use a rotating drum or prism to sweep the beam through a volume. More recently, these techniques have been successfully adapted to deliver whole-volume 3-D Particle Image Velocimetry (PIV) measurements.

Technological advances have also brought the complex and expensive technique of Magnetic Resonance Imaging (MRI). The MRI technique provides detailed cross-sectional images of a flow, and is especially suited to flows devoid of optical access or transparency. The technique provides slow data acquisition rates and is extremely expensive. Some examples are discussed in Miles & Nosenchuck (1989).

Holography has the potential to become the ultimate in 3-D visualization and measurement, given its ability to record and reproduce entire flow volumes with great data compactness and its infinite depth of field. To the best of the authors' knowledge, the applications of holography to visualize 3-D flows have thus far been limited to whole-volume PIV techniques (holographic interferometery is a 2-D technique — see Chapter 9). In holographic particle imaging, a volumetric "snapshot" of a particle-seeded flow is recorded as a hologram. The hologram of the volume is replayed, then interrogated by means of an imager which is physically moved through the (three-dimensional) image. The output of the imager, as a function of spatial location, constitutes the data volume. This method usually requires a large amount of post-processing to reconstruct the sampled volume. However, advantageously, the number and location of data planes can be chosen and varied at the post-processing stage.

Unfortunately, the development of holographic systems has received far less attention than other 3-D techniques such as volume scanning and 3-D PIV, most probably due to the high cost and complexity of the holographic apparatus. In addition, practical limitations to the particle seeding density have, until recently, restricted the spatial resolution to number of vectors available to $O(10^5)$, compared with $O(10^6)$ for light sheet based 3-D PIV systems. This limitation was required to avoid masking and shadowing of neighboring particles in the flow, as well as (for in-line systems) masking of the reference beam. A number of techniques are now available that deliver extraordinary performance, albeit with high complexity and cost. For example, a recent system developed by Zhang *et al.* (1997) has delivered the 3-D velocity distribution within a square duct at $Re = 1.23 \times 10^5$, with a grid of $97 \times 97 \times 87$ vectors. This exceeds the spatial resolution of most 3-D PIV systems, but does not yet offer time-resolved measurements. Another technique called "multiple light sheet holography" is described by Hinsch (1995). The technique uses a holographic recording of a multitude of parallel laser sheets to effect simultaneous multi-plane PIV. The application of time-resolved holographic PIV, or "holocinematographical PIV" has been described by Weinstein & Beeler (1988) and Meng & Hussain (1991). For further information, the reader is directed to recent works by Blackshire *et al.* (1994), Barnhart *et al.* (1994), Hussain *et al.* (1994), Meng & Hussain (1995), Meinhart *et al.* (1994), Zimin *et al.* (1993) and Hinsch (1995).

Optical Tomography is the general name given to a technique for constructing cross-sectional images of a body or a flow from a number of in-plane optical projections. Images are generated by illuminating the plane of interest from a number of in-plane sources, then imaging it from a series of in-plane viewpoints, typically arrayed a semicircle. The result is a series of intensity profiles, or projections, which can be used to reconstruct the intensity profile of the complete planar image. This contrasts with the laser cross-sectioning technique in which the planar image is obtained from scattered light which is viewed by imager outside (typically normal to) the plane of illumination. There are several planar reconstruction algorithms; convolution and Fourier transform methods are most commonly used for fluid mechanical tomographic data: see Eckbreth (1988) and

Hesselink (1988) for details. The imaging method itself can use a number of methods such as shadowgraph, schlieren, interferometry and absorption to obtain distributions of quantities such as density, temperature and concentration.

The primary drawback of optical tomography for the study of fluid flows is the experimental complexity required. To obtain adequate spatial resolution, a large number of projections must be used to construct the volume. This requires either a periodically forced flow field which can be imaged sequentially from a series of viewpoints, or simultaneous imaging from a series of viewpoints. The use of eighteen simultaneous viewpoints was accomplished (Snyder & Hesselink, 1988), but the spatial resolution was low. To achieve the resolution available from other techniques such as planar imaging, the experimental complexity seems unreasonable. Being expensive, both in terms of the imaging equipment necessary and the computations required for the post-processing step, developments in tomography have been limited largely to the medical sciences. A major benefit of these developments is in the volume visualization software that has been developed to reconstruct tomographic volume data sets. This will be discussed below. We must point out that it is quite common for the volumetric laser scanning technique to be described as a "tomographic" technique. Whereas it is true that the sectional data obtained using planar imaging techniques is similar in format to reconstructed tomographic planar images, the method by which the images are obtained is not tomographic (Hesselink 1988).

The use of stereo imaging to obtain a volumetric data set is a natural extension of the binocular nature of human vision. Typically the method uses two imagers aligned at some angle to one-another, imaging the same region of flow. The primary limitation of stereo imaging is the uncertainty in the determination of depth, although this uncertainty can be minimized by the use of two orthogonal viewing axes, or possibly three. To image the entire flow volume, it is necessary to have optical transparency. Thus, the method is better suited to the imaging of particle-seeded flows than dye-marked flows (Praturi & Brodkey 1978). However, in order to minimize aliasing problems, the density of the seeded particles must be quite low. An interesting alternative to particle seeding is mentioned by Miles & Nosenchuck (1989), who propose the possibility of writing grid lines into the flow. In the case of water, hydrogen bubble lines or photochromically excit ed lines can be used. The stereo imaging technique has found much application in resolving the out-of-plane motions in planar PIV techniques, as described by Prasad & Adrian (1993), Hinsch (1995) and Brücker (1995a), and in tracking hydrogen bubble and fluorescent dye markers (Chapter 2 and Chapter 4, respectively).

For a comprehensive discussion of the above techniques, and the associated background theory, the reader should consult Hesselink (1988), Miles & Nosenchuck (1989) and Hinsch (1995). These reviews also discuss a number of less commonly used imaging techniques, such as stereo Schlieren and acoustic imaging, which will not be discussed in this chapter.

10.3   Image Data Types

Three-dimensional imaging techniques can deliver qualitative or quantitative data. It is now commonplace to conduct qualitative dye visualization, quantitative scalar visualization, and PIV, on individual planes within three-dimensional flows. In fact, any technique that can be applied on an individual plane can be used to obtain volumetric information using multiple planes imaged in quick succession. However, the volumetric technique also lends itself to a range of methods that deliver more detailed information such as scalar dissipation rates, strain rates, vorticity and velocity. Some examples of these techniques will now be given.

By far the most common use for 3-D visualization has been in the investigation of the three-dimensional structure of flows. For example, the 3-D dye visualization technique has been applied by Nosenchuck & Lynch (1986) to investigate the changes in flow structure in a perturbed boundary layer. The technique has been further used by Garcia & Hesselink (1986) to investigate the structure of a co-flowing jet; they used iso-surfaces of concentration to assess surface-to-volume ratios and hence entrainment rates. More recently, Goldstein & Smits (1994) and Delo & Smits (1993, 1997) used qualitative data to generate volumetric reconstructions of turbulent boundary layers, and applied conditional sampling and statistical methods to extract information about flow the structure. Other examples include Yoda & Hesselink, (1990) and Kelso *et al.* (1993, 1995). A number of workers have applied 3-D imaging techniques to the measurement of gas concentration through scattering techniques such as molecular Rayleigh scattering. These include Yip & Long (1986), Yip *et al.* (1988), Mantzaras *et al.* (1988) and Sen *et al.* (1989). Filtered Rayleigh scattering (see Chapter 5) has been applied by Forkey *et al.* (1994) to investigate a supersonic flow, using the naturally-occuring ice crystal "fog" as the scattering medium. This provided a series of concentration fields which were reconstructed into a 3-D image of the measured volume.

Examples of volumetric PIV are provided by Brücker (1992, 1995a,b,c, 1996, 1997a,b and 1998) who applied a range of volumetric techniques to "macroscopic" investigations of flows such as vortex breakdown, flow in a T-junction, the near wake of a spherical cap, flows with bubbles, the flow behind a short, surface mounted cylinder and the flow within a cylinder head. Two of these systems performed 2-D PIV on coarsely-spaced discrete planes, using planes of different orientations to provide 3-D velocity information. Three other variants provided three velocity components by either combining the PIV technique with stereoscopic imaging, or using closely-spaced light sheets or color-coded light sheets to measure the out-of-plane velocity component. Other examples include Ushijima & Tanaka (1996) using two orthogonal scanning systems to investigate a rotating flow.

Thus far two groups, consisting of Dahm & Buch and their co-workers, and Dracos & Rys and their co-workers, have applied the 3-D volume scanning technique to "microscopic" investigations of the fine-scale structure of turbulence (see also Chapter 11). These authors used numerous finely-spaced imaging planes to investigate small regions within turbulent jets. Their techniques enabled data such as scalar dissipation rates, strain rates, vorticity and velocity fields to be measured together for the first time.

10.4   Laser Scanner Designs

The following sections describe a range of laser scanning techniques that have been used or can be used in 3-D imaging. The scanning systems all generate a stack of parallel imaged planes, or sheets, which are used to reconstruct images, velocity fields and other volumetric representations of the flow. The list of systems is not exhaustive but does illustrate a range of possibilities and concepts.

Laser scanning systems essentially fall into two categories: those that generate discrete sheets of light and those that generate moving sheets of light. The type of system used depends on the equipment available, the data sought, future expandability and the cost. In some cases it is possible to achieve similar results using different system types, or construct the two system types from the same set of optical components. In nearly every case the optical system achieves the volumetric imaging process using two distinct optical components, or groups of components, to spread, sweep or traverse the laser beam in two orthogonal directions. We will henceforth distinguish these components as the "primary optic" (PO) and "secondary optic" (SO), respectively. Some systems combine both optical stages into one.

In the discussion to follow, we will refer to each complete scan of the flow volume as one time-step. Each time step consists of a number of "slices" or "im-

Discrete sheet systems
===

Drum scanner/Swept beam: Kelso *et al.* (1993, 1995); Brücker (1996); Delo & Smits (1997); Brücker (1997b); Brücker (1998), Guezzennec *et al.* (1996)

Drum scanner/Cyl. lens: Brücker (1995c, 1997a)

Osc. mirror/Osc. mirror: Prenel *et al.* (1986a,b); Dahm *et al.* (1992); Buch & Dahm (1996); Ruck & Pavlovski (1998)

Cyl. lens/Osc. mirror: Brücker (1995b); Ushijima & Tanaka (1996)

Poly. mirror/Osc. mirror: Brücker (1992)

Osc. mirror/trans. mirror: Perry & Lim (1988)

Fixed optics: Mantzaras *et al.* (1988); Yip & Long (1988); Forkey *et al.* (1994); Arndt *et al.* (1998)

Moving sheet systems
===

Osc. mirror/Cyl. lens: Yip *et al.* (1988); Sen *et al.* (1989); Prasad & Sreenivasan (1990); Goldstein & Smits (1994); Merkel *et al.* (1995); Ushijima & Tanaka (1996)

Cyl. lens/Osc. mirror: Yoda & Hesselink (1990)

Poly. mirror/Cyl. lens: Nosenchuck & Lynch(1986)

Galilean transformation/Cyl. lens: Garcia & Hesselink (1986)

Table 10.1. 3D visualization experiments grouped by volume scan technique (Osc. = oscillating, Cyl. = cylindrical, Poly. = polygonal, Trans. = lateral translation).

aged planes," being the individual laser cross-sections of the flow. The various system types and a range of optical arrangements will now be described in turn, with examples. Many examples are summarized in Table 10.1 which provides references for each of a range of scanning system designs.

10.5   Discrete Laser Sheet Systems

These systems generate multiple, discrete, parallel laser light sheets at fixed planes in the flow. The planes are usually illuminated in step-wise fashion.

The main advantages of such systems are that the flow is illuminated in fixed, stationary planes, and the illumination of each plane can be repeated for multi-pulse PIV applications if desired. Such systems can take several forms. The most commonly used form to date is the double-scan laser sweep system where the laser beam is deflected and scanned through the flow by two scanners, usually being oscillating mirrors or rotating prism mirrors. A less common alternative is the drum scanner, which simultaneously generates the light sheet or swept beam and increments it from plane to plane. Other systems include manually traversed light sheets for imaging time-averaged flow structure, scanners that generate non-parallel "designer" laser sheets, and systems that use laser sheets of different colors that exist simultaneously and are discriminated by optical means.

10.6   Double Scan Laser Sweep Systems

Laser sweep systems typically take the form of two oscillating mirrors that to-gether sweep the collimated laser beam through the flow. The image is recorded by keeping the shutter of the camera open during each individual sweep. In this way, the system can generate a raster of parallel laser sheets, very much like the raster scan of the electron beam in a cathode ray tube. A typical system is shown in Fig. 10.1. The primary and secondary optics each consist of a single oscillating mirror. To achieve the raster effect, the frequencies of oscillation of the two mirrors must differ: one mirror sweeping at high frequency to generate the individual laser sheets and the other sweeping at lower frequency to move the beam from plane to plane. At the end of each sweep, each mirror retraces its path, usually rapidly, prior to the next sweep. In general terms, the ratio of the mirror frequencies defines the number of sheets in each volume. The lower frequency mirror can be stepped from level to level, or moved continuously. The latter case causes the raster to be inclined at an angle, since both scanners move simultaneously, which depends upon the ratio of frequencies. For example, equal frequencies generate one sheet at 45°. These systems offer additional flexibility in that any "designer" sweep profile can be achieved, as demonstrated by Prenel *et al.* (1986a,b; 1989), who generated parallel, cylindrical, crossing and radial sheet patterns for different experimental applications.

Figure 10.2 describes the path taken by the laser over each volume scan in two typical applications of the double scan technique. Here the z-coordinate is normal to the imaged planes. The figure illustrates the significance of the "retrace time," being the time required for the beam to retrace its path after

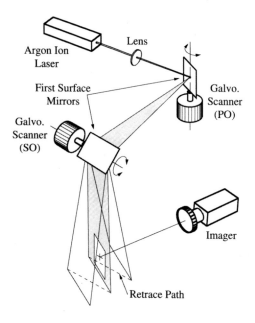

Fig. 10.1. A typical double scan laser sweep system.

each scan. In double-scan systems the retrace times of both scanners affect the amount of light available at each imaged plane and the maximum scanning rate available. For the best performance, the retrace time should be as small as possible. A detailed discussion of this issue is given in Rockwell *et al.* (1993). Figure 10.2 also indicates two of the many possible scanning patterns. The upper graph represents a typical stepwise scanning pattern, whereas the lower graph describes a pattern that delivers pairs of images of each imaged plane, closely spaced in time. This is useful for the collection of PIV data, as discussed in Section 10.11.6. In the case of galvanometer or similar scanners, the latter pattern is possible only if the low frequency scanner moves in a stepwise fashion.

Oscillating mirrors are typically commercial galvanometer-driven units with roll-off frequencies of the order of 100 Hz and linear response. The mirrors and moving components are usually small to minimize inertia and maximize frequency response. Linearity and frequency response of the mirrors ultimately define the maximum speed and position accuracy of such systems and should be

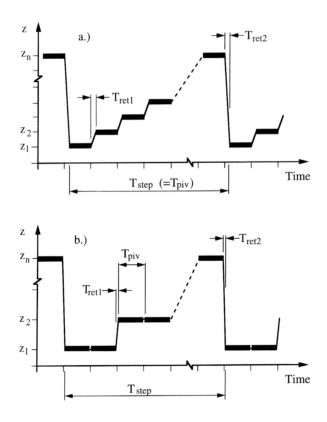

Fig. 10.2. The path taken by the laser over each volume scan in two typical applications of the double scan technique. T_{step} represents the time taken to complete each complete scan of the imaged volume, that is, the length of the time step. T_{ret1} is the retrace time of the high frequency scanner, whereas T_{ret2} is the retrace time of the lower frequency scanner. T_{piv} represents the mimimum time between two successive scans of any given plane and is relevant to PIV applications. (a) Standard step-wise scanning pattern. (b) Double-pulse pattern suitable for PIV in some applications.

chosen with this in mind (Rockwell *et al.*, 1993).

Systems such as this usually rely upon small angular sweeps to ensure that the imaged planes are close to parallel. In many applications a small angular divergence between the laser sheets may be acceptable and the majority of workers have adopted this approach. Angular divergence can be readily corrected optically using plano-convex lenses, as the following example will show.

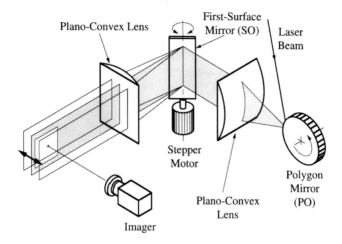

Fig. 10.3. Alternative double scan laser sweep system, similar to that used by Brücker (1992).

An alternative to the dual mirror system is shown in Fig. 10.3. This is similar to the system used by Brücker (1992) for PIV measurements, and incorporates a rotating polygonal mirror as the primary optic and an oscillating mirror as the secondary optic, both driven by stepper motors. Plano-convex (cylindrical) lenses are also incorporated into the system to transform the divergent beams into parallel paths. There are many advantages of such systems over the mirror system above. First, the polygonal mirror is a continuously rotating component driven by a stepper motor at a modest speed, ensuring complete synchronization (frequency and phase) with the imaging system. Second, for the rotating mirror the retrace time is very small (Rockwell *et al.*, 1993). Third, the lower frequency mirror is driven by a stepper motor which turns the mirror through defined angular steps between beam sweeps and then retraces the mirror rapidly before the next time step. Again, the stepper motor is well suited to such a task. Brücker's system comfortably achieved 500 images/sec.

Figure 10.4 describes a suitable replacement for either or both the polygonal mirror and oscillating mirror, with their associated plano-convex lenses, shown in Fig. 10.3. The device is similar to prismatic scanners used by Schluter *et al.* (1995), Deusch *et al.* (1996) and Cutler & Kelso (1997). The device, a rotating prismatic refractor, or "rotating prism," acts essentially as a laser beam translation stage which, when rotated, translates the beam parallel to itself,

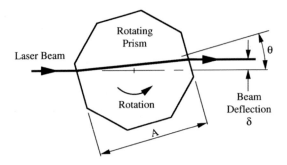

Fig. 10.4. Rotating prism scanner.

sweeping in one direction only. Its retrace time is similar to the polygonal mirror described above. The rotating prism is readily attached to a stepper motor or similar and can operate at constant speed or be rotated in a stepwise fashion. The relationship between the translation δ and the angle of turn θ is nonlinear, but the nonlinearity decreases rapidly as the number of faces increases. An equation relating the deflection distance δ versus the width A and the angle of turn θ_1 is:

$$\delta = A \sin \left(1 - \frac{n_1 \cos \theta}{\left(n_2^2 - n_1^2 \sin^2 \theta\right)^{1/2}} \right)$$

For a prism operating in air, $n_1 = 1.0$.

For correct operation there should be an even number of faces (4, 6, 8, ...) and with a sufficient number of faces, any chosen degree of linearity between δ and θ can be achieved. For example, for an 8-sided acrylic prism ($n_2 = 1.495$) of 100 mm width, the total scan width $2\delta_{max} = 27.6$ mm and the maximum deviation of the δ versus θ curve from linearity is 0.3 mm. The main drawback of this device is that, for a large number of faces, the diameter of the prism becomes large relative to the required sweep distance.

10.7   Single Scan Laser Sweep Systems (Discrete)

These systems can take the form of a single cylindrical lens (see discussion in Section 10.11.4) as the primary optic, with a single oscillating mirror as the secondary optic to generate a sheet of laser light that moves in stepwise fashion. The image is recorded by keeping the shutter of the camera open during each

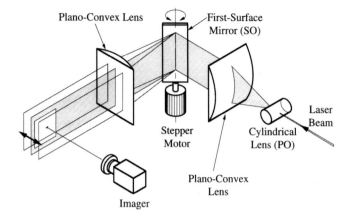

Plano-Convex Lens

First-Surface
Mirror (SO)

Laser
Beam

Stepper
Motor

Cylindrical
Lens (PO)

Plano-Convex
Lens

Imager

Fig. 10.5. Typical single scan laser sweep system, similar to that used by Brücker (1995a,b).

individual step. A typical system, similar to that used by Brücker (1995a,b) for PIV measurements, is shown in Fig. 10.5. Another example is found in Ushijima & Tanaka (1996) who used two orthogonal systems imaged by a high frame-rate camera. The system would also work if the cylindrical lens was used as the secondary optic and the oscillating mirror used as the primary optic, although the angular spread of the laser sheet would vary depending on the angle of incidence between the laser beam and the cylindrical lens. The preferred arrangement may ultimately depend on the size, geometry and quality of the available optical components, especially the cylindrical lens and scanning mirror.

An alternative to the galvanometer-based oscillating mirror is the Bragg cell, or Acousto-Optic Modulator (AOM). These are fast, solid-state devices that deflect the beam at an angle proportional to the voltage applied. A typical maximum deflection angle is of the order of 1 or 2°. The physical arrangement requires that the AOM function as the primary optic, with a cylindrical lens (or scanner) as the secondary optic. The beam can be stepped, swept and modulated to achieve any spatial and/or temporal illumination profile required. The advantage of such a device is low complexity, high operational flexibility and a small retrace time. The main disadvantages are that the loss in light intensity can exceed 50%, and the laser is restricted to single-line operation which reduces the available intensity by at least a further 50%. A detailed discussion of AOMs is provided by Rockwell *et al.* (1993).

The single scan laser sweep system is essentially the same as the moving sheet design discussed below, with one exception: the moving sheet design translates the laser sheet parallel with itself in a continuous motion, but the discrete single scan system moves the sheet in stepwise fashion. Stepwise translation of the light sheet enables multiple laser light pulses at each measurement plane, allowing PIV or Particle Tracking Velocimetry (PTV) measurements to be made.

10.8   Drum Scanners

Figure 10.6 describes a system that embodies an entirely different scanning concept whereby the primary and secondary optics are combined, namely the drum scanner. First developed by Delo & Smits (1993), it consists of a helical array of 45° mirrors, fixed to each of twenty faces of a rotating drum. When used with a continuous wave (CW) laser, the focused laser beam passes parallel to the axis of the drum, and reflects off each mirror as the drum rotates. The motion of the flat mirror face causes the beam to sweep through an angle of 18°, creating a laser sheet similar to those described above. The image is recorded by keeping the shutter of the video camera open during the sweep. As the drum continues to turn, the beam reflects off the next mirror forming another sheet at a different height. Because the location of the sheet is determined by the height of the corresponding mirror, the sheet locations are exactly repeatable and they can be accurately determined from a single static calibration. The retrace time between individual laser sweeps and between one time step and the next is the same, representing approximately 1° of rotation. It should be noted that this design produces a small curvature in each swept plane due to the variation in the height of the point of incidence of the laser beam on the flat mirror. This effect decreases as the number of mirrors increases and the sweep angle per mirror decreases. For the arrangement described by Delo & Smits, this effect is negligible compared with the thickness of the laser sheet. More complete descriptions of this system are given by Kelso *et al.* (1993, 1995), Delo *et al.* (1994), and Delo & Smits (1997).

Drum scanners of this type can readily be applied to PIV. Two examples of such systems have been reported by Brücker (1996, 1997, 1998) who used them to measure velocities in three dimensions. The adaptation of the Delo & Smits system (above) to the measurement of velocities is relatively straightforward and is largely a matter of setting the time separation between successive images (T_{piv}) at each imaged plane so as to provide the required dynamic range in the

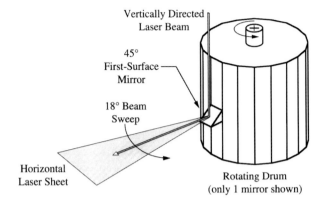

Fig. 10.6. Drum scanner as used by Delo & Smits (1993).

PIV measurements. If necessary, the scanning mirrors on the drum can be set such that each plane is scanned two or more times in each time step (equal to the time to rotate the drum), as shown in Fig. 10.2. Velocity measurements are achieved by the cross-correlating successive images. In systems limited by the frame rate of the imager (as compared with the time-scales of the flow), it may be necessary to locate the mirrors in pairs at each level, as described in Brücker (1996, 1997b) for a similar system. In systems using high frame rate imagers it is possible to achieve a broader dynamic range by cross-correlating images separated by one, two, or more time-steps. Clearly, for an existing drum scanner, the more mirrors at each plane, the fewer planes can be scanned. For example, a 20 mirror drum can image 10 planes using 2 scans, 6 planes using 3 scans, or 5 planes using 4 scans, and so forth.

The drum scanning technique has been further developed by Brücker (1996) to provide 3-D scanning particle image velocimetry using a single scanner. The technique uses a multiline Ar-ion laser and a beam splitter to generate two separate beams, at wavelengths of 488 nm and 514 nm. These are scanned simultaneously using the drum scanning technique, such that the two laser sheets (that is, the two scanning beams) overlap by 50%. The scanned volume is imaged by a 3-CCD color video camera, which separately records the blue and green wavelengths. Thus, the motion of particles from one plane to the other can be determined by a correlation between the "blue" and "green" image pairs from consecutive images (scans). RMS errors in the range of 16% of the mean velocity were achieved. In another system Brücker (1997b) used a "dual plane" method to measure the out-of-plane velocities in a flow with known directional bias in the out-of-plane velocity component. In a third system, Brücker (1998) used two scanners, mounted orthogonally, to measure (2-D) velocities in two sets of orthogonal planes, thus providing 3-D velocity measurements. A split optical system allowed him to image the two sets of scanned planes simultaneously using a single high-speed imager.

A logical variant of the above drum scanner design is also reported by Brücker (1995c, 1997a) who replaced the flat mirrors with curved (conical) ones to achieve the beam height incrementation without any sweep, as described in Fig. 10.7. The spreading of the laser sheet was then achieved using a cylindrical lens, thus illuminating the entire plane at each imaging level. The duration of each light pulse was defined by a shutter plate (plate with holes to pass or block the beam) mounted to the top of the scanner drum. Brücker used two such scanners, mounted orthogonally, to measure velocities in three dimensions (similar to Brücker, 1998). Due to the slow speed of Brücker's S-VHS video recording system, it was necessary to locate the mirrors in pairs at each level. Equally, this could have been achieved using a single mirror at each level and pulsing (for example, using an AOM or a pulsed laser) or shuttering (using a shutter plate) the laser light to produce two or more light pulses per level. Such a system may be advantageous when using a scanner for dye visualization as well as PIV.

10.9   Multiple Fixed Laser Sheets

An alternative to sweeping the beam is to image simultaneously the scattered light from a number of fixed laser sheets. The principle utilizes the high-intensity lines produced by multi-line lasers which are separated according to wavelength, individually aligned and spread into parallel light sheets. Such systems offer the advantage of low mechanical complexity — no moving components - but the disadvantage that a number of laser sources (and/or modifiers) are required for more than two or three simultaneous laser sheets. This technique was attempted successfully by Yip & Long (1986) using two laser sheets of different wavelength and two cameras fitted with optical band-pass filters. The method was later extended by Mantzaras *et al.* (1988) to include four laser sheets (two lasers plus a Raman shifter). The flow was imaged by a single camera, viewing through a quadrupling prism fitted with appropriate filters.

A further alternative to the volume visualization method was adopted by Forkey *et al.* (1994) who imaged the flow field inside a model supersonic inlet

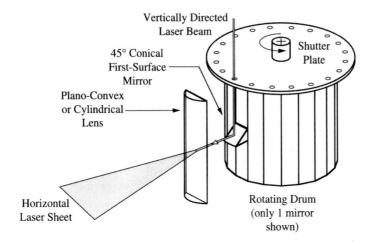

Vertically Directed
Laser Beam

Shutter
Plate

45° Conical
First-Surface
Mirror

Plano-Convex
or Cylindrical
Lens

Horizontal
Laser Sheet

Rotating Drum
(only 1 mirror
shown)

Fig. 10.7. Drum scanner as used by Brücker (1995c, 1997a).

using filtered Rayleigh Scattering from naturally occurring ice particles (see Chapter 5). The camera and light sheet assemblies were manually traversed between each of the 26 imaging planes. Using individual instantaneous images chosen at random from each plane, the system allowed a "time-average" flow pattern to be reconstructed.

Finally, it is important to mention that the phase averaging technique, so common in laser Doppler velocimetry and hot-wire anemometry, appears to be used rarely with imaging techniques. Whereas it has been applied to 2-D PIV techniques, the authors are not aware of any instances where phase averaging has been applied to the volumetric technique. Three-dimensional imaging and, in particular, volumetric PIV, are instances where phase-averaging techniques may offer considerable reductions in complexity and cost. Using forcing to induce perfect periodicity in a flow (eg. Perry & Lim 1978), or in selected highly periodic flows, it is feasible to image the flow on single planes only, using a phase trigger to identify or choose the phase of each image. With a sufficient number of closely-spaced imaged planes and sufficient phases resolved, a high-quality data set can be obtained. It should be pointed out that this technique saves time and cost in using planar PIV (no volume scanning is needed) but requires extra time to collect the data. The processing time is essentially independent of the way in which the data is collected, assuming the data array size and processor are the same. Furthermore, the money saved using the 2-D technique can be invested in a higher resolution imager and faster data processor. Clearly, this technique is less desirable than a 3-D volume imaging system, but in some cases it may be an acceptable, low-cost compromise.

10.10   Moving Laser Sheet Systems

Moving light sheet systems are very similar to the discrete single scan laser sweep system discussed above. Whereas the discrete single scan system moves the sheet in a step-wise fashion, the moving sheet design translates the laser sheet continuously. The main advantage of moving sheet systems is in the speed and flexibility of operation. However, the continuous translation of the light sheet severely restricts the ability of the system to be used for PIV measurements. Hence, most moving light sheet systems have been applied to the visualization of scalar fields.

As before, these systems can take the form of a single cylindrical lens as the primary optic, with a single oscillating mirror as the secondary optic to generate a moving sheet of laser light. However, by far the most common arrangement uses an oscillating mirror as the primary optic, followed by a cylindrical lens as the secondary optic. In each case the image is recorded by fast shuttering of the camera to "freeze" the motion of the flow relative to the moving beam (that is, minimize the distance moved relative to the fluid during each image capture). Three variants of this system have been used successfully. In two of these the oscillating mirror is replaced by a polygonal mirror (Fig. 10.3) and a rotating prism scanner (Fig. 10.4), respectively. The third variant uses the convection of the flow features themselves to provide an effective translation of the sheet. This is akin to the use of Taylor's hypothesis to translate between space and time coordinates. This method was applied by Perry & Lim (1978) to image co-flowing jets and wakes in air, and Garcia & Hesselink (1986) to image co-flowing jets in water. The method is effective in reproducing the flow topology provided that the rate of distortion of the flow features is small. Clearly the spatial growth of the vortical structure cannot be reproduced. This method is viable only in flows such as plane or axi-symmetric jets, wakes and shear layers in regions of monotonic growth.

A significant benefit of the moving laser sheet system is the nearly infinite variability in the number of images per time-step. Unlike many discrete laser sheet systems where the hardware — stepper motors or drum scanners, for example — defines the minimum spacing between images and number of images per time step, moving laser sheets essentially have no such limitation. Furthermore, when a polygonal mirror or prism is used the speed of scanning is relatively unconstrained by the frequency response of the hardware — to scan at higher frequency, the optic can be spun faster or the number of faces of the polygon can be increased. Hence, moving laser sheet systems offer the highest speed and spatial resolution of all the scanning systems. The main system constraints are the speed of the imaging device and the available illumination.

The use of moving laser sheet is more problematic when PIV is attempted, due to the continuous movement of the laser sheet. Where image-to-image time delays of less than the time-step are required, multiple images must then be obtained at each imaged plane in the flow. For multiple images to be obtained, the time-delay must be so small that the displacement of the laser sheet between images is small compared with the thickness of the laser sheet. This limits the time delay to extremely small values. This is a severe limitation which reduces the viability of the moving laser sheet technique for PIV applications. However, one exceptional case where such a system is advantageous, is when the laser sheet moves with the mean convection velocity of the flow structures. Hence, the laser sheet is stationary with respect to the flow and so the normal rules for stationary sheets then apply.

10.11   Imaging Issues and Trade-Offs

In an ideal imaging system where infinite imaging rates, infinite laser power and infinitesimal laser pulse duration are available, many of the following issues would be of little consequence. However, the practicalities are that imaging rates, laser power and pulse duration are limited, sometimes severely, by what we can afford or by the state of the current technology. We must therefore trade the limitations off to achieve the "best" compromise in system performance for the given application. The following sections seek to identify the main issues affecting the performance of volume scanning systems and ways to deal with them.

10.11.1   Position accuracy of laser sheets

Positional accuracy of the laser sheets or scans is a significant issue in volume scanning systems. Problems such as mechanical jitter and resonance as well as roll-off in frequency response can be absent during static or low-speed calibration, yet cause considerable inaccuracy or uncertainty at high-speed. For example, the galvanometer-based, moving laser sheet system used by Goldstein & Smits (1994), presented considerable problems when driven by a 50 Hz triangle wave. As reported by Delo & Smits (1993), these include vibration at the point of maximum acceleration as well as considerable uncertainty as to the location of any given image from one sweep (time step) to the next. Although these problems may have been peculiar to the scanner model used in this instance, they highlight the need to check or calibrate the system at the operating condition. Stepper motor drivers, such as those used by Brücker (1992, 1995a,b) to drive scanning and oscillating mirrors, offer far more certainty in image location, although they can (from the authors' personal experience) suffer from severe resonance if driven by primitive controllers. Again, there is a need to confirm or re-calibrate the system at the operating condition. Finally, perhaps the greatest certainty and positional accuracy is provided by the drum scanner designs of Delo & Smits (1993, 1997) and Brücker (1995c, 1996, 1998). These rugged scanners provide mechanically defined laser scan or sheet locations. They are readily calibrated and maintain their alignment at high speed.

A useful solution to the problem of sheet location and speed when using oscillating mirrors (and any other system for that matter) is reported by Yip *et al.* (1988). The speed and position of their scanning mirror was monitored using a pair of photodiodes triggered by a He-Ne laser reflecting off the back of the mirror. Clearly, the path of the reflected beam, striking a distant wall or screen, would also offer insights into the fidelity of the mirror scan.

10.11.2   Illumination issues

The single, double, and drum scanner systems described above are well suited to CW laser illumination, as evidenced by the frequent use of 5-7 W Argon-ion lasers. However, appropriately synchronized pulsed lasers such as copper vapor lasers can be used provided that a broad enough pulse width is available for a given application. A pulsed laser could, in fact, be advantageous in the double-scan and drum scanner cases because illumination could be avoided during the retrace process between successive scans, minimizing the possibility of signal contamination and stray reflections. In the cases of multiple fixed laser sheets, both CW and pulsed lasers are feasible. Moving laser sheet systems are compatible with both continuous and pulsed laser illumination.

As is often the case with laser-based optical measurement systems, illumination is a significant determinant of the performance of a volume imaging system. It is safe to say that one can never have enough light! As discussed below, the faster a scanning system operates, the better the dye or particle motions are "frozen," but the faster a system operates, the shorter the imaging period. If a CW laser is used, the shorter imaging period results in a smaller amount of light available to the imager. Thus, when all other avenues (such as adjusting the imager's aperture size or reducing the imaged volume size) are exhausted, a speed is ultimately reached at which there is insufficient illumination for the imager to exploit its full image depth (greyscale). When fast-framing CCD imagers are used, this problem is exacerbated since they tend to be significantly less sensitive than conventional CCD imagers. These limitations are somewhat reduced by the use of pulsed lasers such as the copper vapor laser, which will operate at variable frequency and deliver large pulses of light at a constant pulse energy. As discussed in the next section, when using a CW laser, it is highly advantageous to scan the laser beam rather than use pulses of short duration.

When undertaking dye visualization, the intensity of the scattered (and/or fluoresced) light can sometimes be improved by adjusting the concentration of the scalar marker. Care should be taken to ensure that this does not lead to shadowing adjacent to regions of high dye concentration. For example, in visualizating and measuring the scalar field of a turbulent jet, Prasad & Sreenivasan (1989) introduced sodium fluorescein into the jet with a concentration of 10 ppm. Delo & Smits (1997) and Kelso *et al.* (1992, 1995) introduced dye into a boundary layer at a concentration of 500 ppm. This provided exceptional intensity at the expense of some shadowing. In PIV techniques, the use of fluorescent particles, such as rhodamine or fluorescein-filled latex micro-spheres, can similarly improve the intensity of the light reaching the imager.

10.11.3   Sweeps versus sheets for CW lasers

The issue of whether it is preferable to sweep the laser beam to form a laser sheet or to spread the laser beam using a cylindrical lens (or equivalent) is central to the design of a system for any given application using a CW laser. The choice depends essentially on the trade-off between complexity (cost), light intensity and time resolution.

For a continuously illuminated plane, there is essentially no difference in the area-averaged illumination provided by a sweeping beam or a continuous (non-pulsed) sheet, given the same spreading angle and fan origin, and assuming a uniform spread from the cylindrical lens. If the imager integrates light scattered by a passive scalar over one complete sweep of the beam, the average illumination of each pixel or grain on the imager's sensor must be the same as for a uniformly spread sheet over the same time. However, when short duration pulses are required, separated by a specified time delay, the average intensity of the pulsed spread sheet using a CW laser can be significantly less than that of the sweeping beam. For the pulsed case the time delay between pulses represents wasted light, and so the effective illumination is proportional to the ratio between the pulse duration and the time delay, that is, the duty cycle. For the sweeping beam the time delay is equal to the period of each scan and, providing the laser beam diameter is small relative to the width of the imaged plane, no modulation of the beam is required. Thus, the area-averaged intensity of the sweeping beam is unaffected. The sweeping beam therefore provides the maximum illumination achievable from a CW laser (see also Rockwell *et al.*, 1993).

Furthermore, the spread sheet and sweeping beam techniques register different images in each case. In the spread sheet case, the imager integrates the fluid motions over the entire duration of light pulse and every part of the image is integrated simultaneously. In the sweeping beam case, the laser beam illuminates a thin line only, so the integration time at each point in the flow is very small, and different parts of the flow are imaged at different times. Thus, in a typical application, the spread sheet offers truly simultaneous imaging at a lower time resolution, and the swept beam offers non-simultaneous imaging at a significantly better time resolution so far as each pixel is concerned. Clearly, the differences between these two methods depends on the duty cycle of the spread sheet and, in the case of the sweeping beam, the beam diameter relative to the width of the imaged plane. In many applications such as PIV or dye imaging, an acceptable performance compromise can be reached using either method by ensuring that the total imaging time at each imaging plane is small with respect to the smallest time-scale of interest — less than half, according to the Nyquist criterion. For studies where very high spatial and temporal resolution are required, such as in the scalar imaging experiments of Buch & Dahm (1996) and Deusch *et al.* (1996) where length scales below the viscous length scale must be resolved, the benefits of the scanned sheet are clear. To quote Buch & Dahm (1996), "the effective temporal resolution is greatly increased by sweeping the collimated laser beam through the flow, rather than imaging from a fixed laser sheet. This reduced the effective integration time for each pixel to the time spent in the pixel's field of view — a reduction of more than two orders of magnitude."

10.11.4   Optical components

The choice of optical components is very important in the development of volume scanning systems. Beam degradation and losses (due to scattering, absorption and reflection) occur whenever the laser beam meets or passes through a solid surface, be it a mirror, lens, prism or tunnel window. Hence it is important that the number of optical components be minimized and the quality of each component maximized.

For optimum performance, components should be of as high a fidelity as possible and they should be matched to the laser source. Flat optical components should have 10-wave flatness or better wherever possible. Poor surface quality can lead to both scattering of the light and errors in image location. Mirrors should have their reflective coating on the front-surface: rear-surface mirrors should not be used as they generate double images (and waste light). Commercially made mirrors are often coated to improve their resistance to abrasion and corrosion. Lenses should preferably be of "laser" quality and non-reflective coatings are an advantage. Optical glass, reflective coatings and protective or non-reflective coatings are often optimized for specific wavelengths of light, and should be matched to the application where possible.

Fortunately, limited budgets should not necessarily prevent a high-fidelity system from being assembled. Many excellent experimental systems have been constructed from military and industrial surplus and salvaged from redundant, broken or superceded equipment. There are numerous companies that specialize in surplus equipment and many bargains can be found, ranging from front-surface mirrors, prisms, laser printer scanners and lenses, to stepper motors and power supplies. Many of these components are of the highest quality.

One common cost saving strategy is to use a glass or acrylic rod as a cylindrical lens. As a guide, the focal length of a glass rod is approximately equal to its diameter. The main disadvantage of using rods is that a Gaussian intensity profile is produced in the spread laser sheet. Although this can be corrected in image processing, the intensity variation wastes available light and reduces the depth of grey-scales that can be resolved by the imager, particularly at the edges of the fanned light sheet. A uniform illumination profile is ideal and can be generated by appropriately designed, commercially available "line generator" lenses such as the Powell lens. A second disadvantage is that glass and acrylic rods inevitably have surface imperfections which cause striations, or stripes, in the light sheet. These may be eliminated by using ground and polished glass rods or by attaching the rod to a small electric motor and spinning it at high speed. This effectively smears the striations uniformly over the entire sheet. The period of rotation of the rod should ideally be an order of magnitude smaller than the imaging time for each plane.

10.11.5 Methods of control

The effective operation of a volume scanning system relies on the frequency and phase synchronization of each and every component. Modern digitally controlled devices allow systems to be synchronized in a number of ways. When older or less sophisticated components are used, the choices are reduced.

The most difficult way to control the system is to slave the scanning system to the imager. However, this approach is necessary in the case of cine cameras and some analog electronic cameras (such as the analog Kodak or Spin Physics Ektapro 1000), as well as less expensive digital models where the cameras cannot be slaved to external devices. Added to this complication, some cameras generate a frame pulse only when the camera reaches the target operational speed, necessitating quick synchronization of the scanner, especially when recording time is limited. This problem was the primary motivation for the manual system developed by Kelso *et al.* (1993, 1995), which is described below. More recently, a control system has been developed at the University of Adelaide for use with a fast-framing 16 mm cine camera, or any external imaging device from which a frame pulse can be extracted. It uses a programmable logic controller to drive a DC motor, providing fast response and extremely accurate speed control. Phase control, although more difficult to achieve, has been obtained to approximately ±1°.

A far simpler approach, where feasible, is to slave the imager to the scanner itself. Many of the latest digital cameras can readily be slaved to an external frame pulse. Clearly, in such cases the scanning system can be driven by any means (DC, AC, stepper, synchronous motor, etc.) provided that a phase pulse is available to drive the imager. For example, one may choose to fit an optical encoder to the scanning system, thus using the actual position of the scanner for maximum phase accuracy. In such a system, it would be necessary to provide phase adjustment, so as to coordinate the operation of the imager (for example, shuttering) with the pulsing and/or sweeping of the laser beam. Such a system was adopted by Brücker (1997).

A third approach is to slave both the scanner and imager to a common external clock. Again, it will generally be necessary to provide phase adjustment between the clock signals and each device; this may take the form of an adjustable phase delay on the clock input to one of the devices.

10.11.6   Operational considerations

Imaging optics

Three important issues have to be addressed with respect to the imaging optics. The first is the depth of field, given by

$$\delta z = 4 \left(1 + M^{-1}\right)^2 (f^\#)^2 \lambda$$

where M is the magnification of the lens, $f^\#$ is the f-number of the lens and λ is the wavelength of the light. For 3-D volume visualization the depth of field required can be up to two orders of magnitude greater that required for a 2-D planar imaging system. Thus, if a high magnification ($M > 1$) is used to achieve a high spatial resolution, a high f-number will be needed to achieve the required depth of field, meaning that little light will be available to the imager. This will also lead to a larger diffraction limit for the resolution of particle and dye images. On the other hand, if a small f-number is used to achieve adequate saturation of the imager and the same depth of field, a small magnification ($M < 1$) will be required, which delivers poor spatial resolution. The diffraction limit will accordingly be smaller. Clearly, every experimental set-up will have its own unique compromise between these competing factors. Adrian (1991) provides further details.

The second issue relates to variations in magnification due to the differences in distances of the visualized planes from the imager. If the depth of field is sufficient to encompass the entire measurement volume, then this effect is expected to be small. If this is not the case, then each image must be individually calibrated to account for magnification differences, especially if the data is to be used for quantitative purposes. Magnification and indeed image distortion should be systematically checked and accounted for with every new set-up. An accurately drawn grid from a drawing equipment supplier can be used as an effective calibration tool.

Third, when the laser sheet thickness is significant with respect to the distance of the visualized plane from the imager, and/or the angular range of the imaging lens is large, the measurements of in-plane displacements will be contaminated with out-of-plane displacements. In the case of dye visualization, the dye patterns will be distorted around the edges and corners of the image due to the imager viewing the plane somewhat edge-on. In the case of PIV, the in-plane motions will be contaminated by out-of-plane motions, resulting in errors in inferred velocity. This problem is minimized by the use of thin laser sheets and/or long focal length lenses. Further discussion can be found in Adrian (1991).

Spatial and temporal resolution

The spatial and temporal resolution required depend upon what information is sought. The parameter that dictates the temporal (or time) resolution of the system as a whole is the time to perform a complete volume scan, namely the duration of one complete time step, T_{step}. This relates to how well the volume is "frozen," or specifically, the convection and spatial evolution of the pattern between and during successive time-steps (volume scans). The spatial resolution relates to the smallest volume that can be resolved within the region of interest. This is dictated by the dimensions of the region, the resolution of the imager (usually the number of pixels in the array), the characteristics of the optical elements, and the thickness and spacing of the imaged planes. In most turbulent flows of interest it is unlikely that a single imaging device can resolve motions of all relevant length scales within the flow. The spatial and temporal resolution achieved by a volume scanning system represents a compromise between competing factors. There must be a trade-off to achieve the best possible compromise for each experimental application. These issues will now be discussed.

The spatial resolution of a stack of 2-D images is dependent on several parameters. The spatial resolution of each image in the stack depends on the dimensions of the region of interest, the resolving power of the imager (pixel array size) and the optics employed. In addition, imaging a light sheet performs a visual integration along the line of sight of the imager, normally over the thickness of the light sheet itself. It is therefore desirable to use the thinnest light sheets practicable to illuminate the volume, while still providing adequate illumination. The spacing of the sheets introduces another, more stringent restriction on the spatial resolution in the direction normal to the light sheets. Reducing the number of sheets (without changing the spacing) in order to speed up volume acquisition (improve time resolution), reduces the spatial extent of the volume. Furthermore, increasing the size of the volume by spacing the sheets further apart, leaving un-scanned volume between the sheets, decreases the spatial resolution in the direction normal to the sheet, and will usually require the use of an interpolation scheme to "fill-in" the missing information. The method of scanning is also important as it relates to convection of the flow during the imaging of each image plane, as discussed in Section 10.11.3. The retrace time also impacts on the time resolution as it represents a time-delay between individual laser sweeps or between successive volume scans. Imaging systems that minimise the retrace time are generally preferable.

In order to match a 3-D imaging system to a given flow, the temporal and spatial resolution requirements must be established. For example, a system to resolve the large-scale features of a flow, the field of view must encompass the largest length scale within the flow — equal to the jet width or body width (L) or boundary layer thickness (δ) as the case may be. The time between volumes, T_{step}, must also be smaller than the relevant flow time scale — the eddy advection time (L/U) or outer time scale δ/U, also known as the eddy turnover time. The Nyquist criterion requires that T_{step} be less than half of the relevant flow time scale for unambiguous observations to be made. This applies to both PIV and scalar field investigations.

If the smallest scales of a turbulent flow are of interest, the required resolution of a dye-marked flow will depend on the scalar diffusivity of the dye, rather than the momentum diffusivity. If the diffusivity of the scalar marker is significantly lower than the momentum diffusivity, that is, the Schmidt number Sc ($= \nu/D$) is high, the gradients resolvable within the scalar field will be significantly smaller than the Kolmogorov length scale η ($= (\nu^3/\varepsilon)^{1/4}$). Thus, it is necessary to resolve to significantly smaller length scales, known as the diffusion length scale or Batchelor scale λ_D ($= (D/\varepsilon)^{1/2}$). The corresponding time scale for this case is the local molecular diffusion advection time λ_D/U. This concept is discussed in more detail in Chapter 11. In the case of a boundary layer where the inner flow structures are to be resolved, the imaging system should have (at minimum) sufficient spatial resolution to capture the smallest motions, of order $10\nu/u_\tau$, and sufficient time resolution to capture the smallest time scales, of order $\nu/(u_\tau^2)$. In all of these cases the Nyquist criterion sets the maximum spatial or temporal frequency required to avoid aliasing in the sampled data.

In the final analysis the process of system selection and design depends initially on the flow case to be studied, the type of data sought, and the imager and laser source to be used (or money available to buy them). The rest of the choices (scanning system design, number and spacing of light sheets, etc.) must then follow.

When applying the 3-D visualization method to PIV, the situation is more complex. Detailed descriptions of the issues for two different systems are given in Brücker (1995b, 1997b). Two time-scales must be considered for the application of this technique. Firstly, there is the duration of the complete volume scan,

T_{step}, as discussed above. This must be small with respect to the timescale of the overall flow. Secondly, there is the time required between successive images at each image plane, T_{piv}, as dictated by factors such as the spatial resolution of the imaging device, the magnification of the optical system and the velocities to be resolved (see Chapter 6). As pointed out by Brücker (1995b), the ideal system is one where the speed of the scanning system and imager would be sufficiently high such the T_{step} and T_{piv} can be the same. Should this not be possible, it will be necessary to arrange a different imaging pattern to provide T_{piv} less than T_{step}; for example, two images per plane per time step as shown in Figs. 10.2 and 10.5. In some systems, such as those using galvanometer mirrors, the frequency response and retrace time of the optics will define the minimum value of T_{piv}. The use of polygon mirrors, prism scanners and drum scanners overcomes this limitation due to the higher speed capability of the continuously rotating components and the inherently short retrace time.

Given the above, all the principles that apply to single-plane PIV also apply to 3-D whole volume PIV. These issues are discussed in Chapter 6 and by Adrian (1991) and Rockwell *et al.* (1993). For information on flow scales the reader is referred to Landahl & Mollo-Christensen (1992) and Tennekes & Lumley (1973).

10.11.7   Imaging devices

The choice of imaging device is all important in determining the performance of 3-D imaging systems. For a given illumination budget, the imager ultimately dictates the maximum volume acquisition rate, either directly through its maximum frame rate limitation or indirectly through its light sensitivity limitation. With a given cost and technology, the choice of imager usually requires a trade-off between frame rate and resolution, that is, either fast frame rate and low resolution or slow frame rate and high resolution. Among the mainstream technologies, the highest resolution is obtained from photographic film, and high frame rates can be achieved by using sensitive emulsions. The main drawback is the time required for processing the film and digitizing the images. Short duration image sequences are possible at frame rates upwards of 30,000 frames/s (fps) using drum cameras. Longer duration sequences can be obtained at a lower frame rate, typically 100 to 10,000 fps, using fast framing film cameras. The most convenient technology is video, where frame rates between 25 and 3000 fps are available. Standard video (25 or 30 Hz) is of limited use, although it can be improved in time resolution (50 or 60 Hz) at the expense of spatial resolution by separating the interlaced fields (eg. Brücker 1992). At the time of writing, analog or digital high frame rate Charge-Coupled Device (CCD) cameras are typically capable of imaging rates up to 3000 fps at imager array sizes up to 512 × 512 pixels. Depending on the storage technology, these can store images in numbers ranging from 256 to 4096 (digital) or 40,000 (analog). Faster rates are obtained using electronic framing cameras which act as intermediate storage devices, storing a limited number of frames which are then imaged by CCD detectors.

When choosing an imaging device, the following issues should be considered: light sensitivity, resolution (size of pixel array), pixel fill factor (proportion of sensor filled pixels), pixel shape, image (or frame) rate, shutter speed adjustability, number of frames stored, and the availability of a frame sync pulse input and output.

An important issue generally, but especially when slaving the scanning system to the imager, is that of frame rate jitter, hence frame sync pulse jitter. Significant jitter is not unusual in mechanical systems (cine or analog video) and can lead to a number of problems, from the inability to maintain scanner synchronization during operation, through to unacceptably large errors in PIV measurements. Such problems are not expected in digital systems. Unfortunately, manufacturers will not always be aware of the jitter problem, so it is useful to consult with prior purchasers of the imager, arrange for a demonstration, try-before-you-buy or rent before you buy.

10.12 Detailed Example

The following is a detailed discussion of one imaging system developed jointly by the authors at the Gas Dynamics Laboratory, Princeton University. It is hoped that the discussion will illustrate some of the points raised earlier and also provide useful ideas for future developments. The details of the study are given by Delo & Smits (1997).

The purpose of this apparatus was to study a low Reynolds number, nominally zero pressure gradient, turbulent boundary layer. Experiments were conducted in a closed-loop, free-surface apparatus with a full-width flat plate positioned in the test section. The experimental configuration is shown in Fig. 10.8. Flow visualization was accomplished using disodium fluorescein dye introduced from two spanwise dye slots located 39 and 4.7 boundary layer thicknesses upstream of the leading edge of the measurement volume. The concentration of the dye introduced from the slots was 250 and 500 ppm (by weight) respectively. The free stream velocity was $U = 229 \, \text{mm/s}$ and the boundary layer thickness at

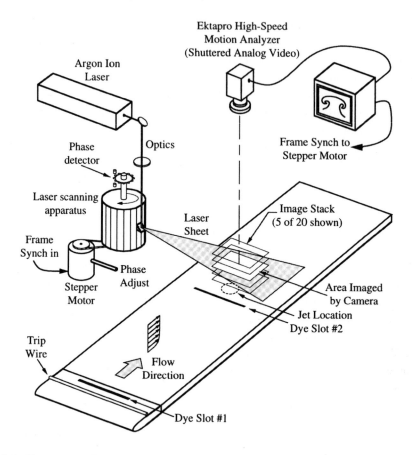

Fig. 10.8. Experimental arrangement used to collect the volumetric data set (Delo & Smits, 1997; Kelso *et al.*, 1993, 1995).

the upstream edge of the volume was $\delta = 26.9$ mm. The Reynolds number based on momentum thickness was 701, the friction velocity was $u_\tau = 11.1$ mm/s, and the Kármán number $(u_\tau \delta / \nu)$ was 299. The useful portion of the interrogation volume had dimensions of $L_x/\delta = 3.53$, $L_y/\delta = 1.49$, and $L_z/\delta = 3.34$ (in viscous units: $L_x^+ = 1054$, $L_y^+ = 444$, $L_z^+ = 999$).

The volume was interrogated using the rotating drum type laser scanning apparatus shown in Fig. 10.6. A stack of twenty laser sheets was formed by sweeping a focused laser beam through the flow volume parallel to the flat plate (in *x-z* planes) at twenty *y*-locations. To sweep the beam, a helical array of

45 mirrors was fixed to twenty faceted faces of the rotating drum. The focused beam of a CW Argon-ion laser operating in single line mode (501 nm, 1.8 W nominal power) was directed parallel to the axis of the drum, and reflected off each mirror as the drum rotated. The rotation of the flat mirror face caused the reflected beam to sweep through an angle of 18°. The resulting laser sheet had uniform intensity, and its y-location (determined by the position of the mirror on the drum) was precisely repeatable. As the drum continued to turn, the beam reflected off the next mirror in the helix, forming another sheet at a different location. To minimize reflections from the flat plate, the bottom x-z laser sheet was set at $y = 2$ mm; to facilitate the volumetric reconstructions, the sheets had uniform separation of 2 mm in the y-direction ($\Delta y/\delta = 0.074$, $\Delta y^+ = 22.2$).

The scanner was driven by a stepper motor synchronized to the frame-rate signal from the analog video imager. The laser sheets were imaged from directly overhead with a Kodak/Spin Physics Ektapro 1000 High-speed Motion Analyzer fitted with a 12.5 to 75 mm $f1.8$ zoom lens. Images were acquired at 500 fps, yielding 25 full volumes per second. The rotation of the drum was such that the top slice of each volume was imaged first, the next lower slice 0.002 s later, and so on. The elapsed time between subsequent volumes (0.04 s) corresponded to approximately one third of a characteristic "eddy turnover time" ($\Delta t U_e/\delta = 0.34$; $\Delta t u_\tau^2/\nu = 4.9$). The time resolution was therefore adequate for the examination of large-scale coherent structures, and no interpolation in time was performed.

The same laser scanning system and water channel facility were used to investigate the structure of the wake of a transverse jet (Kelso *et al.* 1993, 1995). The arrangement is also described in Fig. 10.8. The jet, having a top-hat velocity profile in the case of no cross-flow, discharged normally from the surface of a horizontal flat plate mounted above the channel floor. The experiments were performed with a jet diameter of 25 mm and a free-stream velocity of 150 mm/s. The jet exit was located 1.1 m from the leading edge of a flat plate, giving a laminar boundary layer layer (no trip wire was used) with a thickness $\delta = 13$ mm immediately upstream of the jet. The Reynolds number was 3800 based on the free-stream velocity in the channel and the diameter of the jet. The jet-to-cross-flow velocity ratio was 4.3 based on the average jet velocity. Fluoresceine dye (500 ppm concentration) was injected into the boundary layer of the flat plate from two spanwise slots located upstream of the jet. The imaged volume measured 96 mm in the streamwise direction by 120 mm in the spanwise direction. There were 20 horizontal laser sheets, spaced 2 mm apart with the lowest sheet at 1 mm above the flat plate and the highest 39 mm above the plate. The framing rate of the camera was 250 frames per second, corresponding to 12.5 time steps (drum rotations) per second. The resolution of the imager was 238×192, thus providing spatial resolution to 0.5 mm, sufficient to resolve all large-scale features unambiguously.

The requirements of this experiment were somewhat less stringent than those of the boundary layer experiment. The experiments were designed to investigate the wake vortex roll-up process. This process was known to involve the separation and roll-up of the flat plate boundary layer on the downstream side of the jet. The aim was to track the roll-up and convection of the vortices through space in order to determined how they formed and interacted. The details of the scalar and velocity fields were not sought.

The time scale of separated boundary layer vortex $= \delta/U_e = 0.087$ s, and the time scale of wake vortex $= D/U_e = 0.167$ s. At a scanning rate of 12.5 time-steps per second (250 images/s) the time step was 0.08 s. Thus, it would appear that the scanning rate was just sufficient to resolve the detailed structure of the wake vortices, having a time step of approximately half of the flow time scale, but the scanning rate was insufficient to resolve the details of the boundary layer roll-up process. However, an additional consideration was the average period of the wake, 1.4 s, which was resolved during a separate experiment. If we assume that the period between boundary layer separation events was similar to this, a time-step of 0.08 s seems adequate to track the vortices through space.

10.12.1   Control system design

As mentioned above, the most difficult way to control a scanning system is to slave it to the imager. In the present case where the imaging device (an analog Kodak Ektapro 1000) could not be driven from an external synchronization (sync) pulse, it was necessary to use the imager to drive the scanning system. An additional complication was that the Ektapro system would generate a frame pulse only when the system reached the target operational speed. Thus, it was necessary to develop a system where the scanner was already operating close to the target speed before the imager began. When the imager was operating at the correct speed, the scanner source could be switched over to the imager sync pulse. An additional requirement was that the system should be simple and inexpensive.

The system that was used to provide frequency and phase lock to the imager sync pulse is described in Fig. 10.9. The circuit used a phase-locked-loop with a divide-by-10 in the feedback loop to lock onto and multiply the sync frequency by a factor of 10. The circuit also incorporated a simultaneous electronic switch-over between "dummy" and imager sync signals. In practice, the laser scanning system was first brought up to a speed 5% higher than the required operating speed using a dummy pulse generator. The imager system was then instructed to commence. Once up to full operational speed, the imager output sync pulse was initiated and the sync pulse input of the driver circuit was switched over to the imager sync pulse. The stepper motor speed fell to meet and lock to the new pulse frequency. A mechanical adjustment to the phase of the scanner system was then made to coordinate the beam sweeps with the electronic shutter of the imager. This was achieved by comparing the phase of the imager's sync pulse with the output from a phase detector (optical encoder) on the drum scanner shaft. The adjustment was effected by mechanically rotating the driving stepper motor within its mounts. The process of synchronization took approximately 20 s and consumed approximately 25% of the operating time.

The driver circuit included simple R-C filters to provide damping and slew rate limiting to limit the rate of change of output frequency. These were op-timised to provide stable operation of the stepper motor. The stepper motor itself was a high torque, high inductance unit that was designed for low fre-quency operation. It was also driven by a rudimentary controller. Thus, in order to operate at frequencies up to 12.5 rps it needed to be driven using high voltage (96 V DC) using large series resistors to limit the current. At the full operating speed the motor was susceptible to electronic noise and signal jitter. This was the reason for driving the scanner system to a speed 5% higher than the required operating speed using a dummy sync source prior to switch-over to the imager sync pulse source. If the dummy source was at or below the required frequency, the jitter introduced during the source-switch was sufficient to cause the stepper motor to miss pulses and stall.

It should also be pointed out that the system was successfully driven by a high-speed prism camera operating at 250 fps as well as a standard 30 Hz video camera. In the latter case the sync pulse was obtained from a composite video signal which was low-pass filtered to pass the 60 Hz field pulse. This was buffered, conditioned and frequency halved to provide a 30 Hz frame pulse. This can also be achieved using readily available video signal stripping microchips.

Once recorded onto the Kodak analog video tape, the two-dimensional im-ages were downloaded onto video tape at 1 fps, then transferred to a Personal Iris workstation using a Panasonic AG-6500 editing video cassette recorder and an Imaging Technology Series 151 frame grabber. Wyndham Hannaway image processing software was used to control the frame grabber and to enhance the

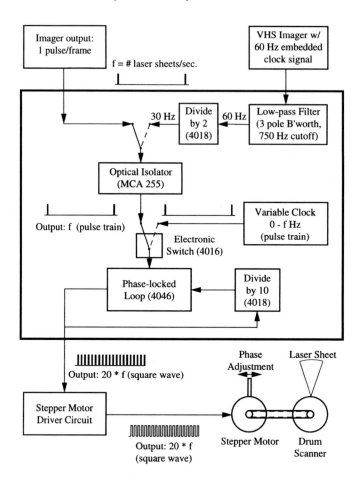

Fig. 10.9. Schematic representation of the circuit used to drive and synchronize the laser scanning system used by Delo & Smits (1997) and Kelso *et al.* (1993, 1995). The semiconductor chips used are indicated in most cases.

images. Reconstruction of the images into 3-D volumetric images was achieved using software described below and in Delo *et al.* (1994).

10.13 Analysis and Display of Data

10.13.1 Processing and analysis of data

The analysis of volumetric data is, for the most part, identical to that of planar data. All of the techniques described above, holography included, involve the representation of the volumetric data as a series of slices. These are enhanced and analysed in much the same way as 2-D data obtained from planar measurements. Many of the techniques used for processing planar concentration and PIV data are described by Hesselink (1988). Detailed discussion of image processing for scalar images is provided in Garcia & Hesselink (1986) and Pratt (1991). A more current treatment can be found in Chapters 6 and 11.

Specific reference should be made here to some additional issues that are not explicitly dealt with in the context of planar measurements. Firstly, there is the issue of the third (or out-of-plane) velocity component in 3-D PIV. This has been dealt with in a number of ways, including the use of the continuity equation (Robinson & Rockwell, 1993), spatial correlation by color coded sheets (Brücker, 1996) and stereoscopy (Brücker, 1995a). Secondly, 3-D scalar measurements have recently been used to calculate velocity field data. These approaches include "Scalar Image Velocimetry" (Dahm *et al.*, 1992) and "Image Correlation Velocimetry" (Tokumaru & Dimotakis, 1995; Deusch *et al.*, 1996). These methods have been used successfully to extract information such as the velocity field, vorticity and rate of strain directly from scalar field.

Having obtained and then processed the data, whether velocity or scalar fields, the data can be re-sampled to generate data fields in any sectional plane. This is akin to visualizing or measuring the flow field itself, within the limitations of the time and spatial resolution of the data set. In this case, interpolation within the data grid may be necessary. It should be noted that individual planes in a volume are imaged at different times, and, if the flow is convecting as a whole, the flow pattern moves between each plane. If a mean velocity gradient exist within the flow, as was the case in the boundary layer investigation of Delo & Smits (1997), the convection velocity will vary with height and will have to be corrected for by measuring the mean velocity profile. Such a velocity profile can, of course, be obtained from the 3-D volumetric data itself by cross-correlating successive images on each plane. Clearly, the magnitude of the convection effect is dependent on the ratio between T_{step} and the appropriate time scale of the flow. Fo r the purposes of qualitative visualization, small corrections for convection effects may not be necessary. For quantitative measurements, however, corrections should be made whenever practical.

Significant convection effects will play an additional role in selecting the orientation of the image planes within the scanned volume. In the boundary layer investigation, the image planes were chosen to be parallel to the wall. The mean velocity was therefore uniform within each plane and correction for convection was straightforward; it was accomplished by a simple downstream translation of image planes, without altering their position relative to the wall. A similar orientation, with image planes parallel to the mean flow and normal to the mean velocity gradient, would be indicated for mixing layers, flow about a long spanwise cylinder, and so forth. For axisymmetric flows (for example, a co-flowing jet, flow about a sphere, etc.), it may be more appropriate to orient the image planes normal to the mean flow. Corrections for mean convection would then be manifested as a normal translation of the image planes, that is, a relatively simple adjustment to the spacing between planes. Clearly the specifics of an individual flow situation will dictate the choice of scanning pattern.

10.13.2   Methods of presentation and display

It is logical to assume that a complete 3-D visualization of a flow field will reveal the complex spatial interrelationships in the flow most effectively. A full volumetric view may be sub-sampled into standard two-dimensional renderings, but knowledge of the entire volume may be necessary in order for the 2-D sampling to effectively reveal the important dynamics. Thus it is necessary to present the data in an appropriate three-dimensional manner. There are a number of ways to achieve this, as will now be described.

Velocity field data are arguably the most difficult to present in a meaningful way. Whereas planar velocity data are meaningfully represented by velocity vectors, streamlines and contour plots, the data becomes obscured when this is attempted for an entire volume. It is therefore more usual, and more effective, to present data as surface contours of quantities such as velocity and vorticity, as streamsurfaces, or as a series of meaningful and indicative streamlines. In the case of concentration data, it is usual to present the data as 3-D iso-surfaces of concentration, or as translucent, cloud-like images using the source-attenuation method. Hesselink (1988), Yoda & Hesselink (1989), Russell & Miles (1987), and Delo *et al.* (1994) provide a useful coverage of these techniques.

Recent advances in the presentation of 3-D volumetric data can be attributed substantially to advances in non-intrusive medical imaging techniques. Many of the computer applications that are typically employed to visualize volumetric data sets had their beginning in the field of medical imaging. An example is "NIH Image," a freeware volumetric image processing package developed at the U.S. National Institutes of Health. NIH Image was designed specifically to visualize and analyze tomographic data. Programs for 3-D display, animation and volume rendering can now be obtained for most computer platforms, ranging from freeware to comprehensive, expensive packages. Furthermore, one should not overlook standard computer aided design packages which offer features such as volume rendering (Ruck & Pavlovski, 1998).

Perhaps the greatest challenge of all is in the final presentation of the data itself. The most tangible of these is the solid model. Perry & Lim (1978) used wire and wooden spoon models to describe their observations. Many recently developed 3-D visualization systems lend themselves to the construction of solid model using slices milled from flat sheet material or direct formation using the rapid prototyping technique.

In the print medium there are four methods that have been successfully employed. The first is the volume rendered image, where perspective, surface texture and cut-away sections provide effective depth cues. Hesselink (1988) provides a good example of such an image. The second method is the hologram, of which Hesselink also provides an example. The third method is one where stereoscopic views are generated in the form of simple stereograms, or stereo pairs, where the images (as seen by each eye) are placed side-by-side. The stereo-pair images are then viewed through simple corrective lenses (or without, after some practice) to provide the stereo effect. Figure 10.10 gives an example of a stereogram of the wake of a jet in cross-flow.

The fourth method is one where the stereo pair views are combined in the form of "anaglyph stereograms." Here, two separate two-dimensional projections of a volume are generated from viewpoints corresponding to human binocular vision. The separate views are colored, combined, and viewed on the computer screen or in print with colored glasses. The glasses act as filters, causing the correct view to be presented to each eye. The optical synthesis of the two views results in the stereoscopic effect. Many examples of anaglyph images can be found in Delo & Smits (1997). The paired projections of the volumetric data used to create the stereoscopic visualizations were calculated with "3Dviewer5.4," a volume rendering program developed by Delo *et al.* (1994) for the creation of stereoscopic visualizations. The program generates true translucent volumetric views of an image stack using a ray-tracing method, using a source-attenuation model (Beer's Law) to create a pair of monochrome projections of a stack of two-dimensional images. It includes a range of variable viewing parameters

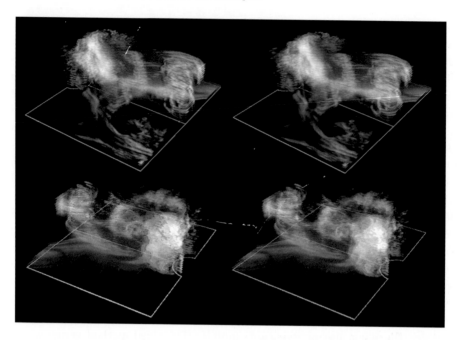

Fig. 10.10. Three-dimensional reconstructions (stereo pairs) of the wake of a jet in cross-flow. The upper and lower pairs represent different views of the same time-step for the case of $Re = 3800$, $\delta/D = 0.5$ and $R = 4$. In the upper pair the cross-flow is towards the bottom right. In the lower pair the cross-flow is towards the top left. The flat wall and rear half of the jet are outlined.

including: opacity, perspective, angle of orientation, projection plane location, viewing distance and binocular parallax angle. The projections from the two viewpoints are calculated separately, then combined to form either stereo pairs or anaglyph stereograms.

Construction of anaglyph stereograms from paired projections is straightforward. A three-color RGB (Red-Green-Blue) image is constructed from the monochrome stereo pair images: the left eye view is copied to the red band of the image, the right eye view is copied to the blue and green bands, resulting in a red/cyan anaglyph image. After the monochrome views are combined, the color stereograms can be enhanced using a "gamma" histogram adjustment, an increase in color saturation and/or an image sharpening routine. Each of these processes improves the perception of depth in the stereograms. It should be pointed out that the generation, or reconstruction of such flow patterns to provide intelligible images requires that the individual planar images first be "cleaned up." This process may involve thresholding the data to remove background noise from un-marked fluid, gray-scale adjustments to ensure that the data fills the dynamic range of the images (for example, 8-bit). Attention to such pre-processing will ensure that the features of interest are not obscured by a "fog" of noise.

The most powerful tool available to visualize 3-D data sets is movement (Russell & Miles, 1987). Animations, be they rotating images of a dye pattern, or a time-series showing the evolution of an eddy, can be the most effective presentation tool of all. When combined with a stereoscopic viewing method, the effect can be stunning. There are currently very few methods that allow this to be achieved using a computer monitor or conventional television. The simplest of these is the anaglyph stereogram method which requires that the screen be viewed using red/cyan colored glasses. A more complex method uses glasses that shutter the viewer's eyes in phase with the screen refresh rate of the computer monitor in such a way that each eye receives a different series of images. A product called "Flow Eyes" is one example of such a system.

10.14   Concluding remarks

We have attempted to provide an overview of the methods used to design, construct and operate an affordable 3-dimensional visualisation system. The design of such a system is clearly a trade-off between illumination, spatial resolution, temporal resolution, experimental requirements and many other factors, most notably, cost. These conflicting requirements have resulted in a wide variety of systems, a small number which have been described here. Rapidly advancing imaging technology will undoubtedly result in a growth in the use of 3-D imaging systems in the future. The authors hope that the present chapter will contribute in a positive way towards this advancement.

10.15   Acknowledgments

The authors acknowledge the support of the Commonwealth Scientific and Industrial research Organisation (Australia), the Australian Research Council, the Fannie and John Hertz Foundation (USA), and from the National Science Foundation (USA). Thanks are due to many supporters of this work including R. B. Miles, J. P. Poggie, P. R. E. Cutler, T. T. Lim, A. J. Smits, F. A. Brake and K. V. McKenzie for their support and encouragement. We also wish to acknowledge the contributions and tolerance of T. T. Lim and A. J. Smits in their roles as the editors of this book.

10.16　References

Arndt S., Heinen C., Hubel M. and Reymann K. 1998. Multi-colour laser light sheet tomography (MLT) for recording and evaluation of three-dimensional turbulent flow structures. *Proc. IMechE International Conference on Optical Methods and Data processing in Heat and Fluid Flow*, London, Paper No. C541/005/98, 481–489.

Adrian R. J. 1991. Particle-imaging techniques for experimental fluid mechanics. *Ann. Rev. Fluid Mech.*, **23**, 261–304.

Barnhart D. H., Adrian R. J. and Papen G. C. 1994. Phase-conjugate holographic system for high-resolution particle-image velocimetry. *Appl. Opt.*, **33**, 7159–7170.

Blackshire J. L., Humphreys W. M. and Bartram S. M. 1994. 3-Dimensional, 3-Component velocity measurements using holographic particle image velocimetry (HPIV). *Proc 18th AIAA Aerospace Ground Testing Conference*, Colorado.

Brücker C. 1992. Study of vortec breakdown by particle tracking velocimetry (PTV). Part 1: Bubble-type vortex breakdown. *Exp. Fluids*, **13**, 339–349.

Brücker C. 1995a. 3D-PIV using stereoscopy and a scanning light sheet: Application to the 3D unsteady sphere wake flow. In *Flow Visualization VII*, ed. J. Crowder, Bengell House, 715–720.

Brücker C. 1995b. Digital-Particle-Image-Velocimetry (DPIV) in a scanning light-sheet: 3D starting flow around a short cylinder. *Exp. Fluids*, **19**, 255–263.

Brücker C. 1995c. Study of the 3-D flow in a T-junction using a dual-scanning method for 3-D Scanning-Particle-Image-Velocimetry (3-D SPIV). In *Turbulent Shear Flows*, **10**, 7-19–24.

Brücker C. 1996. A new method for determination of the out-of-plane component in three-dimensional PIV using a colour-coded light-sheet and spatial correlation: simulation and feasibility study for three-dimensional scanning PIV. *Proc. IMechE International Seminar on Optical Methods and Data processing in Heat and Fluid Flow*, Paper No. C516/014/96, 189–199.

Brücker C. 1997a. Study of the 3-D flow in a T-junction using a dual-scanning method for 3-D Scanning-Particle-Image-Velocimetry (3-D SPIV). *Exp. Thermal Fluid Science*, **14**, 35–44.

Brücker C. 1997b. 3D scanning PIV applied to an air flow in a motored engine using digital high-speed video. *Meas. Sci. Technol.*, **8**, 1480–1492.

Brücker C. 1998.   Time-recording scanning-particle-image-velocimetry (SPIV) technique for the study of bubble-wake interaction in bubbly two-phase flows. *Proc. IMechE International Conference on Optical Methods and Data processing in Heat and Fluid Flow*, London, Paper No. C541/064/98, 31–40.

Buch K. A. and Dahm W. J. A. 1996. Experimental study of the fine-scale structure of conserved scalar mixing in turbulent shear flows. Part 1. Sc ¿¿1. *J. Fluid Mech.*, **317**, 21–71.

Cutler P. R. E. and Kelso R. M. 1997. Private communication.

Dahm W. J. A., Su L. K. and Southerland K. B. 1992. A scalar imaging velocimetry technique for fully resolved four-dimensional vector velocity field measurements in turbulent flows. *Phys. Fluids A*, **4**, 2191–2206.

Delo C., Poggie J. and Smits A. J. 1994. A system for imaging and displaying three-dimensional, time-evolving passive scalar concentration fields in fluid flow. *Tech. Rept. 1992*, Mech. & Aerosp. Eng. Dept., Princeton University.

Delo C. and Smits A. J. 1993. Visualization of the three-dimensional, time-evolving scalar concentration field in a low Reynolds number turbulent boundary layer. In *Near-Wall Turbulent Flows*, ed. C. G. Speziale and B. E. Launder, Elsevier Science Publishers, 573–582.

Delo C. and Smits A. J. 1997. Volumetric visualization of coherent structure in a low Reynolds number turbulent boundary layer. *Int. J. Fluid Dynamics*, **1**, Article 3. Available at: http://elecpress.monash.edu.au/ijfd/index/html

Deusch S., Dracos T. and Rhys P. 1996. Dynamical flow tomography by laser induced fluorescence. In *Three Dimensional Velocity and Vorticity Measuring and Image Analysis Techniques*, Kluwer Academic Publishers, 277–297.

Eckbreth A. C. 1988. *Laser Diagnostics for Combustion Temperature and Species*. Abacus Press, Cambridge, MA.

Forkey J. N., Lempert W. R., Bogdonoff S. M., Miles R. B. and Russell G. 1994. Volumetric imaging of supersonic boundary layers using filtered Rayleigh scattering background suppression. *AIAA 32nd Aerospace Sciences Meeting & Exibit*, Reno, NV.

Garcia J. C. A. and Hesselink L. 1986. 3-D reconstruction of flow visualization images. In *Flow Visualization IV*, ed. C. Veret, Hemisphere, 235–240.

Goldstein J. E. and Smits A. J. 1982. Flow visualization of the three-dimensional, time-evolving structure of a turbulent boundary layer. *Phys. Fluids*, **6**, 577–587.

Guezennec Y. C., Zhao Y. and Gieseke T. 1996. High-speed 3-D scanning particle image velocimetry technique. In *Developments in Laser Techniques and Applications to Fluid Mechanics*, ed. R. J. Adrian, Springer-Verlag, Berlin, 392.

Hesselink L. 1988. Digital Image Processing in flow visualization. *Ann. Rev. Fluid Mech.*, **20**, 421–485.

Hinsch K. D. 1995. Three-dimensional particle velocimetry. *Meas. Sci. Technol.*, **6**, 742–753.

Hussain F., Meng H., Liu D., Zimin V., Simmons S., and Zhou C. 1994. Recent innovations in holographic particle velocimetry. *Proc. 7th ONR Propulsion Meeting*, 233–249.

Kelso R. M., Delo C. and Smits A. J. 1993. Unsteady wake structures in transverse jets. *AGARD CP-534*, Paper No. 4.

Kelso R. M., Delo C. and Smits A. J. 1995. An experimental study of the flow around a transverse jet. In *Flow Visualization VII*, ed. J. Crowder, Bengell House, 452–460.

Landahl M. T. and Mollo-Christensen E. 1992. *Turbulence and Random Processes in Fluid Mechanics*. 2nd ed., CUP.

Mantzaras J., Felton P. G. and Bracco F. V. 1988. Three-dimensional visualization of premixed-charge engine flames: islands of reactants and products; fractal dimensions; and homogeneity. SAE/SP-88/759 *Proc. Intern. Fuels and Lubricants Meeting and Exposition*, Portland Oregon.

Meinhart C. D., Barnhart D. H. and Adrian R. J. 1994. An interrogation and vector validation system for holographic particle image fields. *Proc. 7th Int. Symp. Appl. of Laser Techniques to Fluid Mech.*, Lisbon, 1.4.1–1.4.6.

Meng H. and Hussain F. 1991. Holographic particle velocimetry: a 3D measurement technique for vortex interactions, coherent structures and turbulence. *Fluid. Dyn. Res.*, **8**, 33–52.

Meng H. and Hussain F. 1995. In-line recording and off-axis viewing (IROV) technique for holographic particle velocimetry. *Appl. Opt.*, **34**, 1827–1840.

Merkel G. J., Rys F. S., Rys P. and Dracos T. A. 1995. Concentration and velocity field measurements in turbulent flows using Laser Induced Fluorescence (LIF) tomography. In *Flow Visualization VII*, ed. J. Crowder, Bengell House, 504–509.

Miles R. B. and Nosenchuck D. M. 1989. Three-dimensional quantitative flow diagnostics. In *Advances in Fluid Mechanics Measurements*, ed. M. Gad-el Hak, Springer-Verlag, NY, 33–107.

Nosenchuck D. M. and Lynch M. K. 1986. Three-dimensional flow visualization using laser-sheet scanning. *AGARD CP-413, 18-1–13*.

Perry A. E. and Lim T. T. 1978. Coherent structures in coflowing jets and wakes. J. Fluid Mech., **88**, 451–463.

Porcar R., Prenel J. P., Diemunsch G. and Hamelin P. 1983. Visualizations by means of coherent light sheets; applications to various flows. In *Flow Visualization III*, ed. W. J. Yang, Hemisphere, 123–127.

Prasad R. R. and Adrian R. J. 1993. Stereoscopic particle image velocimetry applied to liquid flows. *Exp. Fluids*, **15**, 49–60.

Prasad R. R. and Sreenivasan K. R. 1989. Scalar interfaces in digital images of turbulent flows. *Exp. Fluids*, **7**, 259–264.

Prasad R. R. and Sreenivasan K. R. 1990. Quantitative three-dimensional imaging and the structure of passive scalar fields in fully turbulent flows. *J. Fluid Mech.*, **216**, 1–34.

Pratt W. K. 1991. *Digital Image Processing*. 2nd ed., Wiley, NY.

Praturi A. K. and Brodkey R. S. 1978. A stereoscopic visual study of coherent structures in turbulent shear flow. *J. Fluid Mech.*, **89**, 251–272.

Prenel J. P., Porcar R. and Diemunsch G. 1986a. Visualizations by means of coherent light sheets; applications to various flows. In *Flow Visualization IV*, ed. C. Veret, Hemisphere, 299–103.

Prenel J. P., Porcar R. and Diemunsch G. 1986b. Visualisations tridimensionnelles d'écoulements non axisymetriques par balayage programme d'un faisceau laser. *Optics Comm.*, **59**, 92–96.

Prenel J. P., Porcar R. and El Rhassouli A. 1989. Three-dimensional flow analysis by means of sequential and volumic laser sheet illumination. *Exp. Fluids*, **7**, 133–137.

Robinson O. and Rockwell D. 1993. Construction of three-dimensional images of flow structure via particle tracking techniques. *Exp. Fluids*, **14**, 257–270.

Rockwell D, Magness C., Towfighi J. and Corcoran T. 1993. High image-density particle image velocimetry using laser scanning techniques. *Exp. Fluids*, **14**, 181–192.

Ruck B. and Pavlovski B. A. 1998. A fast laser-tomography system for flow analysis. *Proc. IMechE International Conference on Optical Methods and Data processing in Heat and Fluid Flow*, London, Paper No. C541/032/98, 465–473.

Russell G. and Miles R. B. 1987. Display and perception of 3-D space-filling data. *App. Opt.*, **26**, (6), 973–982.

Schluter T., Merzkirch W. and Kalkhuler K. 1995. PIV measurements of the velocity field downstream of flow straighteners in a pipe line. In *Flow Visualization VII*, ed. J. Crowder, Bengell House, 604–607.

Sen S., Lyons K., Bennetto J. and Long M. B. 1989. Scalar measurements in two, three and four dimensions. *Proc. International Congress on Applications of Lasers and Electro-Optics*, Orlando FL, 177–184.

Snyder R. and Hesselink, L. 1988. Measurement of mixing fluid flows with optical tomography. *Opt. Lett.*, **13**, 87–89.

Tennekes H. and Lumley J. L. 1973. *A First Course in Turbulence*. The MIT Press.

Tokumaru P. T. and Dimotakis P. E. 1995. Image correlation velocimetry. *Exp. Fluids*, **19**, 1–15.

Ushijima S. and Tanaka N. 1996. Three-Dimensional Particle Tracking Velocimetry with Laser-Light Sheet Scannings. *Trans ASME J. of Fluids Engin.*, **118**, 352–357.

Weinstein L. M. and Beeler G. B. 1986. Flow measurements in a water tunnel using a holocinematographic velocimeter. *AGARD CP-413*, 16-1–7.

Yip B., Schmidt R. L. and Long M. B. 1988. Instantaneous three-dimensional concentration measurements in turbulent jets and flames. *Opt. Lett.*, **13**, 96–98.

Yip B. and Long M. B. 1986. Instantaneous planar measurement of the complete three-dimensional scalar gradient in a turbulent jet. *Opt. Lett.*, **11**, 64–66.

Yoda M. and Hesselink L. 1990. A three-dimensional visualization technique applied to flow around a delta wing. *Phys. Fluids*, **10**, 102–108.

Yoda M. and Hesselink L. 1989. Three-dimensional mesaurement, display, and interpretation of fluid flow datasets. *SPIE Vol.*, **1083**, 112–117.

Zhang J., Tao B. and Katz J. 1997. Turbulent flow measurement in a square duct with hybrid holographic PIV. *Exp. Fluids*, **23**, 373–381.

Zimin V., Meng H. and Hussain F. 1993. Innovative holographic particle velocimeter: a multibeam technique. *Opt. Lett.*, **18**, 1101–1103.

CHAPTER 11

QUANTITATIVE FLOW VISUALIZATION VIA FULLY-RESOLVED FOUR-DIMENSIONAL IMAGING

W. J. A. Dahm and K. B. Southerland[a]

11.1 Introduction

A theme that will become evident to readers of this book is the shift that is occurring in recent years from traditional flow visualizations, which have typically provided qualitative pictures of flow structure and dynamics, to fully quantitative visualizations based on multidimensional imaging measurements, which are offering information of a type and level of detail that was previously associated only with numerical simulations. This shift has in large part been a consequence of advances in computers and related technologies during the same period, and it has produced a revolution of sorts in the ability to experimentally visualize and analyze fluid flows. Flow visualizations can today allow direct experimental access to quantitative information on the three-dimensional spatial structure and temporal dynamics of complex flows at a level of detail that had been almost unimaginable before. Moreover these visualizations are providing such information under conditions that can far exceed those accessible by Direct Numerical Simulations (DNS). Indeed the previously sharp distinctions between numerical simulations, computer visualization, and laboratory experimentation are becoming remarkably irrelevant. Today's experimental visualizer of fluid flows is in certain respects indistinguishable from the numerical simulator, and must have many of the same tools and skills in discrete mathematics, image processing, and scientific visualization.

[a]Laboratory for Turbulence & Combustion (LTC), Dept. of Aerospace Engineering, The University of Michigan, Ann Arbor, MI 48109–2140, U. S. A.

This chapter will describe an approach that has been successfully used for obtaining such quantitative multidimensional flow visualizations, namely fully-resolved, three- and four-dimensional, spatio-temporal imaging of turbulent flows. This is part of the broader field of non-invasive optical measurement techniques, a number of which have been under development for several years to allow quantitative visualization of the velocity and scalar gradient fields in turbulent flows. These techniques all make use of advanced laser diagnostics, high-speed imaging arrays, and high-speed data acquisition systems to facilitate a variety of optically-based measurements providing information over spatial fields of many points. They potentially offer high spatial and temporal resolution, as well as genuine spatial field information in place of classical single-point time-series data.

The most widely used of such methods are particle-based techniques, and a number of these are described elsewhere in this book. At the same time it has become possible to obtain fully-resolved, three- and four-dimensional, spatio-temporal measurements of conserved scalar fields in complex flows (Dahm, Southerland & Buch 1991; Southerland & Dahm 1994; Buch & Dahm 1996). This chapter describes such measurements, having spatial resolution finer than the scalar diffusion length scale and temporal resolution finer than the scalar advection time scale. The resulting conserved scalar field data $\zeta(\mathbf{x}, t)$ simultaneously span all three spatial dimensions and time, and have sufficiently high signal quality to accurately determine the true scalar gradient vector field $\nabla\zeta(\mathbf{x}, t)$. Such four-dimensional data typically are comprised of hundreds of individual three-dimensional spatial data volumes, thousands of two-dimensional planes, and literally billions of single-point measurements.

Moreover, determining velocities then no longer involves finding particle displacements as in particle-based techniques, but is instead based on inversion of the space- and time-evolving conserved scalar field to extract the underlying velocity field $\mathbf{u}(\mathbf{x}, t)$. Such scalar imaging methods have been used to obtain fully-resolved four-dimensional spatio-temporal measurements of the fine scales of turbulent flows. Visualizations of the three-dimensional spatial structure and simultaneous temporal dynamics of the full nine-component velocity gradient tensor field $\nabla\mathbf{u}(\mathbf{x}, t)$ and the conserved scalar gradient field $\nabla\zeta(\mathbf{x}, t)$ at the small scales of turbulent flows are important for developing a more complete understanding of the physics of turbulent flows and physical processes occurring in them. The following sections describe key elements of such quantitative multidimensional flow visualizations.

11.2   Technical Considerations

Visualizations such as these are principally used in two-stream mixing problems, including turbulent shear flows and other flows that involve more than one fluid stream. The conserved scalar field $\zeta(\mathbf{x}, t)$ is obtained from the concentration of an inert, water-soluble, passive, laser fluorescent dye (for example, disodium fluorescein) introduced with one of the free stream fluids, which subsequently mixes with other fluid in the flow of interest. The measurements are based on four-dimensional imaging of the laser induced fluorescence field produced by mixing of the dyed and undyed fluid streams, which is then converted to yield the true space- and time-varying conserved scalar field.

11.2.1   Laser induced fluorescence

The fluorescence properties of disodium fluorescein are well known and can be found in the literature. In visualizations of the present type, an argon-ion laser is used in multiline emission mode to excite the fluorescence. Each photon absorbed by a fluorescein molecule raises an outer electron from its ground state to an excited singlet state. Within a very short time, of the order of 10 ns, the electron falls back to ground state from the lowest vibrational level in the singlet state. The photon emitted has a lower frequency (longer wavelength) than the original photon. The resulting broadband absorption and emission spectra thus span largely different frequency ranges, and can be effectively separated using an optical filter. Typically a filter (for example, HOYA O(G)) is used to block the Mie scattered light from any particles in the flow. The filter above effectively blocks 92% of the light at the longest wavelength (514.5 nm) of the laser emission, and virtually all of the light at the shorter wavelengths of the remaining laser lines in Table 11.1. Near the peak of the dye emission spectrum (at 520 nm) the filter transmits only 19% of the incoming light, but by 540 nm, where the emission is still strong, 78% of the incoming fluorescence intensity is transmitted.

11.2.2   Beam scanning electronics

The laser induced fluorescence intensity pattern from the dye concentration field within the region of interest in the flow is measured using high-speed, high-resolution, successive planar laser induced fluorescence imaging from a collimated laser beam. This beam is rapidly swept through the measurement region in a raster pattern, consisting of fast vertical scans and slower horizontal scans,

Wavelength λ (nm)	Relative Line Strength $\alpha(\lambda)$
514.5	0.392
501.7	0.075
496.5	0.116
488.0	0.262
476.5	0.116
472.7	0.039

Table 11.1. Relative line strengths of the argon-ion laser in multi-line emission mode.

synchronized to the imaging array electronics. On each vertical sweep, a 256×256 imaging array captures the fluorescence intensity field emitted from a single two-dimensional spatial (x-y) plane in this region. A concurrent horizontal sweep effectively steps this x-y measurement plane through a predetermined set of up to 256 increments in the third (z) direction to produce a discrete set of parallel data planes. Collectively, these planes produce a single three-dimensional spatial data volume containing up to 256^3 individual measurement points, as indicated schematically in Fig. 11.1. Within any such volume, the spatial separations ($\Delta x, \Delta y$) between points in each data plane are determined by the size of the photodiode array elements and the effective magnification of the optical system. The effective spatial separation (Δz) between parallel planes is set by the inter-plane spacing and the laser beam diameter. In most cases, the interplane spacing is somewhat smaller than the beam diameter, so that parallel z-planes overlap slightly. A deconvolution is used to reduce the effective Δz to the interplane separation.

Once the desired number (N_z) of parallel planar beam sweeps have been completed, the laser beam executes a fly back to the original position and the process repeats. A temporal sequence of such three-dimensional spatial data volumes is thus sequentially acquired to produce a four-dimensional spatio-temporal data space. The time $\Delta\tau$ during which each element in the photodiode array is illuminated is determined by the laser beam diameter and beam sweep rate. This time interval effectively determines the temporal resolution of each data point. In practice, the $\Delta\tau$ values are typically at least three orders of magnitude smaller than any relevant fluid dynamical time scale. The elapsed time Δt between acquisition of successive parallel spatial data planes within any given

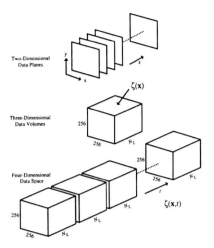

Fig. 11.1. Structure of the four-dimensional spatio-temporal data.

volume is set by the framing rate of the imaging array, since the timing signals
that drive the beam scanners are slaved back to the frame enable (FEN) signal from the array. Thus Δt determines the degree to which any measurement
"freezes" the evolving dye concentration field, and thereby determines in part
the z-differentiability within each three-dimensional spatial data volume. This
Δt is typically an order of magnitude smaller than the shortest relevant fluid
dynamic time scale in the dye concentration field. Lastly, the time ΔT between
acquisition of the same spatial point in temporally successive three-dimensional
spatial volumes is set by the array framing rate and the number of planes per
three-dimensional volume. This intervolume temporal separation is determined
by the number of planes N_z per volume, and effectively determines the time
differentiability of the data. For sufficiently small N_z the resulting data are fully
time- and space-differentiable, yielding a four-dimensional spatio-temporal data
space.

Scanning of the collimated laser beam is accomplished with two fast, low-
inertia, thermally-stabilized, galvanometric mirror scanners and their associated
controllers. Typical examples are General Scanning Inc. Models G120DT and
CX-660. The framing signal FEN from the imaging array formatter triggers
the internal ramp generator in the controller for the fast mirror scanner. This
synchronizes the start of the high speed scan with the frame start of the array,

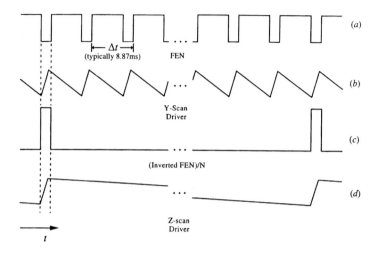

Fig. 11.2. Timing diagram for the camera and scanners, showing (a) frame enable; (b) fast scanner drive; (c) slow scanner trigger; and (d) slow scanner drive.

as indicated by the timing diagram in Fig. 11.2. The ramp period is set to the frame duration of the array. Increasing the integration time of the camera, by changing the number of clock periods before the next readout of the array is begun, allows the required minimum scanner fly back time to be accommodated. A ramp waveform created by a function generator controls the second mirror scanner position. A TTL signal generated from FEN with frequency FEN/N_z triggers this ramp waveform. The full z-scan sweep distance $(N_z - 1)\Delta z$, and thus the interplane separation Δz, is determined by a calibration of the output voltage from the scanner position signal provided by the scanner controller and the measured sweep angle.

11.2.3   Data acquisition system

The fluorescence intensity from dye-containing fluid along the swept laser beam path is typically measured with a 256×256-element photodiode array (for example, EG&G Reticon MC9256/MB9000) having photosites on 40 μm centers. The fluorescence transmitted through the filter is collected by a macro lens (for example, Vivatar 100 mm $f2.8$) operated at full aperture, and projected onto this imaging array. A data acquisition system converts and stores the serial output from the photodiode array in 8-bit digital format to a disk bank. The array formatter controls the sequential (non-interlaced) readout of the array, supplying a sampled-and-held output to the A/D converter. By providing the formatter with an external clock signal, the array can be driven at variable pixel rates up to 11 MHz, corresponding to net framing rates of nearly 120 frames/s, including all overhead cycles needed to accommodate the scanner fly back periods. The formatter uses this signal coupled with a programmable integration time to create the line enable (LEN) and frame enable (FEN) signals used to control the laser beam scanners.

A dual-ported image processor (for example, Recognition Concepts, Inc. Model Trapix 55/256) is used with a set of four 823.9 MB capacity disk drives and a data distribution manager for data acquisition. The overall capacity of the disks is 3.1 GB, allowing storage of nearly 200 individual 256^3 spatial data volumes, or over 50,000 individual 256^2 data planes at the maximum sustained throughput rate of 9.3 MB/s. Control of the data acquisition process is by a separate host computer.

11.2.4   Signal levels

Differentiability of the measured data requires sufficiently high signal quality in the fluorescence intensity measurements. To maximize the overall signal-to-noise ratio, the signal level is increased by operating the laser in multiline mode, by setting the pH of the free stream fluids, and by optimizing the dye concentration.

Multiline laser operation

The CW laser is typically operated in multiline mode to achieve the highest output power. However the resulting multi-spectral nature of the laser excitation complicates the conversion from the measured fluorescence intensity field to the dye concentration field. Table 11.1 gives the relative strengths of each of the argon-ion laser emission lines. The principal excitation wavelengths are 514.5 nm and 488.0 nm, but the beam also contains significant power at wavelengths of 501.7 nm, 496.5 nm, 476.5 nm, and 472.7 nm. For any single wavelength λ, the absorption of beam power by dye is given by

$$dP(\xi) = -\varepsilon(\lambda)c(\xi)P(\xi)\,d\xi. \tag{11.1}$$

Here P is the local beam power, c the local molar concentration, ε the molar extinction coefficient for the excitation wavelength, and ξ the location along the laser beam propagation path. Integrating Eqn. 11.1 gives the classical Beer's Law

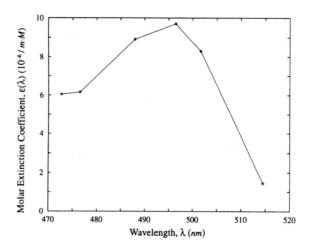

Fig. 11.3. Molar extinction coefficients $\varepsilon(\lambda)$ measured for each laser line in Table 11.1.

for attenuation of the beam power as it propagates through the dye medium. The resulting fluorescence intensity field $F(\xi)$ is then linearly related to the molar extinction coefficient, the local dye concentration, and the local beam power by

$$F(c(\xi)) = \varphi\varepsilon(\lambda)c(\xi)P(\xi). \qquad (11.2)$$

with φ representing the quantum efficiency.

When operating the laser in multiline mode, the combined effects of all excitation wavelengths must be considered. The fluorescence intensity in this case is given by

$$F(c(\xi)) = \sum_i F_i(c(\xi)) \qquad (11.3)$$

$$= \varphi c(\xi)P_0 \sum_i \alpha(\lambda_i)\varepsilon(\lambda_i) \exp\left(-\varepsilon(\lambda_i) \int_0^{\xi} c(\eta)\,d\eta\right). \qquad (11.4)$$

where $\alpha(\lambda_i)$ are the relative line strengths of the laser. The individual measured molar extinction coefficients $\varepsilon(\lambda_i)$ for each of the wavelengths present are given in Fig. 11.3. Note that, of the two principal wavelengths, the shorter one (488.0 nm) is six times more efficient at exciting the fluorescein molecule. For constant beam power, Fig. 11.4 verifies that the fluorescence intensity, even when operating with multiple wavelengths, is linearly related to the dye concentration. In

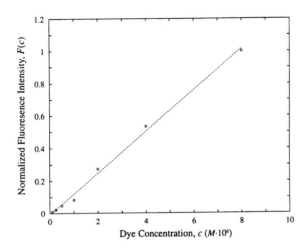

Fig. 11.4. Measured variation of the fluorescence intensity $F(c)$ at a single laser frequency for fixed laser beam power and varying dye concentration c.

addition, the extinction function, a product of the molar extinction coefficient ε and the concentration c, is linearly related to concentration for each wavelength. Fig. 11.5 shows this result for the principle wavelengths of 488.0 nm and 514.5 nm, where the slope of each curve gives the molar extinction associated with that particular wavelength. Fig. 11.6 gives the measured results if a single net extinction function is defined as above for the entire beam. Shown with good agreement is the theoretical result based on the measured line strengths and molar extinction coefficients for the individual wavelengths. This result demonstrates that multiline beam attenuation characteristics can be accurately determined from the characteristics of the individual components in Table 11.1, which is essential for converting the measured fluorescence intensity data to the dye concentration field.

pH effects

The effect of pH on the relative fluorescence intensity is shown in Fig. 11.7. The natural pH of water (typically about 7) lies on the steepest portion of this curve, so a small change in pH can lead to errors in the extinction coefficients. At pH > 8 the curve in Fig. 11.7 is not only flat, ensuring that variations in concentration and beam power alone affect the fluorescence level, but the fluorescence intensity is also maximized. For these reasons, the pH of the free

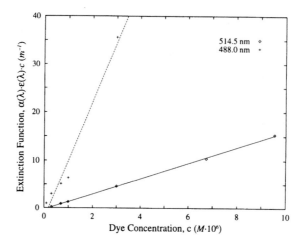

Fig. 11.5. Fluorescence intensity $F(c)$ at 514.5 nm and 488.0 nm, showing strong dependence of molar extinction coefficients on laser line.

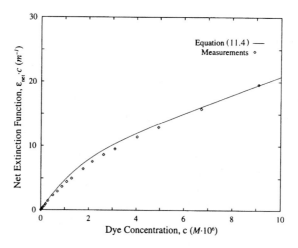

Fig. 11.6. Net extinction function when laser is operated in multi-line mode, showing good agreement of the measured result with Eqn. 11.4 for initial line strengths and coefficients in Fig. 11.3.

stream fluids is typically fixed at 11 by the addition of a small amount of NaOH in both aqueous solutions.

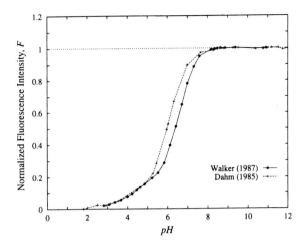

Fig. 11.7. Measured dependence of the normalized fluorescence intensity $F(c)$ on the pH value for multi-line excitation.

Dye concentration

The choice of dye concentration is important in maximizing the fluorescence intensity. Eqn. 11.2 shows that beam power and dye concentration combine to set the fluorescence intensity, however Eqn. 11.4 shows that these two factors also compete against each other. Increasing the dye concentration c also increases the beam absorption along its propagation path, and thus reduces the local power at the measurement location. The beam power decreases via an exponential integral over the dye concentration field up to the measurement location. Thus for very low dye concentrations, attenuation of the beam becomes negligible and the local fluorescence intensity becomes proportional to the local dye concentration, but the fluorescence intensity is weak. On the other hand, for very high concentrations, the dye absorbs most of the laser power as the beam propagates to the measurement volume, thus reducing the fluorescence intensity. From these two competing influences, there is an optimal dye concentration that maximizes the fluorescence intensity at any measurement location, which is found by maximizing $F(c(\xi))$ in Eqn. 11.4 for the theoretical mean dye concentration profile $c(\xi)$.

11.2.5 Signal-to-noise ratio

The noise sources in the imaging array can be classified into two types — those that depend on the incident light intensity and those that do not. The latter, dominated by the "dark" current that results from thermal (Johnson) noise in the photodetector and associated electronics, is independent of the signal level S. Consequently, low light level detection is proportionately more affected by this type of noise than are high light level measurements. When this class of noise sources dominates, the absolute noise level N is constant and thus the resulting signal-to-noise ratio (S/N) increases linearly with the signal level, namely $(S/N) \propto S^1$. On the other hand, photon shot noise increases with the illumination level, producing an rms noise level proportional to the square root of the signal level. Thus when shot noise is dominant, the resulting signal-to-noise ratio increases as $(S/N) \propto S^{1/2}$.

To determine the noise levels and sources in the imaging array, a large number of data planes from a "uniformly" illuminated white sheet are acquired at various f-stops of the imaging lens. The laser power and all gains are the same as in the actual fluorescence intensity measurements so that the results correspond to the true noise characteristics of the experimental data. Fig. 11.8 gives typical distributions of the 8-bit digital signal values obtained for each illumination level from such calibrations. Note that the widths of the distributions increase as expected with the mean signal level. The noise corresponding to each of these distributions is found in Fig. 11.9, where for each illumination level the average signal obtained over all the planes was subtracted from the data to yield the noise distributions shown. The four distributions corresponding to the lowest signal levels collapse quite well to a single curve, for which the width (the *rms* noise level) is constant. The remaining curves, corresponding to higher average signal levels, show noise distributions that increasingly deviate from the above profile by broadening and becoming asymmetric. The broadening of these distributions reflects the increasing noise level. Note that, even in the worst case, the *rms* noise level is less than 1.25 digital signal levels out of the 256 levels discernible with 8-bit measurements.

The *rms* noise level (width) from each of these distributions is given in the log-log plot in Fig. 11.10. This shows the scaling with mean digital signal level of the signal-to-noise ratio (S/N), defined as the mean digital signal level S divided by the *rms* width N of each noise distribution. The result clearly shows the transition from dark noise-limited measurements below digital signal levels of about 50, identifiable by the characteristic $(S/N) \propto S^1$ scaling, to the shot noise-

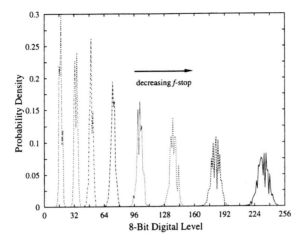

Fig. 11.8. Measured distributions of absolute signal for eight different aperture settings for uniform illumination, giving the associated noise distributions.

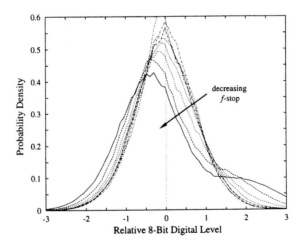

Fig. 11.9. Relative signal levels from Fig. 11.8, showing dependence of measurement noise N on signal level S; see also Fig. 11.10.

limited regime with the characteristic $(S/N) \propto S^{1/2}$ scaling for digital signal levels above about 150. Fluorescence intensity measurements typically span the full 256 digital signal levels under the same operating conditions, and thus span from dark (camera) noise-limited to shot noise-limited. More importantly, the

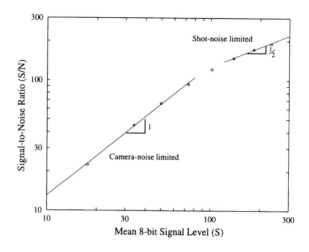

Fig. 11.10. Signal-to-noise ratio (S/N) from Figs. 11.8, 11.9, showing characteristic $(S/N) \propto S^1$ scaling in camera noise-limited regime, and $(S/N) \propto S^{1/2}$ scaling in shot-noise regime.

results in Fig. 11.10 show that when the signal level is maximized, the signal-to-noise ratio is slightly over 200, and even at the mean digital signal level of 50 typical of most measurements, the resulting signal-to-noise ratio is still higher than 65.

11.2.6   Spatial and temporal resolution

A typical goal of flow visualizations such as these is to obtain highly resolved data for the spatio-temporal structure of the scalar energy dissipation rate field $(ReSc)^{-1}\nabla\zeta \cdot \nabla\zeta(\mathbf{x}, t)$ from three- and four-dimensional measurements of the conserved scalar field $\zeta(\mathbf{x}, t)$. In addition to the signal quality noted above, this demands that the scalar field data must be sufficiently highly resolved in both space and time to give accurate values for the individual spatial derivatives underlying the gradient vector field.

From the measured thickness of the imaged portion of the laser beam and the interplane separation, together with the array element size and the image ratio of the measurements, the volume in the flow $(\Delta x \cdot \Delta y \cdot \Delta z)$ imaged onto each pixel can be readily determined. Furthermore, for the clock rates used and the number of planes per volume, the time Δt between acquisition of successive data planes within each spatial volume, and the time ΔT between the same data plane in successive data volumes, can also be determined. These smallest spatial and temporal scales discernible in the data must be compared with the finest local spatial and temporal scales on which gradients can exist in the local conserved scalar field for the flow under consideration in order to assess the resulting relative resolution achieved by the measurements.

Outer scales

In shear-driven turbulent flows, the local outer length and velocity scales u and δ are those characterizing the local mean shear profile. For example, in jets and plumes these are the local mean centerline velocity and the local flow width, while in shear layers the relevant quantities are the free stream velocity difference and the local flow width. All quantities associated with the outer scales are properly normalized by u and δ, and thus the local outer time scale is $\tau_\delta \equiv \delta/u$. The resulting local outer-scale Reynolds number $Re_\delta \equiv u\delta/\nu$ then properly scales the local turbulence properties of the flow, key among which is the relation between the local outer scales and the local inner scales.

Working in local outer scales has several advantages over the more common use of flow-specific source variables, such as the nozzle diameter and exit velocity in the case of jets. Such source variables often have only an indirect influence on the outer scales, as can be seen from the proper momentum-based scaling laws, and thus have an indirect and potentially confusing effect on the local turbulence properties. Moreover, sufficiently small scales of turbulent shear flows at the same local outer-scale Reynolds number Re_δ should have essentially similar structural and statistical properties. Parametrizations and normalizations based on flow-specific variables potentially obscure this quasi-universality and thereby obfuscate one of the strongest organizing principles available in turbulence studies.

Inner scales

The inner scales in turbulent flows characterize the finest length scale and finest Lagrangian time scale on which variations occur in the flow. The finest length scale results from the competing effects of strain, which acts to reduce the gradient length scale, and molecular diffusion, which acts to increase the gradient scale. These reach an equilibrium at the strain-limited viscous diffusion length scale λ_ν in the velocity gradient field, and at the strain-limited scalar diffusion scale λ_D in the scalar gradient field. These inner length scales are related to the local outer scale δ as $\lambda_\nu/\delta = \Lambda \cdot Re_\delta^{-3/4}$ and $\lambda_D/\lambda_\nu = Sc^{-1/2}$. Measurements by Southerland & Dahm (1994) and Buch & Dahm (1998) give $\langle \Lambda \rangle \approx 11.2$; this value is supported by recent measurements of Su & Clemens (1998). As noted above, when working in the local outer-scale Reynolds number Re_δ the value of Λ should be universal; if working in source-based Reynolds numbers it will appear to vary from one flow to another.

The viscous scale λ_ν is directly proportional to the classical Kolmogorov length scale $\lambda_K \equiv (\nu^3/\varepsilon)^{1/4}$ defined in terms of the mean dissipation rate ε. Using dissipation results in turbulent jets and Λ as above gives $\lambda_\nu \approx 5.9\lambda_K$. Although λ_K gives the correct scaling for the finest velocity gradient length scale, it is defined entirely on dimensional grounds and thus does not correspond directly to the resolution requirement. Similarly, the scalar diffusion length scale λ_D is proportional to the Batchelor scale, but it gives the physical size of the smallest structures in the scalar dissipation field in a turbulent flow.

Apart from the inner length scale, viscosity is the only directly relevant physical parameter at the inner scales, and thus the corresponding inner time scale is $\tau_\nu = (\lambda_\nu^2/\nu)$. This gives the shortest time scale on which the underlying vorticity field evolves in a Lagrangian frame. The local outer-scale Reynolds number Re_δ then provides the relation to the local outer time scale as $\tau_\nu/\tau_\delta = \Lambda^2 \cdot Re_\delta^{-1/2}$, where $\tau_\delta \equiv (\delta/u)$. The inner time scale is directly proportional to the classical Kolmogorov time scale $\tau_K \equiv (\nu/\varepsilon)^{1/2}$, where as above $\tau_\nu \approx 35\tau_K$.

When the outer scale Reynolds number Re_δ is sufficiently large, the velocity field $\mathbf{u}(\mathbf{x}, t)$ and scalar field $\zeta(\mathbf{x}, t)$ should, when viewed on the inner scales, be independent of Re_δ. Moreover, since the outer variables enter the governing equations only through Re_δ, the velocity and scalar fields should therefore also be independent of the outer scale variables and, as a further consequence, be independent of the particular shear flow as well. It is in this sense that the fine scale structure of the velocity and scalar fields, when viewed on the inner scales of high Reynolds number turbulent flows, are believed to be largely universal (that is, independent of the Reynolds number and the particular flow).

Advection scales

The inner Lagrangian time scale τ_ν is not, however, the temporal resolution requirement for turbulent flow measurements. The Eulerian nature of measurements obtained at any fixed spatial point introduces the much shorter viscous advection time scale $T_\nu \equiv (\lambda_\nu/u)$ in the velocity gradient field, and the corresponding scalar advection time scale $T_D \equiv (\lambda_D/u)$ in the scalar gradient field. Fully-resolved velocity or scalar field measurements thus need to meet these much more stringent Eulerian resolution requirements. Note that these can be related to the local inner time scale as $\tau_\nu/T_\nu = \Lambda \cdot Re_\delta^{1/4}$, and to the local outer time scale as $T_\nu/\tau_\delta = \Lambda \cdot Re_\delta^{-3/4}$. Note also that statistics of velocity or scalar fields converge on the outer time scale (δ/u), while statistics for velocity gradient and scalar gradient fields converge on the advective time scale T_ν or T_D for Eulerian time-series measurements.

Resolution requirements

At a minimum, the resolution requirements $(\Delta x \cdot \Delta y \cdot \Delta z) \ll \lambda_D$ and $t \ll (\lambda_D/u)$ must be satisfied to permit differentiation in all three directions within each three-dimensional spatial data volume to determine the scalar gradient vector field $\nabla\zeta(\mathbf{x},t)$. If the resulting data are to be time-differentiated as well between successive three-dimensional spatial data volumes, then the additional temporal resolution requirement $\Delta T \ll (\lambda_D/u)$ must also be met. These requirements ultimately place a limit on the highest Re_δ values at which such fully resolved four-dimensional flow visualizations are possible.

While the resolution demands on Δx and Δy can be satisfied by simply reducing the image ratio, the resolution Δz is nominally determined by the laser beam thickness and the interplane spacing. In general, the beam thickness is larger than the desired spatial separation between successive planes, however if the time Δt between planes is small enough that the scalar field is effectively frozen, then the overlap in the measured scalar field represents a convolution of the true scalar field with the laser beam profile. The measured scalar field can then be deconvolved with the measured beam profile to produce an effective resolution Δz comparable to the spatial separation between adjacent planes, which is set by the horizontal scanner and can be made arbitrarily small.

The final issue regarding spatial resolution concerns the depth of field. This can be characterized by measuring the apparent beam diameter at several z-planes ranging from the front-most to the back-most planes in a spatial data volume.

Fully-resolved vs. over-resolved measurements

Fully-resolved scalar field measurements require at least Nyquist sampling relative to λ_D in space and relative to T_D in time, and velocity field measurements require Nyquist sampling relative to λ_ν and T_ν. This resolution allows accurate differentiation in space and time to permit determination of the associated gradient vector fields. While these scales set the minimum resolution required for fully-resolved measurements, it is noteworthy that much higher spatial or temporal resolution is not always desirable. Since data are discretized not only in space and time, but also in digital signal level, it is apparent that there is a finest resolution limit beyond which adjacent points take on the same digital signal level, and thus compromise differentiability of the data. For any field $f(\mathbf{x}, t)$, the finest spatial resolution Δx and temporal resolution Δt occur at critical values of $B_x \equiv |\nabla f| \cdot \Delta x / \Delta f$ and $B_t \equiv |\nabla f| \cdot u\Delta t / \Delta f$, where $|\nabla f|$ characterizes the local gradient magnitude, and Δf is the difference in f between successive digital signal levels. When the B's becomes sufficiently small, spatially or temporally adjacent points will be at the same digital signal level, contributing to an underestimate in the magnitude of the gradient field $\nabla f(\mathbf{x}, t)$ or the time derivative $\partial f(\mathbf{x}, t)/\partial t$.

Resolution verification

The resolution of such quantitative multidimensional flow visualization data can be assessed by a "grid convergence" procedure analogous to that used in numerical studies. The dissipation field $\nabla f \cdot \nabla f(\mathbf{x}, t)$ associated with the energy $\frac{1}{2}f^2(\mathbf{x}, t)$ of any measured quantity $f(\mathbf{x}, t)$ can be integrated over the measured domain, with the procedure repeated as the resolution in the data $f(\mathbf{x}, t)$ is intentionally degraded by successive averaging over adjacent points. If the resulting total dissipation approaches a resolution-independent value, then the data are fully-resolved.

   Figure 11.11 shows the result obtained when such a convergence procedure is applied to fully-resolved four-dimensional scalar field data of the present type. This shows that the resolution achieved essentially reaches the "knee" in the curve, with approximately 80% of the scalar energy dissipation captured by the measurements. Resolution finer by a factor of ten would be needed to capture 98% of the dissipation; resolution coarser by a factor of three would capture less than 15% of the total dissipation.

11.2.7   Data processing

Data processing involves converting the measured fluorescence intensity data $F(\mathbf{x}, t)$ to the true dye concentration field $c(\mathbf{x}, t)$, and then to the conserved

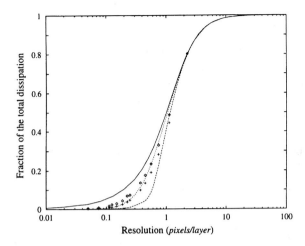

Fig. 11.11. Results from "grid convergence" to determine actual resolution achieved, showing fraction of total dissipation measured at various $\Delta x / \lambda_D$. Lines give theoretical results.

scalar field $\zeta(\mathbf{x}, t)$. Nonidealities in the imaging array and optical system are first collectively removed by dividing the fluorescence intensity data on a frame by frame basis by a measured transfer function $h(x, y)$, obtained by imaging the fluorescence from a uniform dye concentration field and averaging over many frames to remove any effect of noise. Next, the deconvolution decouples the laser beam profile from the measurements to increase the out-of-plane spatial resolution. Conversion of the deconvolved fluorescence intensity field to the dye concentration field then involves integrating as in Eqn. 11.4 along the beam path through the instantaneous dye concentration field to account for the attenuation. Since the attenuation is an integral effect and the path length is typically long relative to the scale λ_D on which variations in the dye concentration field occur, the integrated attenuation up to the imaged region is typically nearly constant.

Figure 11.12 shows typical mean fluorescence intensity fields along the beam path from a few thousand instantaneous data planes in two separate experiments, with the beam propagating from right to left. The same data, after all processing as noted above, are also shown in terms of the true dye concentration field, where they can be seen to agree well with the theoretical mean field, showing only effects of statistical convergence and confirming the efficacy of this procedure for converting the measured fluorescence intensity fields to the true conserved

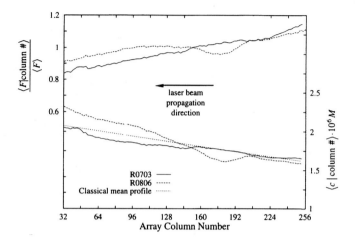

Fig. 11.12. Mean fluorescence intensity across the measurement volume (top), and the fully-corrected dye concentration (bottom). Deviations of the latter from the classical mean are from incomplete statistical convergence.

scalar fields.

11.3   Sample Applications

In this section we give some brief examples of how quantitative multidimensional flow visualizations of this type have been used to visualize and study a number of aspects of the physics associated with turbulent flows.

11.3.1   Fine structure of turbulent scalar fields

Figure 11.13 shows fully-resolved spatio-temporal measurements of the quasi-universal small-scale structure of the conserved scalar field $\zeta(\mathbf{x}, t)$ and the associated scalar energy dissipation rate field $\nabla \zeta \cdot \nabla \zeta(\mathbf{x}, t)$ in turbulent flows. These visualizations were obtained in the self-similar far field of an axisymmetric turbulent jet at outer-scale Reynolds numbers Re_δ from 2,600 to 5,000 and with Taylor-scale Reynolds numbers Re_λ from 38 to 52. These values appear high enough that the basic structure of the scalar field on the inner flow scale λ_ν has attained its asymptotic high Reynolds number form. As a consequence, these quantitative visualizations are largely representative of the small-scale structure of $Sc \gg 1$ scalar mixing in all high Reynolds number turbulent shear flows.

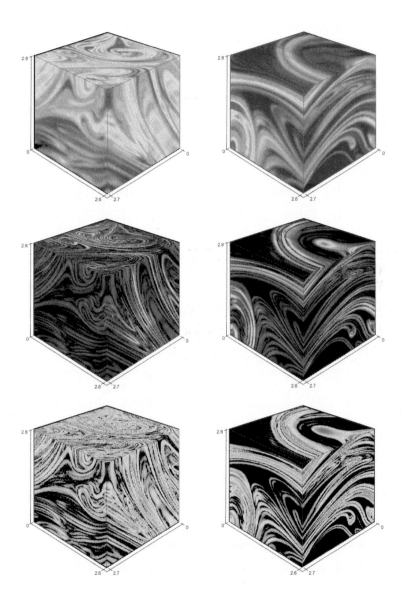

Fig. 11.13. Typical fully-resolved three-dimensional 256^3 spatial data volumes from quantitative visualizations, showing the conserved scalar field $\zeta(\mathbf{x}, t)$ (top row), the scalar energy dissipation rate field $\nabla \zeta \cdot \nabla \zeta(\mathbf{x}, t)$ (middle row), and $\log \nabla \zeta \cdot \nabla \zeta(\mathbf{x}, t)$ (bottom row). Originally from Southerland & Dahm (1994); reproduced with permission from Frederiksen *et al.* (1996). Figure also shown as Color Plate 14.

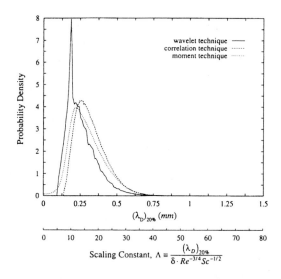

Fig. 11.14. Scalar dissipation layer thickness distributions obtained from quantitative visualizations of the type shown in Fig. 11.13. Thicknesses λ_D are shown in absolute terms as well as scaled on inner variables, where the scaling constant Λ can be determined as $\langle \Lambda \rangle = 11.2$. From Southerland & Dahm (1994).

Such three-dimensional (256^3) spatial data volumes reveal the fundamentally sheet-like physical structure of the scalar dissipation field at the small scales. From such data, probability density functions, spectra, and other quantities describing various structural and statistical features of the mixing process can be obtained (for example, Fig. 11.14). Note that the nature of these visualizations provides detailed spatio-temporal data that in many respects are more like results from direct numerical simulations than from traditional experimental measurements, but unlike DNS are capable of addressing the small scale structure of $Sc \gg 1$ mixing in fully turbulent shear flows.

11.3.2   Assessment of Taylor's hypothesis

Four-dimensional data allow simultaneous evaluation of all three components of the true scalar gradient vector field $\nabla\zeta(\mathbf{x}, t)$ and the time derivative field $(\partial/\partial t)\zeta(\mathbf{x}, t)$ at the small scales of a turbulent shear flow. These can be used to assess the errors made when Taylor's hypothesis is invoked in traditional measurements to estimate spatial derivatives in a turbulent flow. Various such approximations of the scalar energy dissipation rate field are compared in Fig. 11.15

with the true dissipation field at the point of maximum turbulence intensity in a jet. The classical single-point time series approximation yields a correlation of just 0.56 with the true dissipation, while a mixed estimate that combines one spatial derivative and the time derivative gives a correlation of 0.72. An optimal mixed dissipation estimate (Dahm & Southerland 1997) yields a correlation of 0.82.

11.3.3 Scalar imaging velocimetry

It is, furthermore, possible to invert the scalar transport equation using fully-resolved four-dimensional spatio-temporal data of this type to obtain the underlying velocity field $\mathbf{u}(\mathbf{x}, t)$ in a process termed "scalar imaging velocimetry" (Dahm, Su & Southerland 1992; Su & Dahm 1996a,b). The three-dimensional spatial nature of the resulting velocity fields allows all nine components of the velocity gradient tensor to be obtained. This, in turn, permits quantities such as the vorticity vector and strain rate tensor components, as well as higher-order gradient quantities, to be visualized. Figure 11.16 shows typical results for the spatial and temporal structure of various fields of dynamical interest in turbulence studies. These have provided the first fully-resolved, noninvasive measurements of spatio-temporal structure in the velocity gradient fields in turbulent flows. A somewhat different approach, based on pattern matching using optical flow concepts in place of inverting the scalar transport equation to obtain velocity fields from measured scalar field data, has also been examined in a number of studies (for example, Maas 1993; Merkel 1995; Merkel *et al.* 1995; Tokumaru & Dimotakis 1995).

11.3.4 Fractal scaling of turbulent scalar fields

As a final example, Fig. 11.17 shows visualizations of the scale-similarity properties associated with scalar mixing at the small scales of a turbulent flow. In this case, the interest is in the possible fractal structure of the support set on which scalar dissipation rate fields of the type in Figs. 11.13 and 11.15 are concentrated. Owing to the four-dimensional spatio-temporal character of the data involved, it is possible to examine such scale similarity within each spatial data volume and along the temporal direction as well (Frederiksen *et al.* 1996, 1997). This permits, for instance, visualizations of embedded nonfractal inclusions that result from the diffusive cutoff in the repeated stretching and folding action by the strain rate and vorticity fields on the dissipation field. Other investigations

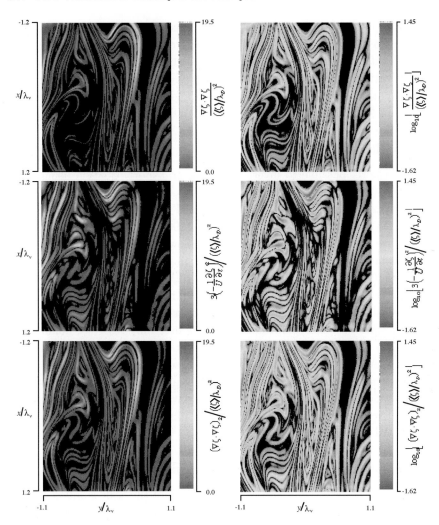

Fig. 11.15. Comparisons of the true scalar dissipation rate field (top) with the single-point Taylor series approximation (middle), and with a two-point mixed approximation (bottom) based on the time-derivative and one spatial derivative. Results in linear form (left) allow comparing relatively high dissipation rates, and in logarithmic form (right) allow comparisons of lower values. Reproduced with permission from Dahm & Southerland (1997). Figure also shown as Color Plate 15.

have also used quantitative multidimensional imaging measurements to examine related scaling processes in turbulent flows (for example, Sreenivasan & Mene-

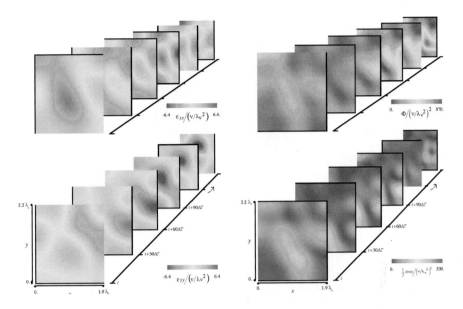

Fig. 11.16. Scalar imaging velocimetry results from fully-resolved four-dimensional spatio-temporal scalar field data, including strain-rate tensor fields (left) showing normal components $\varepsilon_{xx}(\mathbf{x}, t)$ (top) and $\varepsilon_{yy}(\mathbf{x}, t)$ (bottom), and higher-order velocity gradient fields (right) showing kinetic energy dissipation rate (top) and enstrophy (bottom). Reproduced with permission from Su & Dahm (1996b).

veau 1986; Meneveau & Sreenivasan 1991).

11.4 Further Information

This chapter has attempted to show how modern, fully-resolved, quantitative, multidimensional flow visualizations can allow direct experimental access to highly detailed features of complex physical processes in various fields of interest. While necessarily brief, it gives an introduction to the major concepts involved and an indication of the types of results that can be achieved. For more detailed information on the techniques described here, the reader is referred in particular to Buch & Dahm (1996, 1998), Southerland (1994), Dahm & Southerland (1994), and Merkel (1995).

There are a few final points worth noting. Each such visualization typically produces several billion individual, fully-resolved, point measurements of the

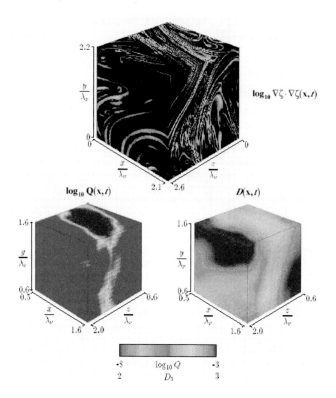

Fig. 11.17. Quantitative visualization of a typical dissipation rate field (top) with the resulting fractal scaling quality $Q(\mathbf{x}, t)$ (left) and local fractal dimension $D(\mathbf{x}, t)$ (right), showing local nonfractal inclusions (blue) within an otherwise fractal background structure (red). Reproduced with permission from Frederiksen *et al.* (1997a).

scalar field values throughout a four-dimensional space-time domain. However owing to the very high resolution of these points, they typically span over only a few local outer time scales (δ/u). As a consequence, while such measurements provide highly detailed information on the spatial structure and temporal dynamics of the flow, long time statistics are inherently more difficult to obtain. These measurements are thus viewed as complementing traditional single-point time series data, from which spatial structure and gradient information are difficult to obtain but which provide long time records suitable for statistics of other types of quantities.

Similarly, while the measurements provide very dense and highly resolved three-dimensional spatial information in data volumes as large as 256^3, the need to resolve the smallest scalar gradients within these volumes presently restricts their physical size to just a few inner flow scales λ_ν in each direction. As a consequence, highly detailed spatial information at the dissipative scales of the flow is available, but no access to the inertial range of spatial scales is currently possible. In this sense as well, these measurements complement traditional time series measurements, which have no access to the three-dimensional spatial spectrum but which can access inertial scales in temporal spectra.

Finally, while the access that this type of quantitative multidimensional visualization makes available to the experimental fluid dynamicist is exciting, readers must be cautioned that it comes at the expense of considerable complexity in comparison with more traditional flow visualization. It is fair to say that these visualizations are, at least at the present time, not appropriate for the casual user. However for situations where the type of information and the level of detail which they make possible are essential, they represent a significant step forward in the ability to visualize and interpret complex processes in fluid flows.

11.5   References

Buch, K.A. and Dahm, W.J.A. 1996. Experimental study of the fine-scale structure of conserved scalar mixing in turbulent flows. Part I. $Sc \gg 1$. *J. Fluid Mech.*, **317**, 21–71.

Buch, K.A. and Dahm, W.J.A. 1998. Experimental study of the fine-scale structure of conserved scalar mixing in turbulent flows. Part II. $Sc \approx 1$. *J. Fluid Mech.*, **364**, 1–29.

Dahm, W.J.A. and Southerland, K.B. 1997. Experimental assessment of Taylor's hypothesis and its applicability to dissipation estimates in turbulent flows. *Phys. Fluids*, **9**, 2101–2107.

Dahm, W.J.A., Southerland, K.B. and Buch, K.A. 1991. Direct, high resolution, four-dimensional measurements of the fine scale structure of $Sc \gg 1$ molecular mixing in turbulent flows. *Phys Fluids A*, **3**, 1115–1127.

Dahm, W.J.A., Su, L.K. and Southerland, K.B. 1992. A scalar imaging velocimetry technique for four-dimensional velocity field measurements in turbulent flows. *Phys. Fluids A*, **4**, 2191–2206.

Frederiksen, R.D., Dahm, W.J.A. and Dowling, D. 1996. Experimental assessment of fractal scale similarity in turbulent flows. Part I: One-dimensional intersections. *J. Fluid Mech.*, **327**, 35–72.

Frederiksen, R.D., Dahm, W.J.A. and Dowling, D. 1997a. Experimental assessment of fractal scale similarity in turbulent flows. Part 2: Higher dimensional intersections and nonfractal inclusions. *J. Fluid Mech.*, **338**, 89–126.

Frederiksen, R.D., Dahm, W.J.A. and Dowling, D. 1997b. Experimental assessment of fractal scale similarity in turbulent flows. Part 3: Multifractal scaling. *J. Fluid Mech.*, **338**, 127–155.

Maas, H.-G. 1993. Determination of velocity field in flow tomography sequences by 3-D least squares matching. *Proc. 2nd Conf. on Optical 3D Measurement Techniques*, Zürich.

Meneveau, C. and Sreenivasan, K.R. 1991. The multifractal nature of turbulent energy dissipation. *J. Fluid Mech.*, **224**, 429–484.

Merkel, G.J. 1995. *Tomographie in einem turbulenten Freistrahl mit Hilfe von pH-abhängiger Laser Induzierter Fluoreszenz.* Ph.D. Thesis No. 11174, Eidgenössische Technische Hochschule Zürich, Zürich.

Merkel, G.J., Rys, P., Rys, F.S. and Dracos, Th.A. 1995. Concentration and velocity field measurements in turbulent flows by Laser Induced Fluorescence Tomography. *Proc. 7th Int'l. Symp. on Flow Visualization*, Seattle.

Southerland, K.B. 1994. *A four-dimensional experimental study of passive scalar mixing in turbulent flows.* Ph.D. Thesis, The University of Michigan, Ann Arbor.

Southerland, K.B. and Dahm, W.J.A. 1994. A four-dimensional experimental study of conserved scalar mixing in turbulent flows. Univ. of Michigan *Report No. 026779-12.*

Sreenivasan, K.R. and Meneveau, C. 1986. The fractal facets of turbulence. *J. Fluid Mech.*, **173**, 357–386.

Su, L.K. and Clemens, N.T. 1998. The structure of the three-dimensional scalar gradient in gas-phase planar turbulent jets. *AIAA Paper 98-0429*, AIAA, Washington, D.C.

Su, L.K. and Dahm, W.J.A. 1996a. Scalar imaging velocimetry measurements of the velocity gradient tensor field at the dissipative scales of turbulent flows. Part I: Validation tests. *Phys. Fluids*, **8**, 1869–1882.

Su, L.K. and Dahm, W.J.A. 1996b. Scalar imaging velocimetry measurements of the velocity gradient tensor field at the dissipative scales of turbulent flows. Part II: Experimental results. *Phys. Fluids*, **8**, 1883–1906.

Tokumaru, P.T. and Dimotakis, P.E. 1995. Image correlation velocimetry. *Expts. Fluids*, **19**, 1–15.

CHAPTER 12

VISUALIZATION, FEATURE EXTRACTION AND QUANTIFICATION OF NUMERICAL VISUALIZATIONS OF HIGH GRADIENT COMPRESSIBLE FLOWS

Ravi Samtaney[a] and Norman J. Zabusky[b]

12.1 Introduction

The inviscid flow of a compressible fluid is governed by a system of hyperbolic conservation laws (which are also called the compressible Euler equations; Courant & Friedrichs, 1948). It is only in exceptional and rather rare circumstances that these nonlinear partial differential equations allow a closed form analytical solution. In most situations, and for almost all problems of practical importance, these equations have to be solved numerically.

It is well-known that for nonlinear systems of hyperbolic conservation laws with C^∞ Cauchy data, the solution may develop discontinuities in a finite time. Examples include the formation of a shock on a wing in transonic flight or the formation and propagation of a shock wave from compressive piston motion. The most common discontinuities which develop in gas dynamics are: (a) shock waves and (b) contact-discontinuities. In the theory of hyperbolic conservation laws, shocks are called genuinely nonlinear waves while contact discontinuities are called linearly degenerate waves. Numerically, the discontinuities are very often handled by "shock-capturing" techniques which typically diffuse or "smear" the discontinuities over transition regions of several grid cells (LeVeque, 1992; LeVeque et al., 1998). We hereafter refer to these near-discontinuities as "discontinuities." Furthermore, with grid refinement, the physical extent of the

[a]MRJ Technology Solutions, Inc., Mail Stop T27A-2, NASA-Ames Research Center, Moffett Field, CA 94035-1000, U. S. A.
[b]Laboratory for Visiometrics and Modeling, Dept. of Mechanical and Aerospace Engineering & CAIP Center, Rutgers University, Piscataway, NJ 08854-8058, U. S. A.

smeared shock reduces in extent, while the number of grid cells over which it is smeared still remains the same for a given numerical method. Consequently, although the derivatives of various field quantities (such as the density or the pressure) are ill-defined, the captured discontinuities in the numerical solution exhibit large gradients over a very small spatial extent, and a numerical evaluation of the derivatives is permitted. A similar discussion of high gradients applies to vorticity bearing contact discontinuities. The reader is reminded that there are other types of waves in compressible flows such as detonation waves which will not be discussed in this chapter.

The earliest scientific work on shock-wave visualization is due to Toepler who developed the schlieren method; followed by Dvorak, one of Mach's assistants who modified the schlieren method to give the shadowgraph method (see Chapter 9). However, there are not many instances of visualizations, extractions and quantifications of flow fields with discontinuities in the scientific visualization literature. Noteworthy efforts in shock wave visualization include the work of Vorozhtsov & Yanenko (1990),who also discuss some issues of quantifying time varying configurations, and Pagendarm & Seitz (1993), Ma *et al.* (1996), and Lovely & Haimes (1998). However, most of the discussion in the literature pertains to shock wave detection in *steady* three-dimensional flow fields. Some of these shock-detection algorithms rely on the gradients of the density field and isosurfaces of unit Mach number. This works because the Mach number (denoted by M) changes from greater than one (supersonic flow) to less than one (subsonic flow) across a shock. However, this criterion ($M = 1$) is not useful for unsteady flows. We note that Lovely & Haimes (1998) have provided correction terms in their algorithm for unsteady flows. Several visualization algorithms which assume at least a continuous field (if not continuity of several derivatives) run into unexpected problems.

In this chapter, we review transformation functions which may be applied to the numerical simulation to generate visual images which correspond to experimental techniques. However, our principal focus is to go beyond generating pictures of flow fields with discontinuities to extract the location and properties of shock waves and contact discontinuities in the flow. Many details of the algorithms are included.

12.1.1   Fundamental configuration

We apply the methods developed in this chapter to two-dimensional simulations of the unsteady interaction of a planar shock wave with a planar inclined

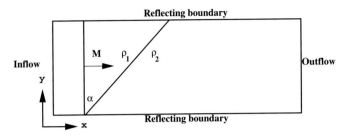

Fig. 12.1. Schematic of the initial conditions for an unsteady two-dimensional shock contact-discontinuity interaction. The physical geometry is a two-dimensional rectangular shock-tube.

contact line, a fundamental configuration in Richtmyer-Meshkov (accelerated inhomogenenous) flows (Samtaney & Pullin, 1996; Zabusky, 1999). This canonical unsteady problem exhibits several interesting features of the discontinuities including interactions and bifurcations of shock waves, triple points with emerging shear layers and rolled-up contact discontinuities. Shown in Fig. 12.1 is a schematic depicting the physical problem. A shock wave of Mach number M translating from left to right encounters an interface initially inclined at an angle α separating two gases. The gas on the left (right) has a density ρ_1 (ρ_2). At the interface, the shock wave refracts and bifurcates into a transmitted shock and a reflected wave which may be a shock or an expansion wave. Further reflections of these waves at the top and bottom boundaries and secondary interactions lead to a complex flow field rich in discontinuities. For convenience, we assume that both gases have the same ideal equation of state with identical ratio of specific heats γ. Thus, a 3-tuple $(M, \rho_2/\rho_1, \alpha)$ defines the principal parameters in this interaction. In this chapter we use parameters $(2.0, 3.0, \pi/4)$ and $\gamma = 1.667$.

The solutions are obtained with two second-order simulation methods, the Godunov and EFM methods (for details see Samtaney & Zabusky (1994) and Pullin (1980). Note that Godunov methods belong to the class of flux difference splitting schemes, whereas EFM belongs to the class of flux vector splitting schemes. The mesh is uniform with square cells. This is not a restriction as our techniques can be extended to body-fitted curvilinear meshes. The domain of simulation is: $[-0.5, 1.5] \times [0, 1.0]$, and it is discretized at two resolutions with 800 and 1600 points in the x-direction, and 400 and 800 points in the y-direction, respectively. See Table 12.1 for the nomenclature of the runs performed.

Unless specified in the figure caption, the images and results in the paper are shown at time $t = 0.72$. Note that time is normalized such that it takes unit

Numerical Method	Low Resolution (400 × 800)	High Resolution (1600 × 800)
Godunov	GL	GH
EFM	EL	EH

Table 12.1. Nomenclature for runs performed

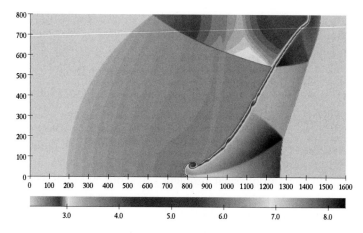

Fig. 12.2. Density field of a two-dimensional shock contact-discontinuity interaction at time $t = 0.72$. Simulation GH. Figure also shown as Color Plate 16.

time for a sound wave in the unshocked incident gas (ρ_1) to traverse the width of the shock-tube.

The density image for GH is shown in Fig. 12.2 at time $t = 0.72$. It exhibits a variety of strong and weak shocks and laminar shear layers. Note the wavy interface is due to the roll-up of deposited vorticity . This effect is not present at this time in simulations GL and EL. The differing results of the Godunov and EFM codes are exhibited, and resolution and convergence issues are discussed. The comparison among methods and resolutions is a first visiometric approach to quantify the quality of codes on this fundamental two-dimensional configuration.

12.2 Visualization Techniques

In this section, we begin by discussing the numerical analog of experimental flow visualization techniques.

12.2.1 Numerical analog of experimental techniques

In an experiment, as a light ray travels through a compressible gas with density variations (density variations are related to the variations in index of refraction via the Gladstone-Dale formula, see Merzkirch, 1974), it undergoes three effects. The first is the angular displacement with respect to an undisturbed path. The second is a displacement from its path which it would have taken in a uniform medium, and the third is a phase shift from the undisturbed light ray (see also Chapter 9). These three effects corresponds to the three main experimental visualization techniques for flows with discontinuities.

Schlieren imaging

Schlieren imaging relies on the angular deflections of light rays that result from a variable refractive index which is a function of the density of the gas. The intensity of a schlieren image corresponds to the gradient of the density (Merzkirch, 1974). In the edge detection literature, the gradient is used in various methods to identify edges. These methods include the Roberts cross, Sobel, Compass, and Prewitt edge detectors (Schalkoff, 1988). Each of these methods uses a different "convolution mask" (convolution mask is the jargon used in image processing for a two-dimensional discrete function or filter). The Roberts cross edge detector, in our notation, is given by:

$$\nabla_x \rho_{i,j} \equiv \frac{\rho_{i+1,j+1} - \rho_{i,j}}{\sqrt{2}h},$$

$$\nabla_y \rho_{i,j} \equiv \frac{\rho_{i,j+1} - \rho_{i+1,j}}{\sqrt{2}h},$$

$$\nabla \rho_{i,j} \equiv [(\nabla_x \rho_{i,j})^2 + (\nabla_y \rho_{i,j})^2]^{\frac{1}{2}}, \tag{12.1}$$

where ρ is the density field, and h is the mesh spacing. The Sobel edge detector, in our notation, is given by:

$$\nabla_x \rho_{i,j} \equiv \frac{1}{8h} [2(\rho_{i+1,j} - \rho_{i-1,j})$$
$$+ \rho_{i+1,j+1} - \rho_{i-1,j+1} + \rho_{i+1,j-1} - \rho_{i-1,j-1}],$$

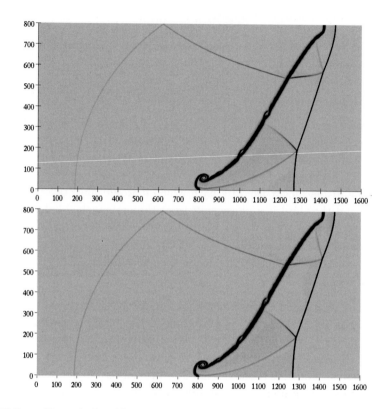

Fig. 12.3.   Numerical schlieren images of a two-dimensional shock contact-discontinuity interaction at $t = 0.72$ from simulation GH. The top image is generated using the Roberts cross edge detector, while the bottom image is generated using the Sobel edge detector.

$$
\begin{aligned}
\nabla_y \rho_{i,j} &\equiv \frac{1}{8h} \left[2(\rho_{i,j+1} - \rho_{i,j-1}) \right. \\
&\qquad \left. + \rho_{i+1,j+1} - \rho_{i+1,j-1} + \rho_{i-1,j+1} - \rho_{i-1,j-1} \right], \\
\nabla \rho_{i,j} &\equiv \left[(\nabla_x \rho_{i,j})^2 + (\nabla_y \rho_{i,j})^2 \right]^{\frac{1}{2}}.
\end{aligned}
\tag{12.2}
$$

The schlieren images, corresponding to the above two methods, are shown in Fig. 12.3. The Sobel edge detector, because of its larger stencil, is smoother and less sensitive to noise. Visually we note that there is little difference between the Sobel and the Roberts cross edge detectors in this example.

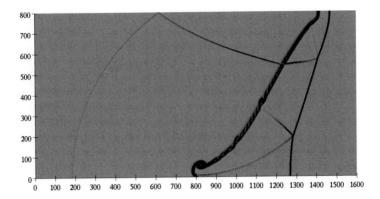

Fig. 12.4. Numerical shadowgraph of a two-dimensional shock contact-discontinuity interaction at $t = 0.72$ from simulation GH.

Shadowgraphs

This technique relies on the displacement of a light ray due to the change in refractive index because of spatial density variations in the gas (Merzkirch, 1974). It can be shown that the displacement experienced by a light ray depends on the second derivative of the density. The numerical equivalent is obtained by taking the the second derivative of the density field. A central difference approximation to calculate this at a point (i, j) on a uniform mesh is the following:

$$\nabla^2 \rho_{i,j} \equiv \frac{\rho_{i+1,j} + \rho_{i-1,j} + \rho_{i,j+1} + \rho_{i,j-1} - 4\rho_{i,j}}{h^2}. \tag{12.3}$$

This formula is applied to the density field in the shock-contact simulation. The shadowgraph image is shown in Fig. 12.4, at time $t = 0.72$. In edge detection literature, this technique of finding edges is sometimes referred to as the "Marr" edge detector (Parker, 1997). Note, in laboratory experiments, shocks and contact discontinuities actually cause a significant amount of light diffraction as opposed to light refraction. Nonetheless, the technique that is proposed highlights the discontinuities in the numerical flow field.

Interferometry

The fringe patterns in an interferogram arise due to the phase shift of light as it moves through a density field (Merzkirch, 1974). Numerically we approximate

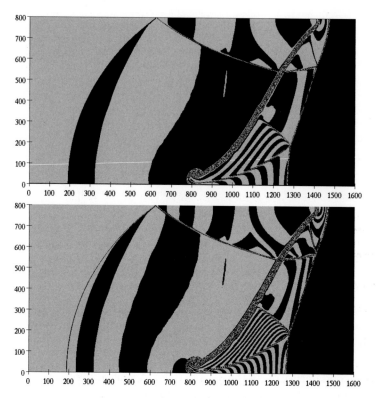

Fig. 12.5. Numerical interferograms of a two-dimensional shock contact-discontinuity interaction at $t = 0.72$ from simulation GH. The top (bottom) image is generated with 64 (128) fringes.

the interferogram as follows:

$$I_{i,j} = 0 \quad \text{(respectively 1)} \tag{12.4}$$

$$\text{if} \quad \mod \left(\text{integer}(N_f \frac{\rho_{i,j} - \rho_{min}}{\rho_{max} - \rho_{min}}), 2 \right) = 0 \quad \text{(respectively 1)},$$

where N_f is the number of fringes in the range $[\rho_{min}, \rho_{max}]$ determined by the user. I is the intensity pattern on the resulting image. Shifts in the fringe patterns occur at the discontinuities (see Fig. 12.5).

The numerical visualization techniques discussed above may be applied to flow variables other than the density field. Note that these three methods are extensively used to visualize experiments wherein the density in one direction is integrated to elicit information about the flow field in two dimensions. Therefore these techniques are useful in visualizing two-dimensional experiments. Experimental techniques to obtain schlieren images and interferograms in color also exist. Furthermore, there are several variants to schlieren and interferometry which we will not discuss here. The reader is referred to Chapter 9 and the book by Merzkirch (1974). For our purposes, the numerical shadowgraph ($\nabla^2\rho$) in particular proves to be useful in isolating discontinuities in unsteady two-dimensional numerical experiments.

12.2.2 Smoothing and noise suppression

Because derivative operations are generally an order less in accuracy than the computed solution, the shadowgraph and the schlieren images are susceptible to error noise. This problem is further exacerbated since most shock-capturing methods reduce to first-order accuracy near discontinuities to maintain monotonicity. To mitigate the effects of noise, the following smoothing techniques have been examined. The first one employs a window around a point (i, j) as follows

$$\tilde{q}_{i,j} = \sum_{k=-n/2}^{n/2} \sum_{l=-n/2}^{n/2} w_{k,l} \, q_{i+k,j+l} \tag{12.5}$$

such that the weights $\sum w_{k,l} = 1$. In the above equation $\tilde{q}$ is the resulting smoothed field. Another smoothing function which has been prominently employed in the image processing literature is to convolve the field with an isotropic Gaussian as

$$\tilde{q}_{i,j} = \sum_{k} \sum_{l} G_{k,l} q_{i+k,j+l}$$

$$G_{k,l} = \frac{1}{2\pi\sigma^2} \exp\left(-\frac{x_{k,l}^2 + y_{k,l}^2}{2\sigma^2}\right) \tag{12.6}$$

where σ is the standard deviation in the Gaussian distribution. It is common practice in edge detection (Parker, 1997) to combine the Laplacian and the Gaussian operations into one convolution mask called the Laplacian of Gaussian or LoG.

12.2.3 Selection of variables for visualization

The selection of the variables and their color maps to highlight discontinuities in the flow is a nontrivial issue. From our experience, gathered by applying the

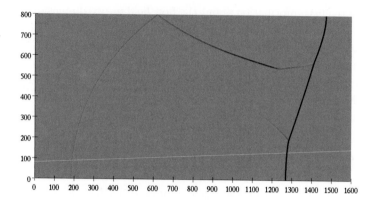

Fig. 12.6. Laplacian of the pressure field in a two-dimensional shock contact-discontinuity interaction at $t = 0.72$ from simulation GH.

edge detection algorithms to various fields such as the density, pressure, entropy, etc., we recommend the following variables:

- *Density.* Magnitude of gradient and Laplacian of the density to visualize shocks and contact discontinuities.

- *Pressure and divergence of velocity.* Magnitude of gradient and Laplacian of the pressure (see Fig. 12.6), as well as the divergence of the velocity field to visualize shocks. Contact discontinuities do not show up in these variables because, in theory, the pressure and normal velocity are both continuous across contact discontinuities. The divergence of the velocity field ($\nabla \cdot \mathbf{V}$) proves to be useful in picking out the shock fronts (see Fig. 12.7). Note that, because shocks in perfect gases are always compressive, $\nabla \cdot \mathbf{V}$ is always negative at the shocks.

- *Entropy.* It has been shown that the entropy jump across a shock wave is a third-order quantity, that is, $\Delta s = O(M - 1)^3$ (Thompson, 1972) where M is the Mach number of the shock. Consequently, entropy gradients are useful to identify only strong shocks in the flow field. Note that entropy is also discontinuous across contacts. The variable $\nabla^2 s$ is shown in Fig. 12.8. The transmitted shock and the primary contact are very clear, while the reflected shocks, which are significantly weaker, are not detected.

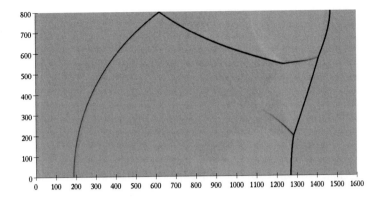

Fig. 12.7. Divergence of the velocity field in a two-dimensional shock contact-discontinuity interaction at $t = 0.72$ from simulation GH.

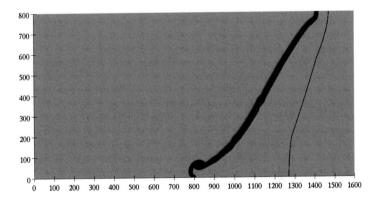

Fig. 12.8. Laplacian of the entropy field in a two-dimensional shock contact-discontinuity interaction at $t = 0.72$ from simulation GH.

12.3 Quantification of Shocks and Contacts

By quantification of discontinuities, we mean the representation of discontinuities in a two-dimensional flow field by curves (and their normals and curvature) along which certain properties are determined. For example, for shocks the properties may include the local strength (pressure jump) or local Mach number (or its jump), and for contacts the local circulation per unit length (tangential velocity jump) (Samtaney & Zabusky, 1994), where the "jumps" are all along the local normal as discussed below. The extracted contours correspond to the zero crossing of the Laplacian of the density field subject to a constraint that the density gradient at the zero crossing be larger than a user specified threshold. In the future, one must be concerned with the topologies, lengths and "widths" of these wave and vortex-transition domains or structures. as well as other measures of their "chaotic" (for example, fractal) complexity.

12.3.1 One-dimensional example

In this section, we will examine the following question: How accurate is the zero crossing of the Laplacian (of density in this example) in quantifying the shock and contact discontinuity locations? This issue is addressed by comparing the analytical solution (Samtaney & Zabusky, 1994) with simulations of a one-dimensional shock contact-discontinuity interaction ($\alpha = 0$ in Fig. 12.1) obtained with the Godunov code. In this interaction, the incident shock bifurcates into a reflected and a transmitted shock. We examine the difference in location of the zero crossing of $\nabla^2 \rho$ in the numerical solution and the analytical solution for various resolutions. The difference is normalized by the mesh spacing and plotted in Fig 12.9 at $t = 0.54$.

We observe that, for all resolutions, the difference in the numerical shock location and the analytical location differs by less than one grid cell. The contact discontinuity, which is typically smeared over a larger extent, is located accurately to within $2h$, twice the mesh spacing at low resolution. However, as the mesh is refined, the zero crossing of $\nabla^2 \rho$ does not converge to the analytical location for a contact discontinuity. An explanation awaits further study.

12.3.2 Algorithm

The details of the algorithm to quantify shocks and contacts in two-dimensional compressible flows follow.

1. *Simplicial decomposition of the mesh*
   The mesh is composed of quadrilaterals and numbered such that quadrilateral (i, j) has four vertices at $\bar{x}(i, j)$, $\bar{x}(i+1, j)$, $\bar{x}(i+1, j+1)$ and $\bar{x}(i, j+1)$ where $i = 0, 1, 2, \cdots, M$, $j = 0, 1, 2, \cdots, N$. Each mesh quadrilateral is decomposed into two triangles. Note that such a decomposition is not unique, but we will not concern ourselves with this issue at this time. Each triangle is given a unique id number which is given by $id = 2(M - 1)j + 2i + k$,

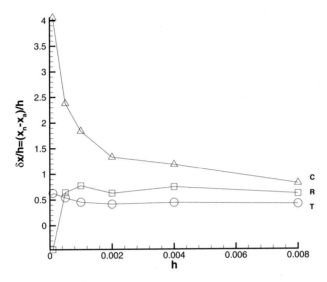

Fig. 12.9. Difference in the numerical location x_n and analytical location x_a of the reflected shock (R), contact discontinuity (C), and transmitted shock (T) at various resolutions in a one dimensional shock contact-discontinuity interaction. The interaction parameters are $(M, \rho_2/\rho_1, \alpha) = (2.0, 3.0, 0.0)$. The mesh spacing is h.

and where $k = 0, 1$. A table of triangles with attributes, called the Triangle Table, is generated. The initial setup by this step in the algorithm is schematically depicted in Fig. 12.10. Then, a global table of edges in the mesh (called the Edge Table) is generated. Each edge is given a unique identification number given by $eid = 3Mj + 3i + k$, $k = 0, 1, 2$. Included in the attributes for each entry in the table of triangles are the unique identification tag for the triangle and three data structures $(E0, E1, E2)$ which contain two pointers. One of these pointers points to the global Edge Table. Since each edge is shared by two triangles or is at the boundary of the domain, the second pointer points to an entry in the Triangle Table which is the neighboring triangle sharing this edge. If the edge is on the boundary, the second pointer is set to NULL. Each entry in the global Edge Table essentially contains pointers to a Vertex Table (not shown in the schematic figure) wherein the coordinates of the vertices are actually stored.

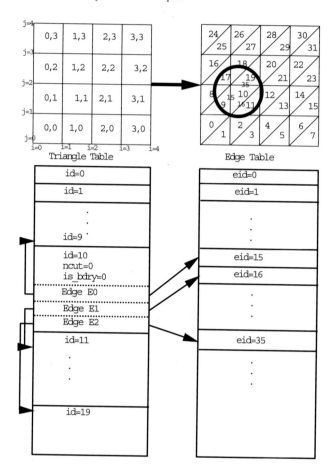

Fig. 12.10. Simplicial decomposition of the mesh and generation of Triangle Table and Edge Table.

2. *Zero crossing of the Laplacian*

   The next step in the process is to compute the Laplacian of the field variable of interest. The edges intersected by the zero contour of the Laplacian are identified. Furthermore, we exclude those edges intersected by the zero contour of the Laplacian where the gradient of the field is below a threshold value. Mathematically, the intersection point is calculated as

$$\bar{x} \;=\; \bar{x}_1 + (\bar{x}_2 - \bar{x}_1)\frac{\nabla^2\rho(\bar{x}_1)}{\nabla^2\rho(\bar{x}_2) - \nabla^2\rho(\bar{x}_1)} \tag{12.7}$$

$$\text{if}\;\; \nabla^2\rho(\bar{x}_2) \cdot \nabla^2\rho(\bar{x}_1) < 0$$

$$\text{and}\;\; |\nabla\rho(\bar{x}_2)| + |\nabla\rho(\bar{x}_1)| \;>\; 2\,|\nabla\rho|_{threshold},$$

where $\bar{x}_1, \bar{x}_2$ are the vertices of the edge. Thus, in the above equation, $\bar{x}$ is the location of the zero crossing of the Laplacian on the edge whose end points are at $\bar{x}_1$ and $\bar{x}_2$, and we choose only those edges for which the average gradient of the field of interest is larger than a user-specified threshold value $|\nabla\rho|_{threshold}$. This is done to eliminate locations where we have a zero Laplacian but which do not lie within a high gradient region. At the end of this step, we have identified all intersection points of the zero contour with all the edges.

3. *Extraction of the discontinuity curve*
   In this step, the intersection points identified in the previous step are connected to form the curves which define the discontinuities in the flow. Each discontinuity is a curve which is stored as a linked list of points. The process of identifying curves is recursive, and the pseudo-code is given in Appendix A.

4. *Spline interpolation*
   In the above step, we have isolated curves which are a list of points. The distribution of these points is, clearly, not uniform. In this step we fit natural cubic splines (Press *et al.* , 1988) to these. The curves are re-meshed so that points along the curve are uniformly distributed.

5. *Shock and contact discontinuity identification and quantification*
   For each curve (this curve is now the fitted spline curve), we identify whether the discontinuity is a shock wave, a contact discontinuity or neither, that is, a spurious discontinuity is identified. The process of identifying shocks is as follows. Recall that the curve is discretized with equally spaced points. At each point along the curve, a line normal to the curve is generated with equally spaced points. For equally spaced points on either side of the curve, flow variables such as the density, pressure and velocity $(\rho, p, \bar{u})$ are calculated using bicubic interpolation. Then we evaluate the normal jump conditions, given below, using equally spaced points on

either side of the curve. The question that arises is: how far must we go in the normal direction so that we are not in the smeared zone of the discontinuity?

For shocks, we travel along the normal direction and find the location where a certain cost function, to be defined later, is minimized. Clearly, the points closest to the shocks do not satisfy the jump conditions because of smearing.

Let W and u_n be the shock and fluid velocity, respectively, in a direction normal to the shock front. The shock speed is calculated using the following jump condition

$$W = \frac{\rho_2 u_{n2} - \rho_1 u_{n1}}{\rho_2 - \rho_1}. \tag{12.8}$$

We define a cost function, $\mathcal{S}$ using the three jump conditions for a planar shock moving with speed W, as

$$S_1 = 1 - \frac{\mu^2 p_r + 1}{(\mu^2 + p_r)\rho_r}, \tag{12.9}$$

$$S_2 = 1 - \frac{p_2 + \rho_2(W - u_{n2})^2}{p_1 + \rho_1(W - u_{n1})^2}, \tag{12.10}$$

$$S_3 = 1 - \frac{h_2 + \frac{1}{2}(W - u_{n2})^2}{h_1 + \frac{1}{2}(W - u_{n1})^2}, \tag{12.11}$$

$$\mathcal{S} = \omega_{s,i} S_i, \quad i = 1, 2, 3, \tag{12.12}$$

where $\rho_r \equiv \rho_2/\rho_1$, $p_r \equiv p_2/p_1$, $\mu^2 \equiv (\gamma + 1)/(\gamma - 1)$, and h is the enthalpy. Ideally, the jump conditions across the shock must be satisfied and therefore $S_i = 0$, $i = 1, 2, 3$ across the shock. Numerically, the jump conditions are not exactly satisfied because shocks are smeared and the contour corresponding to the shock front is only an approximate representation. The final form of the cost function $\mathcal{S}$ is a weighted average of S_i with weights $\omega_{s,i}$. For curves which are not shocks, the jump conditions are obviously not satisfied. We have some simple physically-based constraints which eliminate points as not belonging to shocks. For example, both the density ratio and pressure ratio across shocks in perfect gases have to be greater than unity. Furthermore, the local normal Mach number, computed by using the relative velocities, changes from larger than unity to smaller than unity across the shocks. Therefore, to decrease computational costs, points not satisfying these constraints are eliminated and not processed any further.

For contact discontinuities, we use the fact that the pressure and normal velocities are continuous across the contacts. One difficulty with contacts is that they tend to diffuse more than shocks. The cost function for a contact discontinuity is defined as follows:

$$C_1 = 1 - \frac{p_2}{p_1}, \qquad C_2 = 1 - \frac{u_{n2}}{u_{n1}} \qquad (12.13)$$

$$\mathcal{C} = \omega_{c,i} C_i, \quad i = 1, 2. \qquad (12.14)$$

We find the locations on either side of the discontinuity which minimize the cost functions, ($\mathcal{S}$ for shocks and $\mathcal{C}$ for contact discontinuities). Then the properties at these locations are evaluated, and we can then assign the shock speed, shock strength, etc. along a shock front, and strength of the vortex sheet along a contact discontinuity.

12.3.3   Two-dimensional example

We now apply the above algorithm for extracting curves of discontinuity to the two-dimensional interaction of a shock with an inclined planar contact discontinuity. The threshold used in Eqn. 12.7 is $|\nabla \rho|_{threshold} = 0.008 |\nabla \rho|_{max}$, where $|\nabla \rho|_{max}$ is the maximum gradient magnitude of density. Furthermore, the field $\nabla^2 \rho$ was smoothed four times recursively using Eqn. 12.5 with $n = 2$ and weights $w_{\pm 1, \pm 1} = 1/16$, $w_{0, \pm 1} = 1/8$, $w_{\pm 1, 0} = 1/8$ and $w_{0,0} = 1/4$. The extracted curves are shown in Fig. 12.11. Curves labeled 1, 3, 5, 6, and 7 are shock waves, while the remaining curves 2 and 8 are contact discontinuities (shear layers). A brief explanation of the numbering system follows. The algorithm starts by scanning the (x, y) domain from left to right and bottom to top. Whenever a discontinuity is encountered, a label is generated for it. Thus, in our example, the reflected shock labeled 1 is first encountered, followed by the primary contact discontinuity labeled 2. The next discontinuity that the algorithm encounters is the transmitted shock 3 and so on.

Although not apparent in Fig. 12.11, we find anomalies in a magnification of Fig. 12.11. In regions where the discontinuities approach close to each other, for example the ideal triple point T1, we find that the "weaker" extracted discontinuity curves do not intersect at a point with the "strongest." The neighborhood around the triple point T1 is shown in Fig. 12.12. The weaker extracted curves curves in these regions show an unphysical turning with a high curvature as they

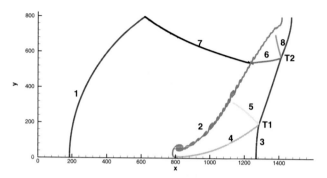

Fig. 12.11. Extracted shocks and contact discontinuities (shear layers) in the two-dimensional shock contact-discontinuity interaction at $t = 0.72$ from simulation GH. Curves labeled 1, 3, 5, 6, and 7 are shock waves, while the remaining curves are contact discontinuities. Labels T1 and T2 are locations of triple points where three shocks and one contact discontinuity meet. The x- and y-axes in the figure are normalized by the mesh spacing. Figure also shown as Color Plate 17.

approach an ideal triple point. The maximum offset is about five grid points. The same is true at T2 and the location where 2, 6 and 7 appear to intersect. We simply caution the reader that this region of non-intersection (unphysical turning) must be ignored in the quantification process until a universal algorithm that resolves this issue is available. A similar disjoint problem with bifurcating skeletal extractions was found in a previous work (Feher & Zabusky, 1996).

Note that the original straight-line contact discontinuity, has evolved into the continuous curve labeled 2, with a many-turn "wall vortex" at lower left and many rolls along its entire length. A discussion of this and related convergence issues is given in the next section.

12.3.4   Contact tracking and convergence of simulations

Some of the convergence issues associated with the contact location were already discussed in the one-dimensional example (Section 12.3.1). To illustrate and better understand the convergence issues associated with interfaces when vorticity exists, we present an additional interface result obtained from solutions of a level set (Sethian, 1996) partial differential equation in two dimensions

$$\frac{\partial \rho \zeta}{\partial t} + \frac{\partial \rho \zeta u}{\partial x} + \frac{\partial \rho \zeta v}{\partial y} = 0, \tag{12.15}$$

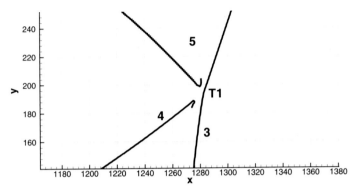

Fig. 12.12. Magnified neighborhood of triple point T1. The extracted shocks are 3 and 5 and contact is 4. The x- and y-axes in the figure are normalized by the mesh spacing.

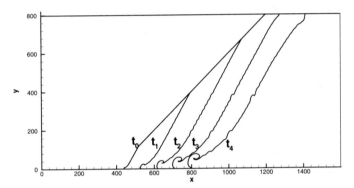

Fig. 12.13. The zero level set at various times in the two-dimensional shock contact-discontinuity interaction, GH. The times shown are $t_0 = 0.0$, $t_1 = 0.20$, $t_2 = 0.38$, $t_3 = 0.54$, $t_4 = 0.72$. The x- and y-axes in the figure are normalized by the mesh spacing.

where ζ is the level-set variable. We initialize $\zeta = \pm 1$ on either side of the interface so that at time t the level set $\zeta(x, y, t) = 0$ defines the interface. Figure 12.13 shows $\zeta(x, y, t) = 0$ at various times for simulation GH, and the rapid appearance of small-scale structure.

As mentioned in the previous section, and as further seen in Fig. 12.13, small-scale structures appear on the contact discontinuity. Ideally, the contact discontinuity is a compressible vortex sheet, while in practice it is a compressible vortex layer. It is well known that a compressible vortex sheet for which the convective Mach number is less than one is unstable to all wave number perturbations (Miles, 1958). Thus linear stability of a plane vortex sheet is ill-posed, and high frequency modes grow rapidly. With mesh refinement, the circulation on the vortex layer converges while the spatial extent of the vortex layer decreases, giving rise to an increase in the local vorticity magnitude. Essentially, the vortex layer develops local regions of high vorticity or "rolls" in the EH and GH higher resolution simulations. For a more detailed discussion of convergence issues in the presence of vortex sheets for numerical solutions of the compressible Euler equations, the reader is referred to Samtaney & Pullin (1996).

In Fig. 12.14, we juxtapose the vorticity (contours or color) with two curves, the zero level set and zero crossing of the Laplacian ($\zeta(t) = 0$ and $\nabla^2\rho = 0$, respectively) in the zoomed domain around the wall vortex for the primary contact discontinuity. The results are from simulations EH and GH with the results EH reflected about the x-axis for convenience.

Several things are to be noted. The $\zeta = 0$ curve (red) is not as tortuous as the $\nabla^2\rho = 0$ curve (black). That is, at minima of vorticity (neighborhoods of vorticity minima are colored dark blue), the former exhibits fewer local rotations or "turns" in comparison to the latter. This is particularly apparent at the dominant wall vortex which is centered on $\nabla^2\rho = 0$ curve but displaced into the heavier fluid domain as delineated by the $\zeta = 0$ curve. Thus we conclude that the level set solution in these simulations is a more diffusive representation of the interface. For the EH simulation, the manifestation of the high-frequency modes is not as dramatic as the GH simulation. Furthermore, the $\nabla^2\rho = 0$ curve shows fewer turns within the wall vortex for the EH simulation as compared with the GH simulation. This is due to the fact that EFM is more diffusive than the Godunov method. We note that for simulations GL and EL no small rolls are evident at this time. Thus our approach provides another approach to quantifying the diffusive nature of codes.

We now focus our attention to the phenomenon of "tip splitting". As evident in the top half of Fig. 12.14, the extreme left end is indented (a "tip split"), a consequence of the near-wall shear layer 4 (red/yellow positive domain near the x-axis), as explained below. We also note that the manifestation of diffusiveness is in the magnitude of the split tip, which is smaller for the $\zeta = 0$ curve. In this figure we readily see the cause of the split tip. Essentially the shear layer arising at T1 is entrained very close to the wall and below the dominant wall vortex. This is a very competitive situation, where the dominant wall vortex, rotating clockwise, wants to move the rolled (localized) end of the shear layer to

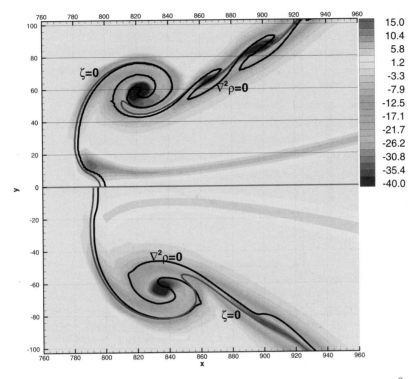

Fig. 12.14. A comparison of vorticity with two interface curves, $\zeta = 0$ and $\nabla^2\rho = 0$ at $t = 0.72$ for the GH (above) and EH (below) simulations. (The results from EH are reflected about the x-axis.) The x and y axes are normalized by the mesh spacing. Figure also shown as Color Plate 18.

the left and up, while the localized region is interacting with its mirror image at close range as a dipolar entity which moves to the right. This right-movement wins out and the dipolar entity entrains the interface causing an indentation. Obviously this very competitive situation is affected by diffusion and in fact is not seen in GL and EL and is only marginally evident in EH.

12.3.5 Quantification of local shock properties

We now apply the quantification part of the algorithm presented in Section 12.3.2. In practice, we find that the cost function which best quantifies the shock is the one with weights $\omega_{s,1} = 1, \omega_{s,2} = \omega_{s,3} = 0$. This cost function, which relates the

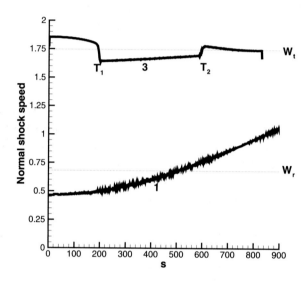

Fig. 12.15. Normal shock speed as a function of arc length s of the shock fronts identified by labels 1 and 3. Labels T1 and T2 on shock number 3 are approximate locations of the triple points on the shock front. The horizontal lines are the speeds of the reflected (labeled W_r) and the transmitted shock (labeled W_t) in a one-dimensional shock-contact interaction, and they are shown for reference.

pressure and density jumps as we move along a direction normal to the shock curve, has a very well-defined minimum. The magnitude of the normal shock velocity for shocks labeled 1 and 3 is plotted in Fig. 12.15 as a function of the length of the shock curve. For reference, we also plot the speeds of the reflected and transmitted shocks in a one-dimensional interaction. Note that shock 3 is really comprised of three different shock fronts. These are the Mach stem extending from the lower boundary to T1, followed by the shock front from T1 to T2, and finally from T2 to the upper boundary. The zero crossing of the Laplacian of density (in fact, even other variables) fails to distinguish between these three shocks and identifies them as a single shock. However, in the quantification of the shock front, we see changes in the normal shock speed.

We also observe that the normal shock speed of shock 1 is a noisier curve. This may be due to several sources of error for this weaker shock wave: the numerical method used to compute the flow, the error in the identification method of the zero crossing (which essentially employs linear interpolation), or the cost function used to quantify the shock front, or a combination of these. A thorough analysis of these errors is beyond the scope of the present study and is left for future work.

12.4 Conclusion

In this chapter, we presented the numerical analog of experimental flow visualization techniques for the compressible flow of a gas with shocks and contact discontinuities. However, we emphasized mainly the extraction, identification and quantification of these "discontinuity" curves and the physical quantities which vary across their normal directions. This allows us a deeper insight into the mathematical and *computational* properties of complex nonlinear fluid phenomena. In particular, we examined aspects of the convergence and errors of the Godunov and EFM codes and the level-set interface tracker.

An algorithm based on the zero crossing of the Laplacian of a field quantity (usually density) was developed to extract the discontinuities. The discontinuities were characterized by curves which were extracted using a recursive technique. Furthermore, we also quantified properties along the extracted curve such as the local strength of the shock or the normal shock speed. This determination is based on the minimum of a cost function in a direction normal to the shock front. The extracted discontinuities with associated properties may be thought of as a means of data reduction. By way of illustration, we applied the methods developed in this paper to the unsteady interaction of a shock wave with a contact discontinuity which yields a flow field rich in bifurcations and discontinuities. An extension of the extraction and quantification algorithm to adaptive meshes in three dimensions is left for the future.

Several other topics were briefly mentioned. These include the selection of appropriate variables. It is recommended that density be used to extract both contact discontinuities and shock; pressure and the divergence of velocity to extract only shocks; and entropy to extract contacts and strong shocks. As far as edge detection techniques is concerned, we recommend the use of the zero crossing of the Laplacian of the density (a Marr edge detector) to extract discontinuities. One-dimensional tests indicate that the location of the centroid of the contact "layer" may be subject to systematic errors. Finally, we remark that there are several sources of error (and noise) in the entire process: from the simulation method, to the discontinuity extraction technique, to the form of the cost function used to quantify the properties along the discontinuity. A judicious application of some smoothing mitigates noise. In the future we look forward to *tracking* the discontinuity curves in time and generalizing all of it to adaptive meshes in three dimensions. Also, it is most important to quantify sources of error and their effect on observable dynamics, including: choice of variables and their thresholds; variation of topologies, lengths and "widths" of the extracted domains about the discontinuity curves and modifying cost functions to be consistent with unsteady motions of shocks, and so on.

The quantification aspects of this paper contribute to the field of post-processing numerical mathematics. This is one cogent approach to the reduction of massive data sets to essential mathematical-physical entities of the numerical solution.

Acknowledgments

The work of R. Samtaney was supported in part by NAS, NASA Contract NAS2-14303. The work of N. J. Zabusky was supported mainly by Dr. Fred Howes and Dr. Daniel Hitchcock of Department of Energy (Grant DE-FG02-98ER25364) and also by grants from SROA and the CAIP center at Rutgers.

Appendix A: Pseudo-code to extract the discontinuity curves

```
Triangle triangle[NTriangles];
Curve curve[];
int n;
int ncurve;
int nedge;
int i;

for(n=0;n<NTriangles;n++){
// If triangle is cut only once
// this is the beginning of a curve.
 if(triangle[n].ncut==1){
// Determine which edge is cut.
  for(i=0;i<3;i++){
   if(triangle[n].edge[i].iscut)
    nedge=i; break;}
 }
```

```
// Get the coordinates of the
// intersection point with the cut edge.
 triangle[n].edge[nedge].
        GetIntersectionPoint(&x, &y);
 triangle[n].edge[nedge].iscut=0;
 if(triangle[n].edge[nedge].
        neighbor_triangle!=NULL){
  nt=triangle[n].edge[nedge].
     neighbor_triangle.id;
  triangle[n].ncut=0;
// Add intersection point to the curve.
  curve[ncurve].AddPoint(x,y);
// Traverse the curve using the
// following recursive routine.
  TraverseCurve(ncurve,nt);
 }
 ncurve++;

// The curve can also start at the boundary.
 if(ncut==2 && triangle[n].IsOnBoundary){

  for(i=0;i<3;i++){
   if(triangle[n].edge[i].iscut &&
      triangle[n].edge[i].
      neighbor_triangle==NULL){
    nt=triangle[n].edge[i].
       neighbor_triangle.id;
    triangle[n].ncut=0;
    triangle[n].edge[i].iscut=0;
    triangle[n].edge[i].
        GetIntersectionPoint(&x, &y);
// Add intersection point to the curve.
    curve[ncurve].AddPoint(x,y);
// Traverse the curve using the
// following recursive routine.
    TraverseCurve(ncurve,nt);
    ncurve++;
    break;
```

```
    }
   }
  }
 }


// Recursive routine to traverse the curve.
TraverseCurve(int ncurve, int n)
{
// Reached end of curve
 if(triangle[n].ncut==1){
  triangle[n].ncut=0;
  return;
 }

// Still on the curve.
 if(triangle[n].ncut==2 ) {

  for(i=0;i<3;i++){
   if(triangle[n].edge[i].iscut &&
      triangle[n].edge[i].
      neighbor_triangle==NULL){
    nt=triangle[n].edge[i].
       neighbor_triangle.id;
    triangle[n].ncut=0;
    triangle[n].edge[i].
       GetIntersectionPoint(&x, &y);
    triangle[n].edge[i].iscut=0;
// Add intersection point to the curve.
    curve[ncurve].AddPoint(x,y);
// Traverse the curve using the
// following recursive routine.
    TraverseCurve(ncurve,nt);
    break;
   }
  }

 }
// Should never reach here.
```

```
    return;
}
```

12.5   References

Courant, R. and Friedrichs, K. O. 1948. *Supersonic Flow and Shock Waves.* Springer-Verlag.

Feher, A. and Zabusky, N. J. 1996.   An interactive imaging environment for scientific visualization and quantification. *International Journal of Imaging Systems and Technology,* **7**, 121–130.

Krehl, P. and Engemann, E. 1995. August Toepler – the first who visualized shock waves. *Shock Waves,* **5**, 1–18.

LeVeque, R. J. 1992. *Numerical Methods for Conservation Laws.* Birkhauser Verlag.

LeVeque, R. J., Mihalas, D., Dorfi, E.A. and Müller, E. 1998. *Computational Methods for Astrophysical Fluid Flow.* Springer Verlag.

Lovely, D. and Haimes, R. 1998. Shock detection from the results of computational fluid dynamics. *Preprint..*

Ma, K.-L., Van Rosendale, J. and Vermeer, W. 1996.   3D Shock wave visualization on unstructured grids. In *Proceeding of the 1996 Symposium on volume visualization, San Francisco, California, October 28–29.,* 87–94. ACM SIGGRAPH.

Merzkirch, W. 1974. *Flow Visualization.* Academic Press.

Miles, J. W. 1958.   On the disturbed motion of a plane vortex sheet.   *J. Fluid Mech.,* **3**, 538–552.

Pagendarm, H.-G. and Seitz, B. 1993.   An algorithm for detection and visualization of discontinuities in scientific data fields applied to flow data with shock waves.   In P. Palamidese, editor, *Visualization in Scientific Computing.* Ellis Horwood Workshop Series.

Parker, J. R. 1997. *Algorithms for Image Processing and Computer Vision.* John Wiley and Sons, Inc..

Press, W. H., Flannery, B. P., Teukolsky, S. A. and Vetterling, W. T. 1988. *Numerical Recipes in C.* Cambridge University Press.

Pullin, D.I. 1980. Direct simulation methods for compressible ideal gas flow. *J. Comput. Phys.,* **34**, 231–244.

Samtaney, R. and Pullin, D.I. 1996. On initial-value and self-similar solutions of the compressible Euler equations. *Phys. Fluids.,* **8** (10), 2650–2655.

Samtaney, R. and Zabusky, N. J. 1994.   Circulation deposition on shock-accelerated planar and curved density-stratified interfaces: models and scaling laws. *J. Fluid Mech.*, **269**, 45–78.

Schalkoff, R. J. 1988. *Digital Image Processing and Computer Vision.* John Wiley and Sons, Inc..

Sethian, J. A. 1996. *Level Set Methods: Evolving Interfaces in Geometry, Fluid Mechanics, Computer Vision, and Material Science.* Cambridge University Press.

Thompson, P. A. 1972. *Compressible Fluid Dynamics.* McGraw Hill, New York.

Vorozhtsov, E. N. and Yanenko, N. N. 1990. *Method for the localization of solutions of gas dynamic problems.* Springer-Verlag.

Zabusky, N. J. 1999.   Vortex paradigm for accelerated inhomogeneous flows: Visiometrics for the Rayleigh-Taylor and Richtmyer-Meshkov environments. *Ann. Rev. Fluid Mech.*, **31**, 495–526.

Color Plates

And

Flow Gallery

346

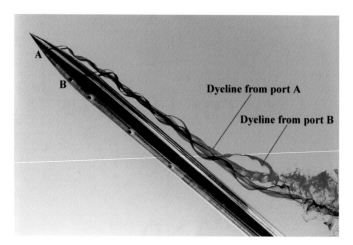

Plate 1. Picture showing dye lines of flow past a tangent ogive cylinder at high angle of attack. The flow is from left to right. Note the dependence of the streakline pattern on the location where the dye is released (Luo *et al.*, 1998).

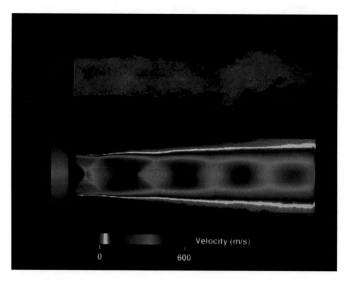

Plate 2. Instantaneous and time-averaged velocity fields of an over-expanded supersonic jet (Smith & Northam, 1995).

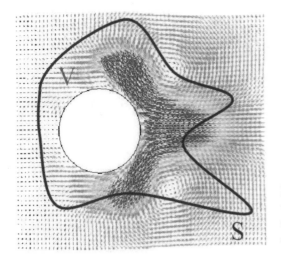

Plate 3. The control surface and volume
used with the instantaneous velocity field
of an oscillating cylinder.(Figure courtesy
of F. Noca.)

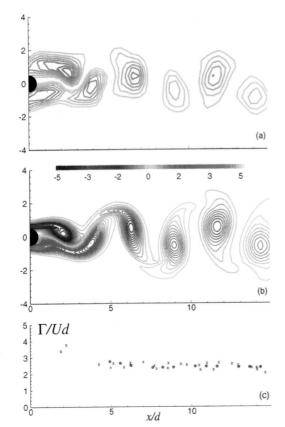

Plate 4. Vorticity in the wake of a circular cylinder, Re =100: (a) DPIV measurements (b) 2-D numerical simulations (c)
computed values of the circulation of wake
vorticies from experiments, ×, and simulations, ●.

348

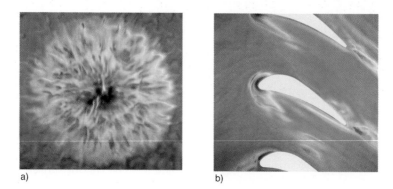

a) b)

Plate 5. Images of *instantaneous* liquid crystal surface temperature patterns generated by: (a) a jet impinging perpendicular to a heated surface; and (b) a turbulent juncture end-wall in a linear turbine blade cascade (Sabatino & Praisner, 1998).

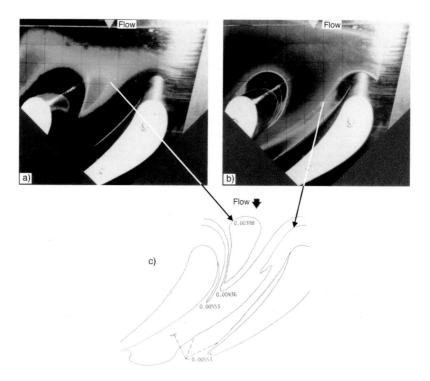

Plate 6. Photographic images illustrating the employment of a constant heat flux surface in conjunction with a narrow-band calibration technique. Images in (a) and (b) were used to determine the contours indicated in (c). After Hippensteele & Russell (1998).

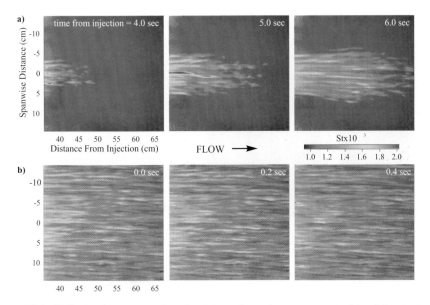

Plate 7. Instantaneous surface heat transfer patterns generated by (a) a turbulent spot (produced by an upstream injection into a laminar boundary layer), passing over a constant heat flux surface at $Re_x = 2 \times 10^5$; (b) a fully turbulent boundary at $Re_\theta = 10,000$. After Sabatino (1997).

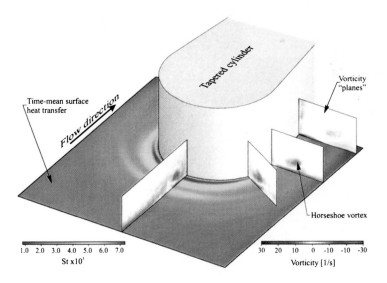

Plate 8. Composite image of time-mean vorticity and end-wall heat transfer for a turbulent end-wall juncture. Image is to scale except for the height of the cylinder which was twice the diameter. After Praisner (1998).

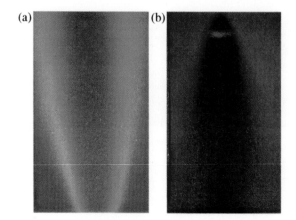

Plate 9. Color-change response of liquid crystal coating to tangential jet flow, $\alpha_L = 90°$, $\alpha_C = 35°$. (a) Flow away from, and (b) flow toward observer.

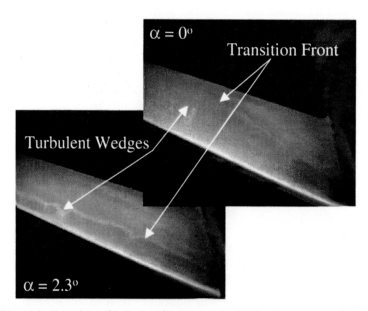

Plate 10. Transition-front visualizations recorded by downstream-facing camera at $M = 0.4$ and $Re = 8.2 \times 10^6$/m.

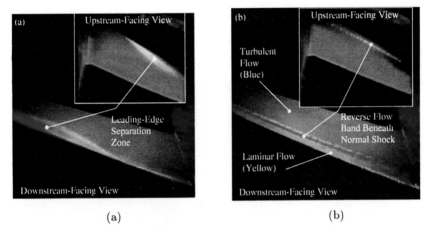

(a)   (b)

Plate 11. Color-change response as recorded by opposing-view cameras: (a) leading-edge separation, $\alpha = 8°$, $M = 0.4$, $Re = 8.2 \times 10^6$/m; (b) normal-shock/boundary-layer interaction, $\alpha = 5°$, $M = 0.8$, $Re = 11.2 \times 10^6$/m.

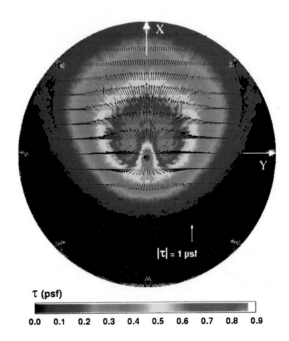

Plate 12. Measured surface shear stress vector field beneath inclined, impinging jet: colors show shear magnitudes and vector profiles every $\triangle X/D = 1$ show shear orientations.

352

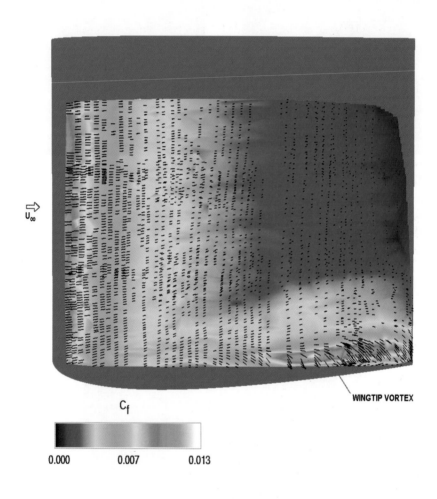

Plate 13. Measured wingtip skin friction distribution.

353

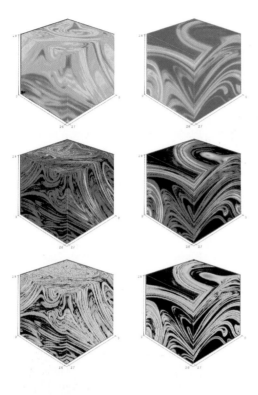

Plate 14. Typical fully-resolved three-dimensional 256^3 spatial data volumes from quantitative visualizations, showing the conserved scalar field $\zeta(\mathbf{x}, t)$ (top row), the scalar energy dissipation rate field $\nabla \zeta \cdot \nabla \zeta(\mathbf{x}, t)$ (middle row), and $\log \nabla \zeta \cdot \nabla \zeta(\mathbf{x}, t)$ (bottom row). Originally from Southerland & Dahm (1994); reproduced with permission from Frederiksen *et al.* (1996).

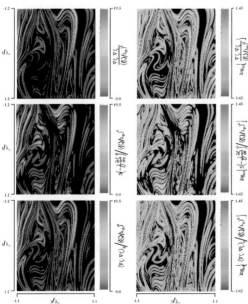

Plate 15. Comparisons of the true scalar dissipation rate field (top) with the single-point Taylor series approximation (middle), and with a two-point mixed approximation (bottom) based on the time-derivative and one spatial derivative. Results in linear form (left) allow comparing relatively high dissipation rates, and in logarithmic form (right) allow comparisons of lower values. Reproduced with permission from Dahm & Southerland (1997).

354

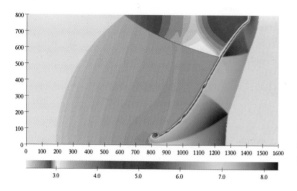

Plate 16. Density field of a two-dimensional shock contact-discontinuity interaction at time $t = 0.72$. Simulation GH.

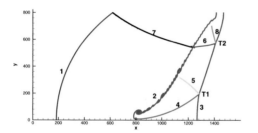

Plate 17. Extracted shocks and contact discontinuities (shear layers) in the two-dimensional shock contact-discontinuity interaction at $t = 0.72$ from simulation GH. Curves labeled 1, 3, 5, 6, and 7 are shock waves, while the remaining curves are contact discontinuities. Labels T1 and T2 are locations of triple points where three shocks and one contact discontinuity meet. The x- and y-axes in the figure are normalized by the mesh spacing.

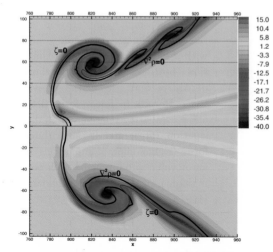

Plate 18. A comparison of vorticity with two interface curves, $\zeta = 0$ and $\nabla^2 \rho = 0$ at $t = 0.72$ for the GH (above) and EH (below) simulations. (The results from EH are reflected about the x-axis.) The x and y axes are normalized by the mesh spacing.

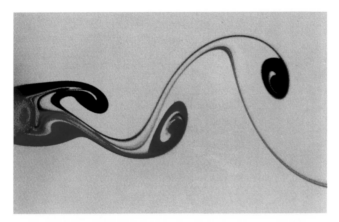

Dye traces in a Kármán vortex street behind a circular cylinder. Blue dye corresponds to positive vorticity and red dye corresponds to negative vorticity. Reynolds number based on the diameter of the cylinder is approximately 80 (A.E.Perry *et. al.*, 1982).

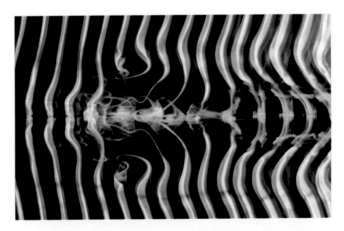

A forced two-sided or symmetric vortex dislocations in the wake of a circular cylinder, in the laminar vortex shedding regime (Re=120). The flow is from left to right. The experiment is conducted by towing a cylinder (with a small ring located at mid-span of the cylinder) along the length of the towing tank, and the shed vortices (vertical green lines) were visualized using laser light, which excited flourescein and Rhodamine dye washed of the surface of the body. Here, it can be seen that the spanwise extent of the symmetric dislocation is far larger than the width of the small ring-disturbance (shown as the yellow dye). Also, the structures are remarkably symmetric on either side of the ring disturbance, even including similar vortex linking and 'wisps' of stretched vortex tubes. Interestingly, these large two-sided structures occur naturally in wake transition, but have not yet been fully simulated in computation (C.H.K. Williamson, 1992).

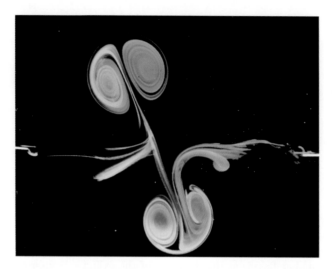

Dipoles collision in a stratified fluid. In this experiment, two "identical" dipoles are produced by injecting a fixed volume of fluid simultaneously through two nozzles of equal diameter placed opposite each other at some distance apart. The Reynolds number based on the nozzle diameter is of the order of 1000. The photograph is obtained 225s after the injection has stopped. When two dipolar vortices collide frontally, the so-called 'partner exchange' is observed: each dipole splits into two, and two new dipoles are formed that move away along straight trajectories. Here, the partner exchange is visualized by using two different dyes (the original dipoles were green and orange). The collision is not exactly symmetric and slightly mis-aligned (G.J.F. van Heijst & J.B. Flor, 1989).

The photograph shows a tripolar vortex which is produced from an unstable cyclonic vortex in a homogeneous rotating fluid. It consists of cyclonic motion in the core and anticyclonic motion concentrated in the two satellite vortices (G.J.F. van Heijst & R.C. Kloosterziel, 1989).

A round jet discharges normal to a cross-stream in a water flow. The jet Reynolds number is 3,800 and the jet/cross-stream velocity ratio is approximately 5. Visualization is by coloured dyes released from a small hole below the lip of the circular pipe and from a dye probe far upstream. The dye marks the jet shear layer, which rolls up naturally to produce distorted ring vortices. The shear layer roll-up leading to the counter-rotating trailing vortices in the jet can also be seen on the downstream side. These vortices are observed to break down a short distance downstream of the jet exit (R.M. Kelso *et. al.*, 1992,1996).

The photograph shows the flow past a tangent ogive cylinder at high angle of attack. The Reynolds number based on the diameter of the cylinder is approximately 2400, and the angle of attack is about 50°. Dye was released from selected ports close to the nose tip. The asymmetry of the vortex pattern can be clearly seen, with the starboard-side vortex lifted off from the cylinder earlier than the port-side vortex. The flow asymmetry is partly responsible for the generation of side-force acting on the cylinder (S.C.Luo *et. al.*, 1998).

358

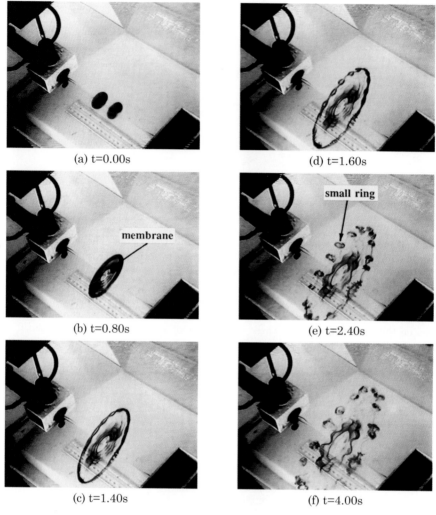

(a) t=0.00s

(b) t=0.80s

membrane

(c) t=1.40s

(d) t=1.60s

(e) t=2.40s

small ring

(f) t=4.00s

A sequence of photographs showing different stages of head-on collision between two identical vortex rings generated by simultaneously ejecting fluid through two nozzles of equal diameter placed opposite each other at 220mm apart. The Reynolds number based on the initial translation velocity and the diameter of the ring is approximately 1000. The rings were made visible by blue and red dye released around the circumference of the nozzle. The first photograph of the sequence has been arbitrarily assigned as t=0.00s. The elapsed time for the subsequent stages of the collision is shown below photograph. During collision, the rings develop azimuthal waves which grow until they touch. At the point of contact, "vortex reconnection" takes place, which subsequently leads to the formation of smaller rings(T.T. Lim & T.B. Nickels, 1992).

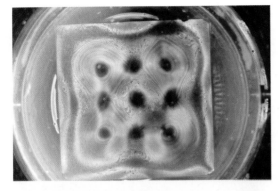

Mode patterns produced by horizontal soap films stretched over a square frame 7.0cm on a side and a circular frame 8.0cm in diameter undergoing periodic transverse oscillations at different frequencies and accelerations. The soap solution consists typically of 94% of water, 5% glycerin and 1% liquid soap. The square cell exhibits a shadowgraph image of flexure mode pattern for a relatively new film at f=70Hz and g/g_o=13.7, where g_o is Earth's gravity. The circular film exhibits interference fringes displaying vortex motion in a relatively old fim(thin, by evaporation)at f=40.4 Hz and g/g_o=17.1 (Afenchenko *et. al.*, 1998).

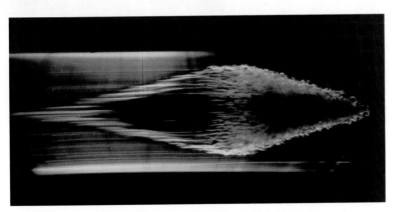

A plan view of a turbulent spot created when a small amount of fluid is ejected from a 0.5mm diameter hole located at 33cm downstream of the leading edge of a flat plate. The turbulent spot was made visible by a sheet of laser which excited the flourescent dye released uniformly from the spanwise slot. The Reynolds number based upon a towing speed of the plate of 40 cm/s and the displacement thickness at the injection hole was about 625 (M. Gad-el-Hak *et. al.*, 1981).

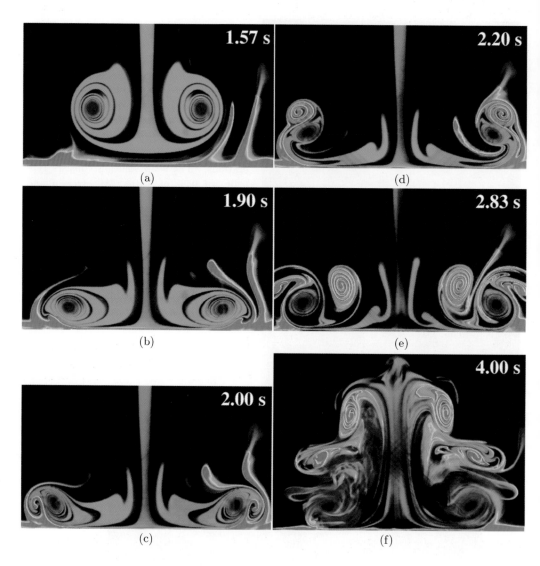

When a vortex ring approaches a wall at normal incidence, the unsteady adverse pressure gradient on the wall causes boundary layer separation and the formation of a secondary vortex of opposite sign to that of the primary ring. In the case shown here, as primary and secondary vortices interact, a tertiary vortex is formed and later in time the secondary vortex ring moves quickly away from the wall. The flow visualization images shown here were obtained using two color LIF, where the vortex ring and boundary layer fluids are labelled by green and red remitting laser dyes respectively (C.P. Gendrich *et. al.*, 1997).

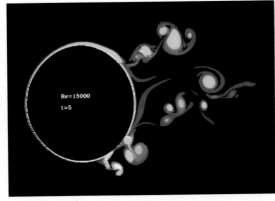

The figure presents the vorticity field computed with viscous vortex method developed by the authors for 2D, incompressible, Re=15000 flow over a circular cylinder. The cylinder is impulsively started and is performing a time dependent rotation of $\Omega(t)=2\sin(\pi t)$. The rotation induces the ejection of multipole vortex structures from the boundary layer. As a result, the time-averaged flow shows a significant separation delay that yields drag reduction (D. Shiels & A. Leonard, 1998).

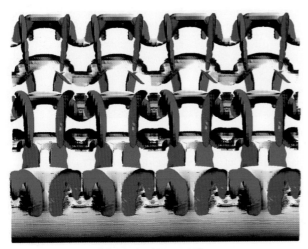

(a) Mode A instability

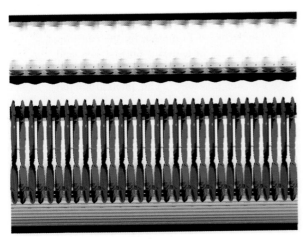

(b) Mode B instability

Numerical simulations of Mode A and Mode B secondary instability in the wake of a circular cylinder. The flows were computed by solving the full three-dimensional Navier-Stokes equations using the linear instability mode as an initial condition and periodic boundary conditions that exclude the more unstable long-wavelength modes of the wake. As presented, the flow direction is to the top. (a) Mode A at Re=195. The image shows isosurfaces of vorticity with values of $\xi_x= +0.75$ (red), $\xi_x =$-0.75 (blue), and $|\xi_z| =1$ (silver). This is a relatively long-wavelength instability ($\lambda_z=$ 3.96d, where d is the cylinder diameter) that scales on the distance between Kármán vortices in the wake. (b) Mode B at Re=265. The image shows isosurfaces of vorticity with values of $\xi_x = +0.5$ (red), $\xi_x = $-0.5 (blue), and $|\xi_z| =1$ (silver). This is a relatively short-wavelength instability (λ_z = 0.822d, where d is the cylinder diameter) that scales on the thickness of the separating shear layer (R. D. Henderson, 1997).

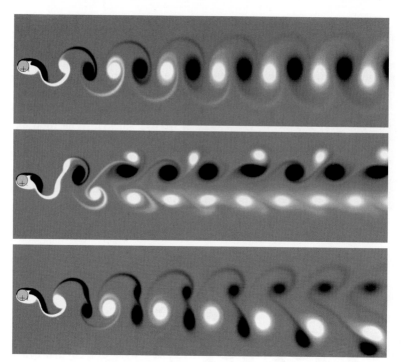

Vorticity fields for two-dimensional incompressible flows past oscillating cylinders computed using a spectral element discretisation. In each case, Re=500, and the cross-flow oscillation amplitude is a quarter of the cylinder diameter. The first plot shows a conventional Kármán street wake, while the two lower plots illustrate alternative time-periodic wakes obtained on solution branches that bifurcate from the Kármán street branch. These pictures are flooded contour versions of figures 11(a, b) and 14 in (H.M. Blackburn & R.D. Henderson 1999).

When the fluid inside a completely filled cylinder is set in motion by the constant rotation of an endwall, both steady and unsteady axisymmetric vortex breakdown flow are possible. The photograph on the left is a perspective view of a snapshot of a dye sheet that was released from an axisymmetric ring co-centric with the axis, near the stationary endwall, when the flow is time-periodic. Evident is the filling and emptying through the downstream end of the vortex breakdown structure. The Reynolds number of the flow is Re=2765 and the cylinder aspect ratio is H/R=2.5 (J.M.Lopez & A.D. Perry, 1992).

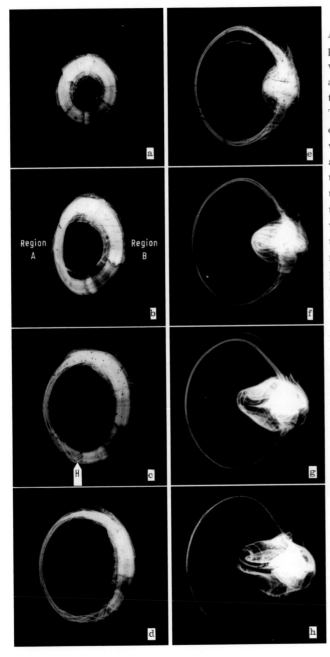

A sequence of front-view photographs showing a vortex ring approaching an inclined wall made of transparent glass plate. The experiment was conducted in a water tank with the plate inclined at 51.5° to the direction of vortex ring motion. As presented, the ring is moving towards the reader and the left-hand side of the ring first touches the plate at region A (see (b)). The time interval between successive photographs is 0.42s. The ring was made visible by milk/alcohol mixture and the interaction was recorded using a 16mm motion picture camera. The Reynolds number of the flow based on the initial velocity and the diameter of the nozzle is about 600. Note how differential vortex stretching has led to the formation of bi-helical vortex lines (see (c)) which are constantly being displaced along the circumferential axis and towards the region of the ring furthest away from the wall (T. T. Lim, 1989).

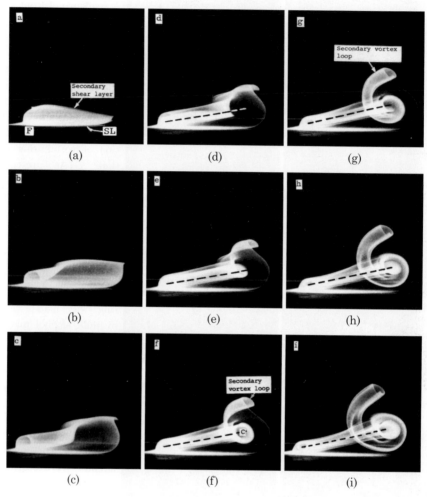

A sequence of side-view photographs showing the development of boundary layer material when a vortex ring interacts obliquely with a solid boundary. The ring, which is not visible, is travelling from top left-hand corner towards the bottom of the picture. The interaction is similar to that shown in the previous figure except that *dye was placed on the floor only*. The time interval between successive photographs is 1s. Dashed line in (d)-(i) indicate the approximate position of the circumferential axis of the primary vortex ring (T.T. Lim, 1989).

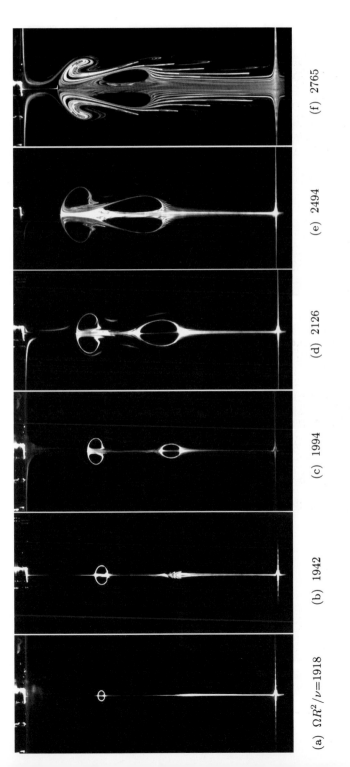

(a) $\Omega R^2/\nu = 1918$ (b) 1942   (c) 1994   (d) 2126   (e) 2494   (f) 2765

Swirling flow: The steady-state flowfield produced in a closed cylindrical container by rotation of one endwall is determined by the aspect ratio H/R and Reynolds number $\Omega R^2/\nu$. H=cylinder height, R=radius, Ω=angular velcity of endwall, and ν=viscosity of fluid. The above photographs show changes in vortex structure with increasing Reynolds number for H/R=2.5. Here, the rotating wall is located at the bottom of each picture. Flow visualization is carried out with the aids 5 W Argon laser and flourescein dye which is introduced from a hypodermic syringe through the center of the non-rotating endwall (M.P. Escudier, 1984).

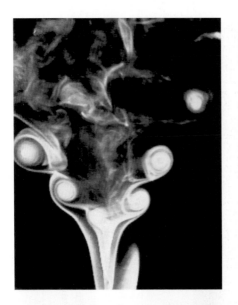

Unconfined vortex breakdowns: An unconfined vortex is referred to a vortex that is not directly influenced by the boundary layer formed on the vertical apparatus wall. The vortex is generated from an axisymmetric flowfield in rotation above a fixed horizontal surface, with vertical volumetric suction at the center. The pressure gradient associated with the flow field causes the fluid to flow radially toward the center, and through the effects of viscosity a vortex boundary layer is formed next to the fixed surface. The radially inflowing fluid, on reaching the center, effuses vertically upward, forming the core of the vortex (effusing core). (a) Laser cross-section of a spiral breakdown. The Reynolds number (Re) which is defined $(\Gamma/2\pi\nu)\approx1000$, and the swirl number (S)≈9.0. (b) Double helix vortex disruption $Re < 750$, $1.5 < S < 3$. (c) Closed bubble breakdown. Re$\approx$2500, S=2.5 (Khoo *et. al.*, 1997).

367

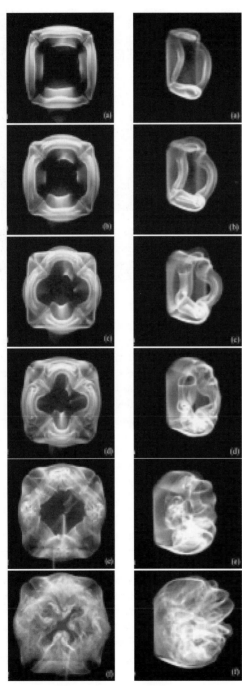

End View   Oblique View

Three-dimensional wake formation behind a square plate. The experiment was conducted in a low speed water channel. Flourescein dye was injected at a fixed upstream location of the plate, and the flow structure was illuminated with an ultraviolet light. The photographs show two sequences of the wake development behind a square plate obtained from two camera angles. (a) t=1.5s after the flow has started, (b) t=2.0s, (c) t=2.5s, (d)=t=3.0s, (e) t=4.0s and (f) t=6.0s (H. Higuchi *et. al.*, 1996).

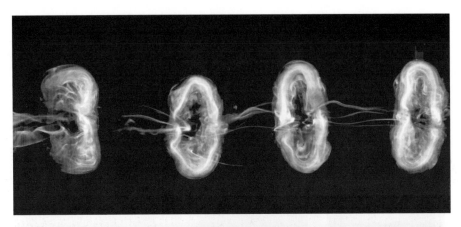

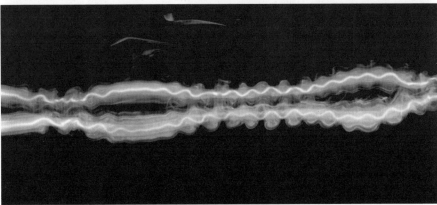

The long and short of vortex pair instability. Here, the vortex pair is generated at the sharpened edges of two flat plates, hinged to a common base and moved in a prescribed symmetric way. Visualization is achieved using flourescent dye. The evolution of the vortex pair was found to depend strongly on the vortex velocity profiles, which are determined by the motion history of the plates. The photographs show two of the three different length scales that have been identified. The upper image shows a plan view of the late stage of the long-wave instability whose axial wavelength is several times the (initial) distance between vortex centers. The initially straight vortices develop a waviness (similar to the long-wavelength deformations in the lower picture), which is amplified until they touch, breakup, and reconnect to form periodic vortex rings, which then elongate in the transverse direction. The lower image shows the development of a short-wave instability (wavelength less than one vortex separation) superimposed on the long waves. The remarkably clear visualization of the vortex core reveals its complicated internal structure, and the observed phase relationships show that the symmetry of the flow with respect to the mid-plane between the vortices is lost (T. Leweke & C.H.K. Williamson, 1996).

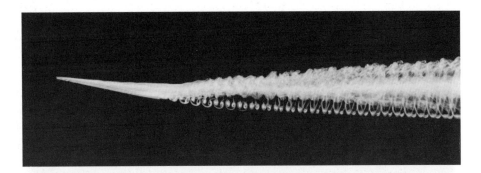

Free flight of a delta wing in water. The photographs show, in side view, the development of the trailing vortex pair as it travels downstream. The fluorescent dye, illuminated by a laser, indicates that the near wake comprises an interaction between the strong primary streamwise vortex pair with the weak "braid" wakes vortices between the pair. Far downstream (64 chordlengths behind the wing) the primary vortex pair have reconnected and become large-scale rings (lower picture), although with a distinctly smaller normalized length scale than predicted from Crow's instability (G.D. Miller & C.H.K. Williamson, 1995).

370

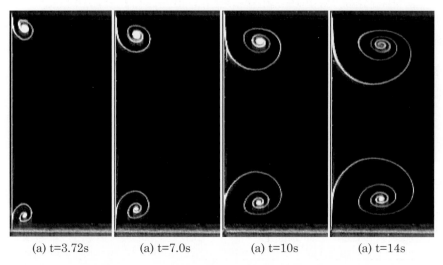

| (a) t=3.72s | (a) t=7.0s | (a) t=10s | (a) t=14s |

The photographs show the formation of a vortex ring in front of a piston as it moves through a cylinder. The piston is located at the left-hand side of each picture. The Reynolds number of the flow based on piston speed and piston diameter is about 3,164. Flow visualization is conducted using flourescent dye illuminated with a thin sheet of Argon ion laser. The mechanism for the formation of this vortex is the removal of the boundary layer forming over the stationary surface in front of the advancing piston. The size of the vortex is solely the function of viscosity and time, and appears to follow similarity scaling (J.J. Allen & M.S. Chong, 1999).

The particle pathline of the flow around one of the vortices generated in front of an advancing piston. The velocity of the piston follows a power law given by $U=At^m$, where m=0.69, A=2.4 cm/sec. The Reynolds number based on the piston velocity and diameter is 8632. The photograph shows an instant, 4.8 s after the initiation of the piston motion (J.J. Allen & M.S. Chong, 1999).

371

Streamlines and integrated streaklines in the wake behind a cylinder at a Reynolds number of 100. The cylinder has a diameter of 2.2 cm and is moving through still water at 0.5 cm/s. The flow is visualized using aluminum dust method and the electrolytic precipitation method simultaneously. Aluminum dust suspension in water shows the "instantaneous" streamlines, and the white smoke produced on the surface of the cylinder by electrolyzing the water shows the integrated streaklines. (S. Taneda, 1985).

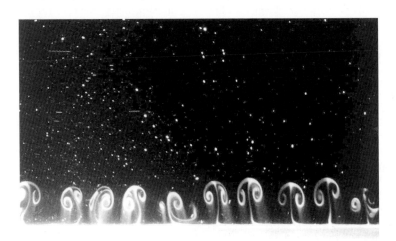

Cross section of the boundary layer on a circular cylinder oscillating rotationally about its center axis in still water. The diameter of the cylinder is 3.2 cm, the oscillation frequency = 0.1 Hz, the angular amplitude = 270°, and the time which has elapsed since the start of oscillation = 13.9s. The boundary layer is made visible by the electrolytic precipitation method (S. Taneda, 1977).

372

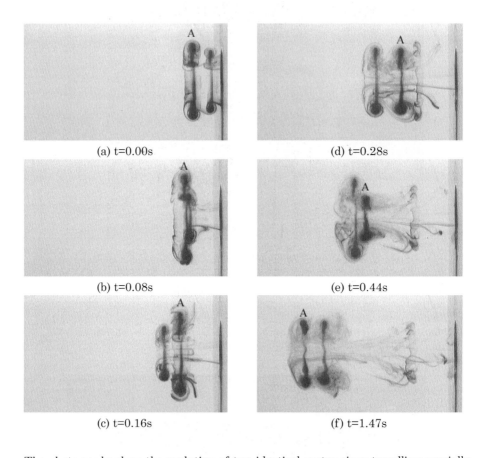

(a) t=0.00s

(b) t=0.08s

(c) t=0.16s

(d) t=0.28s

(e) t=0.44s

(f) t=1.47s

The photographs show the evolution of two identical vortex rings travelling coaxially in water. The rings were generated in quick succession using a piston-cylinder arrangement (see Lim ,1997). The Reynolds number based on the translation velocity and the diameter of the ring is about 2077. The time in the first photograph has been assigned arbitrarily as t=0.00s to indicate the beginning of the leapfrogging process. Rickett "bluo" was used as a tracer. During leapfrogging, the induced velocity of the front ring caused the rear ring to contract and accelerate. In contrast, the rear ring caused the front ring to expand and slow down. The rear ring finally caught up with the front ring, and slipped through the center of the front ring and emerged ahead of it. When this happened, the role of the rings was reversed, and the process repeated itself as can be seen in the figure (d) to (f). In this experiment, the flow was captured by using Sony DXC-930P video camera and SVO-9620 S-VHS recorder (Lim,1997).

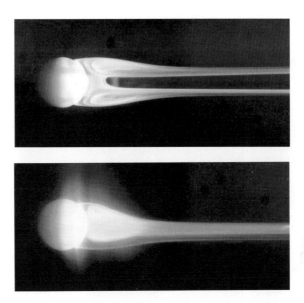

The wake structure of a sphere placed in a uniform flow at Reynolds
number of 270. The flow is made visible using fluorescent dye which is
introduced through a small hole located at the rear stagnation point of
the sphere. A 5W argon ion laser is used to illuminate the flow. The pair
of photographs shows the wake structure viewed from two perpendicular
direction. The top photograph depicts the recirculation region splitting
into two vortex filament while the bottom one shows the asymmetry of
the wake (T. Leweke, 1999).

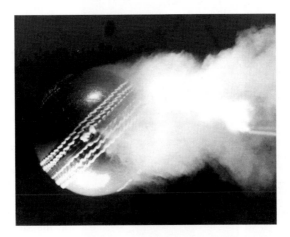

The above picture shows a cricket ball held stationary in a wind tunnel
with the seam set at an incidence angle of 40° to the airflow. Smoke is
injected into the separated region behind the ball where it is entrained
right up to the separation points. The Reynolds number of the flow is
about 0.85×10^5 (R. D. Mehta et. al., 1983).

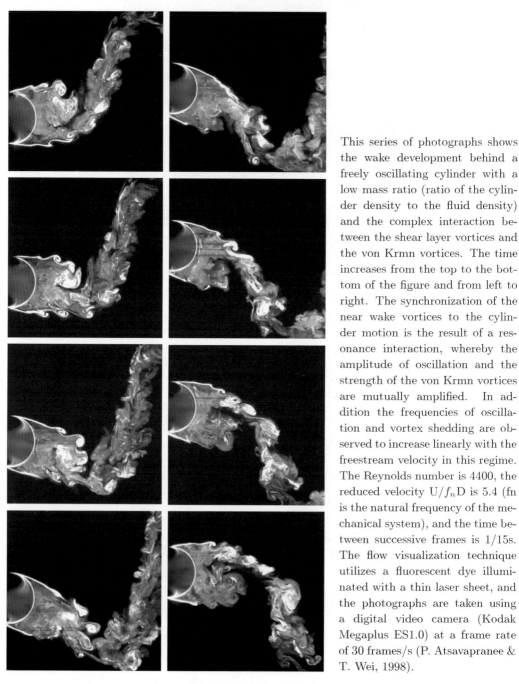

This series of photographs shows the wake development behind a freely oscillating cylinder with a low mass ratio (ratio of the cylinder density to the fluid density) and the complex interaction between the shear layer vortices and the von Krmn vortices. The time increases from the top to the bottom of the figure and from left to right. The synchronization of the near wake vortices to the cylinder motion is the result of a resonance interaction, whereby the amplitude of oscillation and the strength of the von Krmn vortices are mutually amplified. In addition the frequencies of oscillation and vortex shedding are observed to increase linearly with the freestream velocity in this regime. The Reynolds number is 4400, the reduced velocity $U/f_n D$ is 5.4 (fn is the natural frequency of the mechanical system), and the time between successive frames is $1/15$s. The flow visualization technique utilizes a fluorescent dye illuminated with a thin laser sheet, and the photographs are taken using a digital video camera (Kodak Megaplus ES1.0) at a frame rate of 30 frames/s (P. Atsavapranee & T. Wei, 1998).

The picture shows the negatively buoyant wake structures, produced when a small low frequency oscillation is applied to a glass tube from which the smoke is issuing. The flow is from left to right. The inner flow of the tube is lower than the surrounding outer flow in the wind tunnel. The structures resemble a 'daisy chain' of interlocking loops, which are similar to those for the wake behind a sphere. The Reynolds number based on outer flow velocity and the tube diameter is about 350, and the frequency of vibration is about 8 Hz (A.E. Perry & T.T. Lim, 1978).

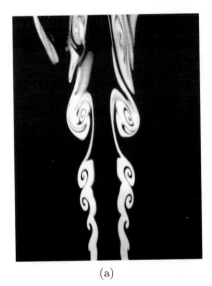
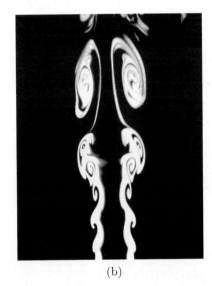

(a)   (b)

Cross-sectional view of coflowing jet structures issuing from a circular tube located along the centerline of a wind tunnel. The flow is to the top. The Reynolds number based on the tube diameter is of the order of 500. Smoke is introduced uniformly around the circumference of the tube only, and a thin laser sheet is used to illuminate the flow. (a) Depicts the formation of the shear layer vortices, and (b) shows the process of vortex tripling (T. T. Lim).

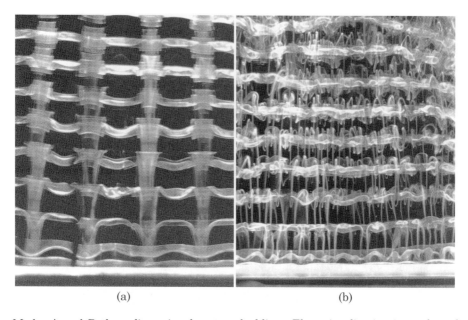

(a)   (b)

Modes A and B three-dimensional vortex shedding. Flow visualization is conducted using flourescent dye. (a) Mode A represents the inception of streamwise vortex loops, for Re=180 and above. Spanwise lengthscale is around 3-4. (b) Mode B represents the formation of finer-scale streamwise vortex pairs, for Re=230 and above, at a lengthscale of around 1 diameter. Experimental evidence shows that Mode A is a vortex core instability, and Mode B is an instability of the "braid" vorticity lying between primary vortices. It is suggested that Mode A is an "elliptic" instability of the cores, and Mode B is a "hyperbolic" instability of the braid vorticity. Note that both of these photographs are to the same scale (T. Leweke & C.H.K.Williamson, 1998; C.H.K. Williamson, 1988, 1996).

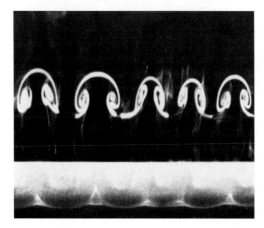

Cross-sectional view of Mode B streamwise vortex structure obtained using the Laser-Induced Fluorescence (L.I.F.) technique. The flow direction is from the bottom to the top. Here, the smaller-scale "mushroom" vortex pair can be clearly seen. $\lambda/D=0.98$; Re=300 (C.H.K.Williamson, 1996).

377

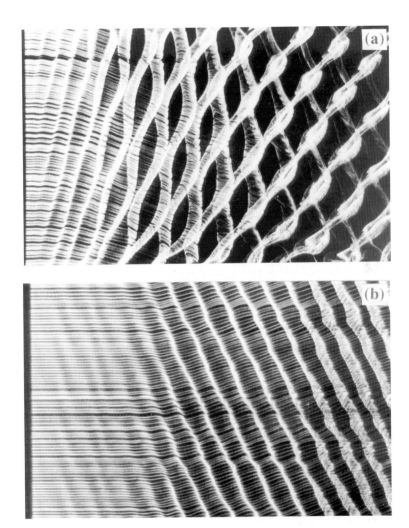

A mechanism of "oblique wave resonance" in the far wake of a circular cylinder. In both cases, the flow is to the right. The flows were visualized using the smoke wire technique. With the smoke wire located downstream of the cylinder at x/D=50, in (a), one can observe at the left the oblique (vortex) shedding waves. These waves interact with two-dimensional instability waves, in the middle of the picture, that are amplified as the flow travels downstream, to produce the large-angle "oblique resonance waves" to the right. The lower photograph in (b) shows that, if the smoke is introduced further downstream at x/D=100, then one observes almost wholly the strong oblique resonance waves. One lesson from the technique, also pointed out clearly by Cimbala, Nagib & Roshko (1988), is that the visualized flow at a particular downstream location is strongly influenced by the upstream point of introduction of the smoke, and is a "history" effect of the method (C.H.K. Williamson & A. Prasad, 1993a,b).

378

A traverse jet in a crossflow is visualised by using smoke-wire technique. The crossflow is from left to right. The wire is oriented at the center-plane of the crossflow and located upstream of the jet. The photograph shows that the initial portion of the jet is dominated by shear-layer vortices, which are a result of the Kelvin-Helmholtz instability of the annular shear layer that separates from the edge of the jet orifice. The jet-crossflow velocity ratio (VR) is 2, and the Reynolds number based on the crossflow velocity (Re) is 3800 (T. F. Fric & A. Roshko, 1994).

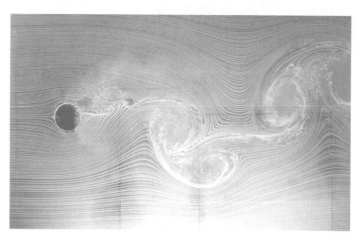

Cross-sectional view of the flow around the traverse jet which is issuing toward the viewer. VR= 4 and Re=11400. Here, the smoke wire is aligned with the $Z_{sw}/D = 0.5$ plane, where Z_{sw} is the vertical distance from the wall and D is the jet diameter (T. F. Fric & A. Roshko 1994).

Vortex/mixing layer interaction. The smoke flow visualization photograph shows a streamwise vortex embedded in a two-stream mixing layer with a velocity ratio of 0.5. The vortex was generated by a half-delta wing mounted in the wind tunnel settling chamber (R.D. Mehta, 1984).

This vortex ring was generated by a smoke-filled bursting soap bubble resting on a liquid surface. This is similar to the simpler case of arising bubble penetrating a free surface, the "flip" experiment of a falling water drop impacting a free surface (Peck and Sigurdson, Phys. Fluids A 3, 2032 (1991)). In both cases a vortex ring is often created. This is a triple exposure illuminated by: the spark used to break the bubble, a strobe showing the original bubble shape, and the strobe again at a later time showing the resulting vortex ring. The base width of the bubble is 15 mm. The vertical white stripe in the middle is strobe light reflected from the upper electrode (Buchholz, Sigurdson & Peck, 1995).

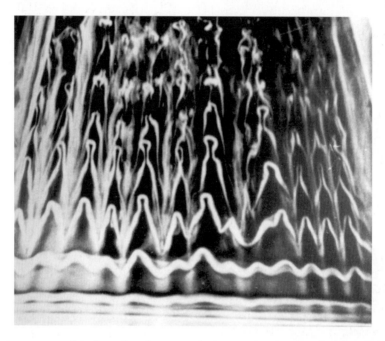

Smoke pattern of Λ-shaped or hairpin vortices generated by an oscillating steel rod of 4.15mm diameter submerged in the smoke layer. The flow is to the top of the picture. The rod was located at where the boundary layer was approximately 2mm thick. The non-dimensional frequency of oscillation is $2.85x10^{-5}$, and the Reynolds number based on the displacement thickness is approximately 320. The picture clearly shows that the vortex filaments generated have a strong tendency to develop a three dimensionalility giving a longitudinal component of vorticity (A.E.Perry *et. al.*, 1981).

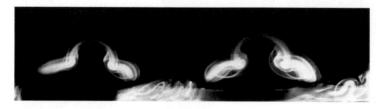

Front view of two of the Λ-shaped or hairpin vortices with Ω-shaped secondary vortices (A.E.Perry *et. al.*, 1981).

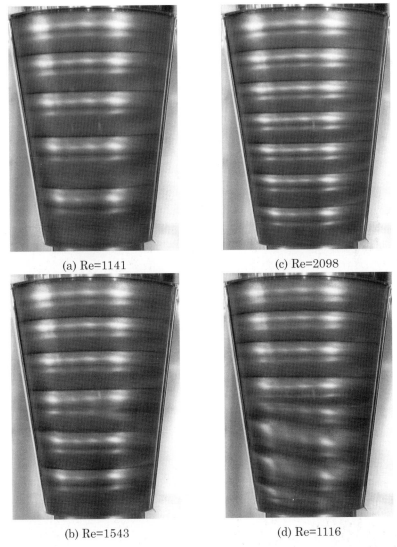

(a) Re=1141   (c) Re=2098

(b) Re=1543   (d) Re=1116

Taylor vortex flow between conical cylinders with the inner cone rotating and the outer one at rest. Both the cones have the apex angle of $16.03°$ giving a constant width for co-axially rotating bodies. The base radius of the outer cone is 50mm while the inner cone is 40mm, thus giving a gap size (s) of 10mm. The length of the fluid column is fixed at L=125mm. Silicone oils are used as working fluids. The flow is made visible by a small amount of aluminium flakes with a typical dimension of $30\mu m$. Photograph (a) is obtained under quasi-steady condition, while (b) and (c) are subjected to different acceleration. The number of vortex pairs in (a), (b) and (c) is five, six and seven, respectively. The flow pattern in (d) is obtained under different initial condition. Here, unsteady helical vortices are formed below the toroidal vortices (M. Wimmer, 1995).

(a)	(b)

Taylor vortex flow between concentric prolate ellipsoids with the inner ellipsoid rotating and the outer one at rest. Axis ratio is B : A = 2 : 1. The vertical longer axis of the outer ellipsoid B = 80mm is the axis of rotation. The gap width s = A - a = 4.9mm, with A and a denoting the shorter axes of the outer and inner ellipsoids, respectively. Flow visualization is achieved by aluminium flakes suspended in Silicon oils. (a) Regular Taylor vortices in the equatorial region, Re = 4650. (b) Wavy Taylor vortices in the whole gap, Re = 28800 (M. Wimmer, 1989).

(a)	(b)

Taylor vortex flow between concentric rotating spheres. Flow visualization is achieved by suspending a small amount of aluminium flakes (with a typical mean dimension of 50 μm) in Silicon oil. (a) $\sigma = s/R_1 = 0.0133$, Re=27,000, where s= gap width, and R_1=radius of the inner sphere. (b) s =0.046, Re=7600 (M. Wimmer, 1976).

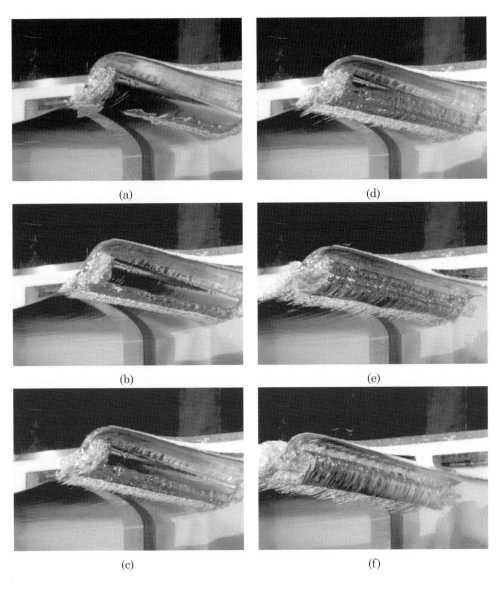

(a) (d)

(b) (e)

(c) (f)

An elliptical air tube is formed when the plunging jet touches the wave front. During the initial stages, the surface of the tube is relatively smooth except near the contact region between the plunging jet and the wave front. In connection with wave propagation, the air tube rolls forward in the direction of wave propagation, consistent with the direction of wave plunging. Instability sets as a result of perturbations at the contact region. As the air tube rolls forward, the surface of the air tube develops into a wavy surface. (J.H.L. Kway & E.S. Chan, 1998).

384

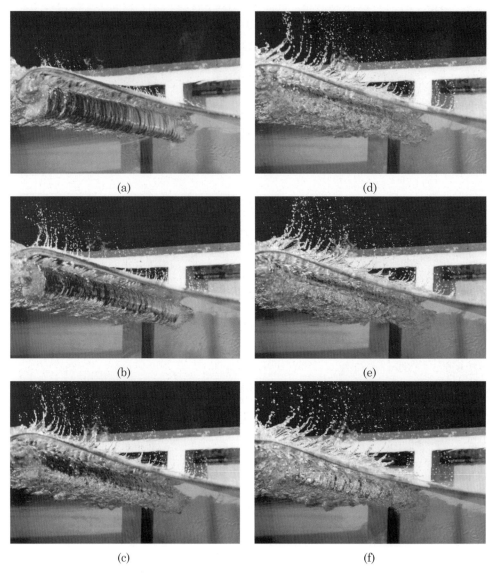

(a)

(d)

(b)

(e)

(c)

(f)

The pictures show the breakdown of entrapped air tube. The continuous rolling action of
the entrapped air tube coupled with the buoyancy of the tube is highly unstable. As wave
plunging progresses, water is sprayed upwards near the crest and backwards relative to the
direction of wave propagation. The backward spray suggests that the air tube is bursting
near the crest. Some air remains trapped and breaks down into entrained air bubbles that
are eventually dispersed (J.H.L. Kway & E.S. Chan, 1998).

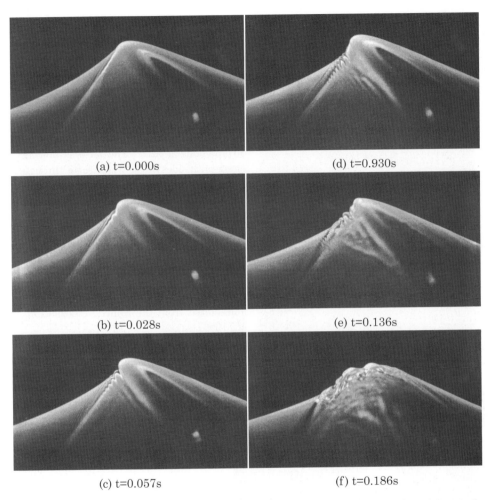

(a) t=0.000s

(d) t=0.930s

(b) t=0.028s

(e) t=0.136s

(c) t=0.057s

(f) t=0.186s

Photographs showing the surface profile histories of gentle spilling breakers generated mechanically with a dispersive focusing technique. The average frequency of the wave is (f_o)=1.42, nominal wavelength (λ_o)=77.43cm, and amplitude to wavelength ratio (A/λ_o) =0.0487. The beginning of the breaking process is marked by the formation of a bulge in the profile at the crest on the forward face of the wave (see (a)). The leading edge of this bulge is called the toe. As the breaking process continues, the bulge becomes more pronounced while the toe remains in nearly a fixed position relative to the crest. Capillary waves formed ahead of the toe (see(b)). At a time of about $0.1/f$ after the bulge first becomes visible, the toe begins to move down the face of the wave and very quickly accelerates to a constant velocity which scales with the wave crest speed (see (c)). During this phase of the breaker evolution, the surface profile between the toe and the crest develops ripples which eventually are left behind the wave crest (see (e)) (J.H. Duncan *et. al.*, 1999).

This image corresponds to a front view of an oscillating planar liquid sheet interacting with coflowing air streams. The liquid (water) was exiting through a 0.9 mm wide 80 mm long nozzle. The air was flowing along both sides of the water sheet through 1 cm apertures. Here, the water exit velocity was 2.4 m/s, and air velocity was 18 m/s. Under these conditions the sheet oscillates with a mixture of sinusoidal and dilatational waves, and atomization is rather poor. To obtain instantaneous images, a 0.5 ms flash lamp was used to freeze the water motion. In this case, the sheet was illuminated from the front (A. Lozano *et. al.*, 1996).

For the same experimental facility, water exit velocity was reduced to 1 m/s, and air exit velocity increased to 30 m/s. In this situation, the sheet oscillates in a dominant sinusoidal mode, the wave amplitude growth rate is higher, and the atomization process is more efficient. This photograph was obtained with back flash illumination. Although the sheet was flowing downwards, the image displayed upside down shows a curious resemblance to a night landscape (A. Lozano *et. al.*, 1994).

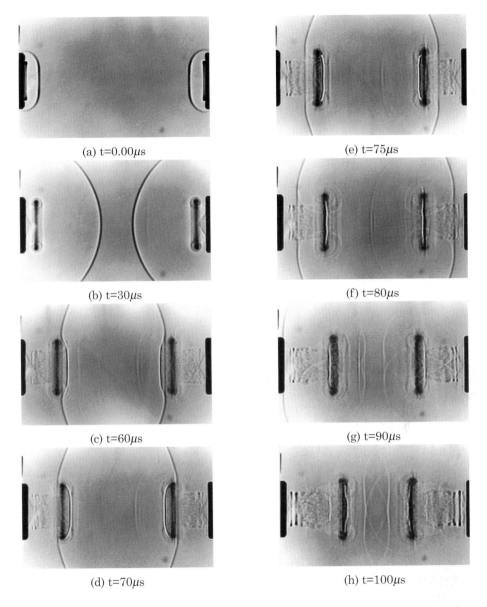

(a) t=0.00μs

(b) t=30μs

(c) t=60μs

(d) t=70μs

(e) t=75μs

(f) t=80μs

(g) t=90μs

(h) t=100μs

Optical shadowgraphs showing the time sequence of the flow field during the interaction between a vortex ring and a shock wave. The Mach number of the shock inside the nozzle is 1.34. In (a) and (b), the shock waves emitted from the nozzle diffract at the edges of the nozzles, evolve into a spherical form, and travel toward each other. The vortex ring is generated at the nozzle exit by the rolling up of the shear layer, and moves toward the other at self-induced velocity. The shock wave travels through the vortex ring, but the portion of the shock wave which comes into the head-on collision with the ring is retarded (2(c)), and is captured inside the vortex ring (2(d)-(f)) (T. Minota *et. al.*, 1997).

388

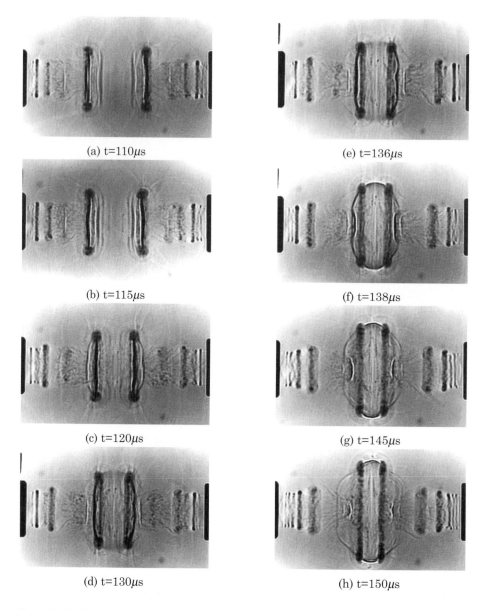

(a) t=110μs

(b) t=115μs

(c) t=120μs

(d) t=130μs

(e) t=136μs

(f) t=138μs

(g) t=145μs

(h) t=150μs

Optical shadowgraphs showing the interaction of two vortex rings. In (a), the hitting diffracted shock fronts are intensified and compressed against the ring surface. The diffracted shock ((a)-(c)) passing over the ring becomes very weak. As the vortex rings come close to each other, the forward motion is blocked and the radial motion is accelerated. In (d) and (e), the front surface of the vortex ring comes into contact and a dark curve, meaning low density, is seen between the cores of the two rings. This becomes an inward-facing shock (f), and moves together with the vortex core ((f)-(h)) (T. Minota *et. al.*, 1998).

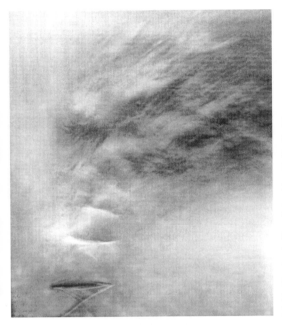

Retroreflective Focusing Schlieren (RFS) image of jet in wind tunnel. The photograph shows a high pressure air jet from a table top is blown upwards into a cross flow. The jet Mach number was 1.07, the wind tunnel velocity was 170 miles per hour, wind tunnel temperature: 78.9°F. The retroreflective material and source grid are placed outside the test section behind a window. The light source, camera, beam splitter and cutoff grid are mounted on the other side of the wind tunnel. Light source was a xenon flash with a duration of 1 msec (J.T. Heineck & Steven Jaeger, 1997).

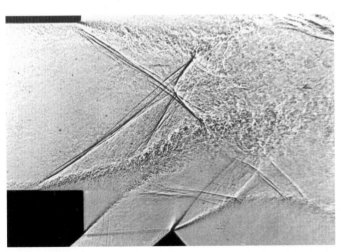

This shadowgraph shows the planar, two dimensional flowfield which consists of an upper Mach 2.5 stream with unit Reynolds number (Re) of 48.9×10^6/m, and a lower Mach 1.5 stream with unit (Re) of 36.2×10^6/m converging at a $40°$ angle past a 12.7mm high base plane. The spanwise width of the flowfield and the height of the upper stream are 50.8 mm. The upper stream is analogous to the supersonic freestream surrounding a rocket afterbody while the lower stream is analogous to an underexpanded exhaust plume. The shadowgaph was produced using a 25ns pulse from Xenon model 437B Nanopulser at a jet static pressure ratio of 2.35 between the two streams (R.J. Shaw, 1995).

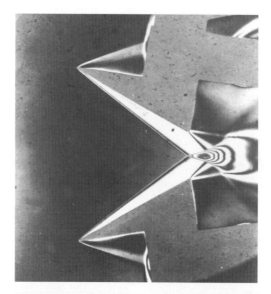

$a = 36.2^{\circ}$,
g/w=0.37

$a = 36.2^{\circ}$,
g/w=0.37

Interferograms of steady-flow shock reflexion in dissociating carbon dioxide.
U_{∞}=3.6 km s^{-1}, ρ_{∞} =3.8x10^{-6} g cm^{-3}, M_{∞}=5.5. g=gap between the trailing
edges of the wedges, w=streamwise length of the wedge face, a=angle between the
wedge face and the horizontal axis. Free-stream composition: C, 10^{-11} mole/g;
O, 6x10^{-6} mole/g; CO_2, 0.0089 mole/g; CO, 0.0138 mole/g, O_2, 0.0069 mole/g
(H.G. Hornung $et.$ $al.$, 1979).

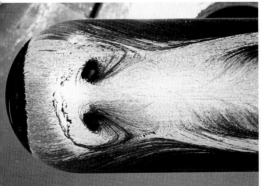

A surface-oil pattern of a hemispherical-cylinder model at an angle of attack of 25°. The flow is from left to right. $M_\infty = 0.55$ and $Re_D = 1.6 \times 10^6$. The boundary layer transition is artificially fixed via a band of carborundum grit visible close to the nose. The picture shows the familiar "owl-face" flow-pattern over the windward side of the cylinder. The dark areas (the owl's "eyes") are areas of high shear, delineating the source of the body vortices which are shed into the flow. (B.D. Fairlie, 1980).

The flow around a circular cylinder mounted normal to a flat plate. The flow is from left to right. $M_\infty = 0.55$ and $Re_D = 1.0 \times 10^6$. The approaching boundary layer is turbulent and is approximately twice the height of the cylinder. The photograph shows the classic formation of the necklace vortex upstream of the cylinder, and the two three-dimensional separation lines clearly visible upstream of the cylinder merging into one some distance downstream. The formation of two vortical structures is also visible behind the cylinder. (B.D. Fairlie, 1980).

Vortex/separated boundary layer interaction: Surface oil-flow pattern showing the qualitative effect of a streamwise vortex on a separated boundary layer. The flow is from left to right, and the Mach number is about 0.8 which is just below the critical Mach number). Here, two foci are generated in the region of the interaction. The flow pattern changed drastically as the critical Mach number was crossed during testing (R.D. Mehta , 1988).

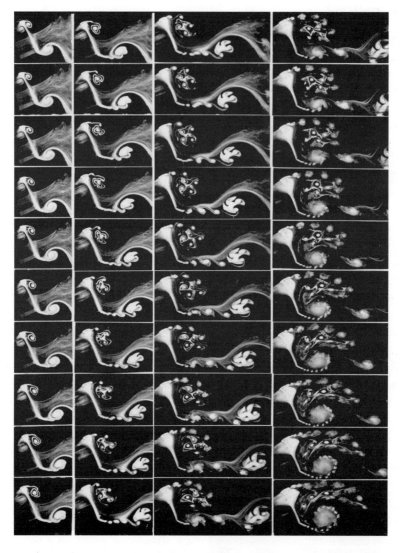

A movie sequence showing the vortical pattern development during dynamic separation in accelerating flow around NACA 0015 airfoil at an angle of attack of 60°. The flow is accelerating from left to right at $2.4 m/s^2$. The Reynolds number based on the airfoil chordlength is 5200. The time from flow start-up to the first frame (t_1) is 26/64s, and increases from top to bottom and then across columns from left to right with (Δt) of 1/64s between consecutive frames. The flow pattern was visualized by using titanium tetrachloride technique. Here, the leading edge starting vortex forms a spiral which under the influence of filament instability takes on a triangular shape in the upper part of column 2. The triangle undergoes a metamorphosis into a triarm in the lower half of column 2 which by incorporation of more vortices transform into a fourarm in column 3. Turbulence and splitting set in in column 4 (P. Freymuth, 1985).

References

Afenchenko, V.O., Ezersky, A.B., Kiyashko, S.V., Rabinovich, M.I. and Weidman, P.D. 1998. The generation of two-dimensional vortices by transverse oscillation of a soap film. *Phys. of Fluids*,**10**(2), 390–399.

Allen, J.J. and Chong, M.S. 1999. Experimental study of the vortex formed in front of a piston as it moves through a cylinder and a rectangular duct (submitted to JFM).

Atsavapranee, P. and Wei, T. 1998. *Bulletin of American Physical Society/Division of Fluid Dynamics*.

Blackburn, H.M. and Henderson, R.D. 1999. A study of two-dimensional flow past an oscillating cylinder *J. Fluid Mech.***385**, 255–286.

Buchholz, J., Sigurdson, L. and Peck, W. 1995. Bursting soap bubble. In Gallery of fluid motion, *Phys. of Fluids*, **7**, S3.

Duncan,J.H., Qiao, H., Philomin, V. and Wenz, A. 1999. Gentle spilling breakers: crest profile evolution. *J. Fluid Mech.*,**352**, 191–222.

Escudier, M.P. (1984). Observations of the flow produced in a cylindrical container by a rotating end wall, *Expts. in Fluids*, **2**, 189–196.

Fairlie, B.D. (1980). Flow separation on bodies of revolution at incidence, 7th Australiasian conference on hydraulics and Fluid Mechanics, 18-22 August, Brisbane, Australia.

Freymuth, P. 1985. The vortex patterns of dynamic separation: A parametric and comparative study, *Progress in Aerospace Sci.*, **22**, 161–208.

Fric, T.F. and Roshko, A. 1994. Vortical structure in the wake of a transverse jet, *J. Fluid Mech.*, **279**, 1–47.

Gad-El-Hak, M., Blackwelder, R.F. and Riley, J.J. 1981. On the growth of turbulent regions in laminar boundary layers. *J. Fluid Mech.*, **110**, 73–96.

Gendrich, C.P., Koochesfahani, M.M. and Nocera, D.G. 1997. Molecular tagging velocimetry and other novel applications of a new phosphorescent supramolecule. *Expts in Fluids*, **23**, 261–372.

Henderson, R.D. 1997. Nonlinear dynamics and pattern formation in turbulent wake transition. *J. Fluid Mech.*,**352**, 65–112.

Heineck, J.T. and Jaeger, S. 1997. One-sided focusing schlieren system with reflective grid, *NASA Technical Briefs* **21**(7).

Higuchi, H., Anderson, R.W. and Zhang, J. 1996. Three-dimensional wake formations behind a family of regular polygonal plates. *AIAA Journal.*, **34**, 1138–1145.

Hornung, H.G., Oertel. H. and Sandeman R.J. 1978. Transition to Mach

394

reflexison of shock waves in steady and pseudosteady flow with and without relaxation. *J. Fluid Mech.*, **90**, 541–560.

Kelso, R. M., Lim, T.T. and Perry, A.E. 1996. An experimental study of a round jet in cross-flow, *J. Fluid Mech.*, **306**, 111–144.

Kelso, R. M., Lim, T.T. and Perry, A.E. 1992. A round jet in cross-flow, *Album of Visualization.*, **9**, 30.

Khoo, B.C., Yeo, K.S., Lim, D.F. and He, X. 1997. Vortex breakdown in an unconfined vortical flow. *Expt Thermal and Fluid Sci.*, **14**, 131–148.

Kway J.H.L. and Chan E.S. 1998. Air entrainment and bubble breakdown in plunging waves. *Technical Report GR6414-6-96*, 1–22.

Lozano, A., Call, C.J. and Dopazo, C. 1994. Atomization of a Planar Liquid Sheet. *Phys. of Fluids*, **6**(9), S5.

Lozano,A., Call, C.J., Dopazo, C. and Garcia-Olivares, A. 1996. Atomization of a Planar Liquid Sheet, *Atomization and Sprays*,**6**, 77–94.

Leweke, T. and Williamson, C.H.K. 1996. The long and short of vortex pair instability, *Phys. of Fluids*,**8**, S5.

Leweke, T. and Williamson, C.H.K. 1998. Three-dimensional instabilities in wake transition, *European J. Mech. B - Fluid*,**17**(4), 571-586.

Leweke, T. 1999. The wake structure of a sphere placed in a uniform flow (private communication).

Lopez, J.M. and Perry A.D. 1992. Periodic axisymmetric vortex breakdown in a cylinder with a rotating end wall. *Phys. of Fluids*, **4**(9), pp 187.

Lim, T.T. 1989. An experimental study of a vortex ring interacting with an inclined wall. *Expts. in Fluids*, **7**(7), 453–463.

Lim, T.T. and Nickels, T.B. 1992. Instability and reconnection in head-on collision of two vortex rings, *Nature*, **357**, 225–227.

Lim, T.T. 1997. A note on the leapfrogging between two coaxial vortex rings at low Reynolds numbers, *Phys. of Fluids*, **9**, 239–241.

Luo, S.C., Lim, T.T., Lua, K.B., Chia, H.T., Goh, E.K.R. and Ho, Q.W. 1998. Flowfield around ogive/elliptic-tip cylinder at high angle of attack, *AIAA Journal.*, **36**, 1778–1787.

Mehta, R.D., Bentley, K., Proudlove, M. and Varty, P. 1983. Factors affecting cricket ball swing. *Nature*, **30**, 787–788.

Mehta, R.D. 1984. An experimental study of a vortex/mixing layer interaction, *Paper # 84-1543*, presented at the AIAA 17th Fluid Dynamics, Plasma Dynamics and Lasers Conference, Snowmass, Colorado, June 25-27.

Mehta, R.D.1988. Vortex/separated boundary-layer interactions at transonic Mach numbers, *AIAA Journal.*, *26*, 15–26.

Miller, G.D. and Williamson, C.H.K. 1995. Free flight of a delta wing. *Phys. of Fluids*, **7**, S9.

Minota T., Nishida M., and Lee M.G. 1997. Shock formation by compressible vortex ring impinging on a wall, *Fluid Dyn Research*,**21**(3), 139–157.

Minota T., Nishida M., and Lee M.G. 1998. Head-on collision of two compressible vortex rings, *Fluid Dyn Research*,**22**(1), 43–60.

Perry, A.E. and Lim, T.T. 1978. Coherent structures in coflowing jets and wakes. *J. Fluid Mech.*, **88**, 451–463.

Perry, A.E., Lim, T.T. and Teh, E.W.1981. A visual study of turbulent spots. *J. Fluid Mech.*, **104**, 387–405.

Perry, A.E., Chong, M.S. and Lim, T.T. 1982. Two vortex shedding process behind two-dimensional blunt bodies. *J. Fluid Mech.*, **116**, 77–90.

Shaw, R.J. 1995. An Experimental investigation of usteady separation shock wave motion in a plume-induced, separated flowfield. *Ph.D. thesis*, University of Illinois at Urbana-Champaign, Illinois, USA.

Shiels, D. and Leonard, A. 1998. Investigation of drag reduction on a circular cylinder in rotary oscillation(submitted to JFM).

Taneda, S. 1977. Visual study of unsteady separated flows around bodies. *Progress in Aerospace Sci.*, **17**, 287–348.

Taneda, S. 1985. Flow field visualization. *Proceedings of the XVIth International Congress of Theoretical and Applied Mechanics.*, Lyngby, Denmark, August 19-25, 399–410.

van Heijst, G.J.F. and Kloosterziel, R.C 1989. Tripolar vortices in a rotating fluid. *Nature*, **338**, 567–571.

van Heijst, G.J.F. and Flor, J.B. 1989. Dipole formation and collision in a stratified fluid. *Nature*, **340**, 212–215.

Williamson, C.H.K. 1988. The existence of two stages in the transition to three-dimensionality of a cylinder wake. *Phys. of Fluids*, **31**(11), 3165–3168.

Williamson, C.H.K. 1992. The natural and forced formation of spot-like 'vortex dislocations' in the transition of a wake. *J. Fluid Mech.* **243**, 393–441.

Williamson, C.H.K. and Prasad, A. 1993. A new mechanism for oblique wave resonance in the "natural" far wake, *J. Fluid Mech.* **256**, 269–313.

Williamson, C.H.K. and Prasad, A. 1993. Acoustic forcing of oblique wave resonance in the far wake, *J. Fluid Mech.* **256**, 313–341.

Williamson, C.H.K. 1996. Three-dimensional wake transition. *J. Fluid Mech.* **306**, 345–407.

Wimmer, M. 1976. Experiments on a viscous fluid flow between concentric rotating spheres. *J. Fluid Mech.*,**78**, 317–335.

Wimmer, M. 1989. Strömungen zwischen rotierenden ellipsen. *Z. agnew Math. Mech.*,**69**, T616-T619.

Wimmer, M. 1995. An experimental investigation of Taylor vortex flow between conical cylinders. *J. Fluid Mech.*,**292**, 205–227.